Digital Logic Design Principles

Norman Balabanian
University of Florida

Bradley Carlson
Symbol Technologies, Inc.

WILEY

John Wiley & Sons, Inc.

New York • Chichester • Weinheim • Brisbane • Toronto • Singapore

AQUISITIONS EDITOR	Bill Zobrist
MARKETING MANAGER	Katherine Hepburn
PRODUCTION SERVICES MANAGER	Jeanine M. Furino
COVER DESIGNER	Dawn L. Stanley
ILLUSTRATION EDITOR	Gene Aiello
PRODUCTION MANAGEMENT SERVICES	Publication Services

This book was set in Times Roman by Publication Services and printed and bound by Courier Companies. The cover was printed by Phoenix Color Corporation.

This book is printed on acid-free paper. ♾

ISBN 0–471–29351-2

Printed in the United States of America

10 9 8 7 6 5 4 3 2 1

Preface

THE BOOK

This in an introductory-level book on the *principles* of digital logic design. It is intended for use by first- or second-year students of electrical engineering, computer engineering, or computer science. No previous knowledge of electrical circuits or of electronics is assumed. Others who need a first exposure to, or a refresher on, digital design principles may also find it useful.

Pedagogical Issues

The deductive process—applying general principles to specific cases—is usually well illustrated in textbooks. Often, on the authority of the author, a general concept or an approach to a topic, or some result, is stated, followed by examples of how to apply the concept. As students begin a topic, it is unclear to them what the motivation is for introducing a particular definition or general procedure. Students don't have a clear idea why this particular topic may be useful or interesting, or why and how anybody thought it up in the first place.

In this book, we include an inductive approach in the development of subject matter. This approach involves the development of a generally valid result from an examination of specific cases, the way a research investigation proceeds. An investigator reaches a generally valid result by carrying out a number of specific experiments or calculations. Sometimes the study of one or more specific cases leads to a conjecture about something generally valid. The conjecture is then explored and justified using previously established results.

In a similar vein, we introduce most topics in an exploratory spirit, rather than dropping them on the reader without any preceding justification. The tenor of the text is that we are conducting an investigative exploration, almost like a research project, for the purpose of discovering and assimilating knowledge about the subject under study. When a topic is introduced, a considerable effort is made to help students understand why we ought to devote time to it. After a particular topic has been exhausted (that is, when we are faced with the need for taking a further step) alternatives are explored. "We could do this or we could do that," the commentary might go; "let's first try this, for the following reasons." *Why* to pursue a particular thread and *how* a particular procedure might come about are just as important to clarify for the learner as the details of following that specific procedure or of applying some particular algorithm.

When a subject such as digital circuits reaches a degree of maturity, there is a tendency for a textbook to acquire some of the characteristics of an encyclopedia: every conceivable topic is "covered." This approach robs the learner of all the joys of discovery. The learner is given the complete story and told to learn it, mainly by practicing on exercises and problems like those that the book has just worked out. In this book, we try to avoid the pitfall of cataloguing for students all that we know on a subject. In the form of problems, we leave for students the pleasure of developing (with guidance) some results that are not essential for going on with the subject matter being developed, and so need not be part of the exposition.

Students learn best if they are engaged. There isn't much that authors can do to keep them engaged, but we do remind them to participate in the derivation of an equation by carrying out the missing steps, to observe the relevant features of a diagram or table by describing it carefully to themselves, or to think through a proposed plan before carrying it out in detail. We do this often.

Level of Presentation

The presentation of material in this book is at the introductory level, first or second year of college. However, the level of a book should not dictate the degree of rigor in the presentation. Everything treated in this book is treated with rigor.

Topic Selection

There is nothing unusual in the selection of subject matter. The selection and ordering of topics has been carried out to facilitate use of the book at institutions with different calendars and a variety of emphases. The book can be used in courses spanning an academic year, either two semesters or three quarters, especially if attention is given to the laboratory component. (See the description of the laboratory manual in what follows.) By proper selection of chapters and topics within chapters, a one-semester course can also be accommodated. Several "enrichment" topics are introduced in sections that instructors can omit without incurring a subsequent penalty. Later sections or problems based on this material can be similarly omitted. The inclusion of such material permits students having more time or interest to benefit without penalizing others.

The choice of ABEL to introduce hardware description language (HDL) as a tool for design minimizes the effort of students to learn the language, thus enabling them to concentrate on the concepts behind designing with an HDL. All concepts of HDL specification, simulation, and synthesis can be taught using ABEL, and the student is not burdened with the task of learning the syntax and semantics of a complex language such as VHDL or Verilog.

Numbering Scheme for Equations and Figures

Some schemes for numbering sections, equations, and figures, and the manner of referring to them, can cause students to be distracted as they engage unproductively in reading the numbers and searching for them. In this book a sequential numbering system, starting fresh in each chapter, is used for both equations and figures. (On the few occasions when reference might be made to an equation in an earlier chapter, the chapter number is also given.) Similarly, major sections within a chapter are numbered consecutively, without a chapter identifier, but subsections and sub-subsections are not numbered, thus obviating the unproductive reading of such section numbers as 4.3-5 that might identify subsection 5 of section 3 in Chapter 4. Subsequent reference to such a particular subsection is rarely, if ever made in *any* book; hence, there is no reference value to such a numbering scheme. Not all equations, only significant ones or those to which reference might be made later, are numbered. When referring to an equation or figure, we spell out the name: "Equation" or "Figure."

Illustrations, Examples, Exercises, and Problems

When a particular topic is being developed, illustrations are used to illuminate it. Indeed, an illustration might precede the development of the topic as part of the process of induction. *Illustrations* are thus incorporated into the development of the material. There are also numbered *examples,* separated from the text and easily distinguished, which are worked out using the concepts just developed, together with other recently assimilated ideas.

Scattered throughout the development, but in a format that distinguishes them from the text, are numbered *exercises* for students to work out at the time they are studying the relevant sections. The purpose of these exercises is to provide reinforcement for the concepts under study by having students carry out some simple calculations and apply results then under discussion. They form part of the "research project" idea. The excitation requirements for one type of flip-flop might be developed within the text, for example; the excitation requirements for other types of flip-flops are then left as an exercise for the students to work out. Where useful, answers are provided so that students can confirm the results of their efforts. (Most of the time, especially if answers are brief and thus easy for students to glance at within the text, they are provided as footnotes.) The exercises do not simply call for repeating the steps of a just-worked-out example using changed values or circuit configurations. Hence, there is no need to provide worked-out examples before asking students to perform an exercise.

At the end of each chapter is a set of *problems*. The problems in each set range from a simple application of procedures developed in the book to a challenging solution of a more complex problem. Sometimes a problem requires students to apply a specified technique. At other times they are asked to solve a problem using two or more specified approaches and to compare the ease or difficulty. In both cases, they are practicing specific techniques and reinforcing their understanding of them. Sometimes the problem is open-ended so that students have to make decisions about the methods to use, and then apply them.

Text Supplements

There are two packages of supplements. One is provided to instructors who adopt the book for use in their courses and is not available to students. It includes a solutions manual that contains full solutions of all the problems in the book. It also includes a set of transparencies of appropriate figures from the book. The figures are enlarged so that instructors can use them in the classroom.

The other package consists of a laboratory manual. In the book itself, although specific families of digital circuits are referred to from time to time (e.g., 74LS02), major stress is on design *principles*. The laboratory manual is intended to engage students in the *practice* of digital design, using the latest in currently available technology. We show how specific design projects from the manual can be incorporated at specific points in the book. Even though some students may be learning about digital design from other texts, they too can use this laboratory manual to gain experience with digital design practice. For additional information concerning the laboratory manual, please review the text web site (http://www.wiley.com/college/elec/balabanian293512).

SOFTWARE

We recommend the use of schematic entry and timing and functional simulation in the laboratory from the beginning (even with simple experiments or labs). The Xilinx WebPack software can be used, and is available for free from the Xilinx web site (http://www.xilinx.com). This software supports the most recent version of ABEL, so students will be familiar with the user interface by the time Chapter 8 is reached.

ACKNOWLEDGMENTS

We would like to acknowledge our indebtedness to several people who have contributed to the development of this book in many ways. The first author wants to thank specifically Dr. Vijay Pitchumani (now with Intel) and Dr. Dikran Meliksetian (now with IBM), both formerly at Syracuse University. At different times they were to be coauthors of this book, and they made important contributions to the development of the text.

Individuals who made invaluable comments and observations when evaluating the manuscript at different stages in its development include:

Yu Hen Hu, University of Wisconsin–Madison
David R. Kaeli, Northeastern University
Juanita DeLoach, University of Wisconsin–Milwaukee
Mehmet Celenk, Ohio University
James G. Harris, California Polytechnic State University, San Luis Obispo
Sotirios G. Ziavras, New Jersey Institute of Technology
James H. Aylor, University of Virginia
Ward D. Getty, University of Michigan, Ann Arbor
Alexandros Eleftheriadis, Columbia University in the City of New York
Ike Evans, The University of Iowa and Evolutionary Heuristics
Shahram Latifi, University of Nevada, Las Vegas
Gregory B. Lush, University of Texas at El Paso

Finally, we want to thank Ko-Chi Kuo, who produced the solutions for the chapter-end problems and created the solutions manual.

Contents

Chapter 1. NUMBER REPRESENTATION, CODES, AND CODE CONVERSION 1

Chapter 2.   SWITCHING ALGEBRA AND LOGIC GATES   34

Chapter 3.   REPRESENTATION AND IMPLEMENTATION
OF LOGIC FUNCTIONS   81

Chapter 4. COMBINATIONAL LOGIC DESIGN 132

Chapter 5. SEQUENTIAL CIRCUIT COMPONENTS 168

Chapter 6.   SYNCHRONOUS SEQUENTIAL

MACHINES   198

Chapter 7.   ASYNCHRONOUS SEQUENTIAL

MACHINES   254

Chapter 8.   DESIGN USING HARDWARE

DESCRIPTION LANGUAGES   290

Chapter 9. COMPUTER ORGANIZATION 325

Chapter 1

Number Representation, Codes, and Code Conversion

It may appear odd to start this book with a chapter title that is about number representations and codes. But doing so is not dissimilar to starting with the alphabet when learning a language. Bear with us until the relevance becomes obvious.

1 SYSTEMS: DIGITAL AND ANALOG

This book is concerned with the study of one aspect of what are called *digital systems*. Such systems have been gradually replacing earlier ones called *analog systems*. Digital systems can be found in many areas of modern technology, the most common of which is a *digital computer,* but the methods and components described in this book are used in many other digital systems as well, including

- Systems that control operations in a process

  - Traffic signals
  - The flow of chemicals in a chemical plant
  - The temperature in many different processes
  - The operation of many parts of an automobile engine
  - The keeping of time

- Machines that dispense products (vending machines) or allow access (toll gates in parking lots, subway stations, toll roads)
- Electronic machines to record and play music, voice, and video
- Medical instrumentation and machines
- The telephone system: touch-tone phones, answering machines, central switching office, mobile systems
- Computer peripherals: parts of the display and printer, keyboard interface, scanner
- Aircraft control, radar signal processing

1

- Electronic instrumentation: digital oscilloscope, logic analyzer
- Video games

For scientific or technical purposes, quantitative processes in the natural world, such as air pressure, velocity, voltage, and position of a gear, are described by means of variables. Relationships among variables are described in terms of *laws* or *formulas:* Newton's laws, Ohm's law, Hooke's law, and others. Variables are often thought of as *signals,* some of which are taken as causes, or *inputs,* and others as effects, or *outputs,* produced by these causes through the action of appropriate laws. An unbalanced force (input), for example, applied to a physical object results in motion of the object, with an acceleration (output) whose value is specified by Newton's law.[1]

In normal experience, physical signals seem to change in a continuous manner. In a typical day, the ambient temperature rises continuously to some high for the day, and then drops gradually to the low at night. An electric current, although consisting of the flow of millions of discrete charged particles (usually electrons), seems to vary in a continuous manner. However, any continuous curve can be approximated by a function consisting of discrete steps, as shown in Figure 1. The approximation can be made as close as desired by proper choice of the intervals into which the independent variable is broken up. Although the approximation illustrated in Figure 1 is graphical, such approximations and the solutions of any mathematical relationships among variables can also be carried out by numerical methods.

A process in which signals vary in a continuous fashion is called an *analog* system; one in which the signals are discrete is said to be *digital.*[2] As described above, continuous signals can be discretized, or *digitized.* (The converse is also possible; digital signals can be converted to analog form.)[3] All variables processed in a digital system must be digital. We will henceforth assume that all of the signals treated in this book are digital; they are either inherently digital or have been digitized.[4]

An example in which analog signals are digitized is digital audio recordings: compact disks (CDs) and digital audiotapes (DATs). The analog music waveform is sampled every few microseconds, and each sample is represented by a digital number ranging from 0 to the digital equivalent of about 4100. This

---

[1]Qualities such as truth, beauty, and justice are obviously not included in this description.

[2]The term *analog* used for *continuous* originates from the use of a specific electrical instrument, or computing machine, whose output is a continuous function and whose internal structure simulates mechanical, aerodynamic, or other processes by *analogy* between such systems and electrical systems. The instrument is hence called an *analog computer.*

[3]Devices that perform these functions are called A/D (analog-to-digital) converters and D/A (digital-to-analog) converters, respectively.

[4]Some quantities, such as the number of students in a class, the face value of government bonds and money in coins or bills, are inherently discrete. Think of a vending machine from which you want to purchase a soft drink or candy bar. Depending on the prices of products carried, some machines accept only certain coins as input. Other machines (e.g., those vending stamps or fares on a rapid-transit system) accept bills only. The money "signal" in this case is inherently digital; of course, the "product" purchased is also "digital" since it comes from the machine in specific sizes. Liquid products come in arbitrarily chosen discrete-size containers. (As an aside, by specifying the size of beer containers permitted within a specific governmental jurisdiction, authorities can discriminate against vendors whose product comes in different sizes — for example, beer from microbreweries or in containers based on the metric system.)

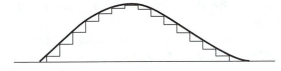

Figure 1 Continuous function approximated by a set of discrete values.

frequency and volume information is stored on the CD or DAT. The CD or DAT player then reconverts the digital information to analog form to actuate the loudspeaker for the listener's enjoyment.

2   HARDWARE, SOFTWARE, AND FIRMWARE

Two broad categories define the field of computer engineering: *hardware* and *software.* Hardware refers to the physical objects, components, circuits, and subassemblies that are physically and electronically coupled to form large and small digital systems. (As described in the preceding section, the adjective *digital* means that the signals processed by such systems are not *continuous* signals, like those that make up music or speech or the variation of power system voltage over time, but are *discrete,* taking on only a specific number of values.)

Software, in a broad sense, denotes the instructions that specify those tasks that the hardware is to carry out. Two general categories of software are usually distinguished: *operating systems* and *application programs*. A computer operating system carries detailed instructions as to how the computer is to operate in order to carry out its function. Examples are DOS (disk operating system), UNIX, and Windows. An application program, on the other hand, specifies the instructions that a computer is to carry out in order to implement a specific task. A simple case consists of the steps that a computer should carry out to find the roots of a polynomial. A more general example is a program for carrying out a wide range of mathematical operations (the program MATLAB, for example), a word-processing program, or a game played by one or more individuals through a computer. Software—either operating systems or application programs—consists of programs that are installed in the memory of a computer after it has been manufactured. (Operating systems are usually installed by the computer vendor; application programs are often installed by the user.)

There is also another category, designated *firmware*. It consists of software that is permanently installed in the computer during manufacture. It is said to be *hard-wired*. Firmware, thus, is a hybrid between software and hardware that is an integral part of the hardware. Its function is to control the operation of the rest of the hardware.

This book does not deal with software as such, or with firmware; it is concerned with hardware at a fundamental level. (However, we do spend some time with the hardware description language labeled ABEL, used in the design of hardware.) That is, it deals mainly with the components, circuits, and subassemblies that make up a functioning computer or other digital device. Later chapters discuss larger units; the control of such large units involves some aspects of programming.

A glance at the rest of this chapter and Chapter 2 reveals that the subject matter contains a substantial dose of mathematics. There are some dangers in starting a book on computer engineering with topics in mathematics. One is that many of you are interested in getting on with the design of hardware and might feel impatient at becoming sidetracked into what seems to be unrelated to this goal. Another danger is that some of you may have some previous exposure to what is discussed here and might think that you already know the subject and, thus, have no need to concern yourself with it here.

The Boolean algebra presented here, however, is essential for understanding the operation of digital systems and, hence, for carrying out their design. Furthermore, this material is useful for introducing the terminology and notation to be used in the rest of the book. Hence, even if you *think* you know the material, you are urged to study it. Read it rapidly if you are very familiar with the subject; if you are only partially familiar with it, read it in a more intense fashion, allowing sufficient time to work out some exercises and problems. You may need to return to this chapter for review when later use is made of signed number representation or codes.

Computers are sometimes affectionately called "number crunchers." The numbers that they crunch, however, are not *decimal* but *binary* numbers and numbers in other number systems related to the binary. If you want to understand the operation of computers, you must be familiar with number representation. If you already have knowledge of the subject, test yourself by solving the relevant problems at the end of this chapter. If you don't have such a background, you will be well rewarded by mastering the contents of the following section. The material in sections 3 and 4, on codes and code conversions, is not needed until Chapter 4. If you are unfamiliar with the subject, you should study those sections and solve the relevant problems at the end of this chapter before you undertake Chapter 4.

The present-day building blocks of computers and other digital systems are electronic circuits. However, this book does not require a knowledge of electronics. At times, we will make reference to certain aspects of transistors and integrated circuits for ease of explaining the subject under discussion, but such reference is meant to be descriptive and does not require foreknowledge. Some elements of transistors and electronics are covered in a brief appendix.

3   NUMBER SYSTEMS

As noted in section 1, digital systems include many other devices besides the preeminent ones: computers. The operation of such devices might not require the manipulation of numbers. Computers, however, operate by processing vast quantities of numbers and other data. It is therefore important to become familiar with those number systems by which numbers can be expressed and with the manner in which arithmetic operations are carried out by computers using these systems. The number system most familiar to contemporary humans is the decimal number system. It utilizes a familiar *alphabet* of 10 digits denoted by 0,

1, 2, 3, 4, 5, 6, 7, 8, 9. The decimal representation of a number, such as the integer 3756, is a shorthand notation for writing a polynomial in base 10:

$$3756 = 3 \times 10^3 + 7 \times 10^2 + 5 \times 10^1 + 6 \times 10^0$$

The number 10 is the *base* of the decimal system. In the usual way of expressing the integer 3756, the powers of 10 are implied by the positions occupied by the digits, starting with the 0 power on the right and increasing to the left. Thus, decimal is a *positional* number system.

Any positional number system can be created by specifying two things: its base b and the b digits that make up its alphabet. When b is 10 or less, it is customary to use the digits 0 to $b - 1$ for the alphabet. When b is larger than 10, new symbols are needed to represent the extra digits. We indicate that a number N is expressed in base b by writing it as $(N)_b$. If the digits in the alphabet are designated a_i, then, by analogy with the decimal system, N can be written as a polynomial in b as follows:

$$(N)_b = a_{n-1}b^{n-1} + a_{n-2}b^{n-2} + \cdots + a_1b^1 + a_0b^0 + a_{-1}b^{-1} + \cdots + a_{-m}b^{-m} \qquad (1)$$

The given number has an *integral* part $a_{n-1}a_{n-2}...a_1a_0$ and a *fractional* part $a_{-1}a_{-2}...a_{-m}$. The digit a_{n-1} multiplying the highest power of b is the *most-significant digit* (msb); the digit a_{-m} multiplying the lowest power of b is the *least-significant digit* (lsb).[5]

Binary and Other Number Systems

The positional number system of greatest importance in digital computers is the *binary* number system. Its base is 2, and its alphabet consists of two digits: 0 and 1. Like a decimal number, a binary number is written as a juxtaposed sequence of digits, binary in this case—for example, 1101001. In accordance with (1), this is shorthand for a polynomial in 2.

Each digit in the binary alphabet conveys a minimum of information, such as whether a switch is *on* or *off*. This unit of information has been called a *bit*. For this reason, it is customary to call a binary digit a bit. ("Bit" can also be viewed as an abbreviation of *b*inary dig*it*.) Thus, msb and lsb are abbreviations for "most (least) significant bit."

Other number systems that find useful applications are

Base 8: *octal* system
Base 16: *hexadecimal* system

Since the latter system needs an alphabet with more than 10 digits, it is customary to use the 10 decimal digits plus the first six (capital) letters of the English alphabet. The representation of decimal integers 0 to 15 in various number systems is shown in Figure 2. Study this table and familiarize yourself with the patterns of digits in the last three columns.

There is a certain lack of consistency in this table. In all but the decimal column each number has the same number of digits. To be consistent, we should append the digit 0 in front of the first 10 decimal numbers, that is, in the tens position. But it isn't customary to do this for decimal numbers, so we didn't. If we had done so, and done so also for the hexadecimal numbers, then each of the

---

[5]Why *msb* and *lsb,* and not *msd* and *lsd,* will be explained shortly.

Number System			
Decimal (Base 10)	Binary (Base 2)	Octal (Base 8)	Hexadecimal (Base 16)
0	0000	00	0
1	0001	01	1
2	0010	02	2
3	0011	03	3
4	0100	04	4
5	0101	05	5
6	0110	06	6
7	0111	07	7
8	1000	10	8
9	1001	11	9
10	1010	12	A
11	1011	13	B
12	1100	14	C
13	1101	15	D
14	1110	16	E
15	1111	17	F

Figure 2 Representations of integer numbers.

first eight numbers in the table would have been represented by the same string of digits in all but the binary system. A second observation is that the same string of digits can represent different numbers depending on the base of the number system. Thus, the string 12 represents decimal twelve, octal ten, and hexadecimal eighteen. (You'll have to extend the table to check the last one.)

You may wonder why all this concentration on different number systems is needed; why not just use the decimal system in computing, or why bring in other number systems if the binary system is what computers now utilize? The earliest physical devices used to carry out modern computing were switches and relays.[6] A switch inherently has two positions: on and off. These two positions can easily be used to represent the digits in a binary system. Later electronic devices also have two easily distinguished stable positions. If the decimal system were used for carrying out computer arithmetic, physical devices with 10 stable positions to represent the 10 digits would be needed. Although it is conceivable to design electronic components with 10 distinct positions, they would likely be quite expensive. Furthermore, systems utilizing such components would be unreliable, since ensuring the stability of all 10 positions would be difficult.

OK, the case *for* binary and *against* decimal has been made; but, you might say, what about the octal and hexadecimal systems? A glance at Figure 2 shows that even for these rather low integers, the number of binary digits needed to represent a number is quite large. (A 6-digit decimal number would require 18 binary digits.) A number represented in the octal system needs about three times fewer digits than the same number represented in binary, and a hexadec-

---

[6]Claude E. Shannon, "A Symbolic Analysis of Relay and Switching Circuits," *Trans AIEE,* 57 (1938), 713–723. This is a highly readable paper; no one who has not read it should claim to be computer literate.

imal system has approximately a 4-to-1 advantage in this respect. The next section describes how easy it is to convert from binary to one of these systems, and vice versa. So, even though physical devices with 8 or 16 stable positions are not expected to be utilized in computers, the convenience of fewer digits in the number representation and easy conversion to and from binary make the octal and hexadecimal systems appropriate subjects for brief study.

Base Conversions

A problem that occurs often is to convert a number expressed in one particular base to a different base. Because of our familiarity with decimal arithmetic, two cases are distinguished:

- Converting from some other number system to decimal
- Converting from the decimal system to one with another base

Converting to the Decimal System

Suppose, for example, that a binary number is to be converted to a decimal number. This is easily done by forming, in accordance with (1), the polynomial in 2 corresponding to the binary number and then expressing each power of 2 as a decimal number. To illustrate,

$$(101011.11)_2 = 2^5 + 2^3 + 2^1 + 2^0 + 2^{-1} + 2^{-2}$$
$$= 32 + 8 + 2 + 1 + 0.5 + 0.25 = 43.75$$

Notice that decimal arithmetic is used in this conversion.

Exercise 1   In a decimal number, the point separating the integral part and the fractional part is called the *decimal point*. What would be a reasonable name for the point separating the two parts of a binary number, as in the preceding illustration? How about the corresponding point in the octal and hexadecimal systems? ◆

Exercise 2   The following is an octal number: 2034. Using the previous illustration of converting from binary to decimal as a model, describe how to convert an octal number to a decimal number. Then convert octal number 2034 to decimal. (Do it first, then look at the answer.)
Answer[7]

Converting from the Decimal System

The reverse process, converting from decimal to another number system, is somewhat more complicated. Let's use decimal to binary as a model for any such conversion. Decimal arithmetic is again used. The integral and fractional parts are handled separately.

---

[7]The base is 8. Each digit of an octal number is multiplied by 8 raised to a power based on the position of the digit:

$$(2034)_8 = 2 \times 8^3 + 0 \times 8^2 + 3 \times 8^1 + 4 \times 8^0 = 1024 + 24 + 4 = 1052 \qquad ◆$$

The decimal integer N is to be converted to a binary number. (The designation of the base 10 has been omitted for convenience.) Note in (1) with $b = 2$ that the last digit in the integral part is 1 if the number is odd and 0 if it is even. Leaving the last digit aside, a 2 can be factored from the rest. The number remaining after the 2 is factored can be treated in exactly the same way; its last digit will be 1 only if that number is odd, and a 2 can be factored from the remaining part. Here's what it will look like after these two steps:

$$N = a_{n-1}2^{n-1} + a_{n-2}2^{n-2} + \cdots + 2a_1 + a_0$$
$$= 2(a_{n-1}2^{n-2} + a_{n-2}2^{n-3} + \cdots + 2a_2 + a_1) + a_0$$
$$= 2[2(a_{n-1}2^{n-3} + a_{n-2}2^{n-4} + \cdots + 2a_3 + a_2) + a_1] + a_0 \tag{2}$$

This process, which gives a blueprint for how to proceed, can be continued until only one digit remains within the innermost parentheses.

Thus, we start with a decimal number N and work in the opposite sense from the preceding development. Dividing the given decimal number by 2 will result in a quotient Q_0 and a remainder R_0. Hence, the number can be written

$$N = 2Q_0 + R_0$$

The remainder, R_0, is 1 if N is odd, and 0 if N is even. Now the quotient Q_0 can be treated as the original number N was. Dividing Q_0 by 2 results in a new quotient Q_1 and a new remainder R_1. The latter is 1 or 0 depending on whether the first quotient, Q_0, is odd or even. Again, $Q_0 = 2Q_1 + R_1$. The original number can now be written

$$N = 2(2Q_1 + R_1) + R_0$$

It is not hard to see the pattern as each quotient is divided by 2. Note that at some point in the process of dividing by 2, the quotient will become 3 or 2. One more division by 2 will yield a quotient of 1; no further division by 2 is possible, and the process terminates. The result of the step after the one just preceding and the final result after k divisions by 2 are

$$N = 2[2(2Q_2 + R_2) + R_1] + R_0 = 2^3Q_2 + 2^2R_2 + 2R_1 + R_0$$
$$N = 2^k + R_{k-1}2^{k-1} + R_{k-2}2^{k-2} + \cdots + R_12^1 + R_0 \tag{3}$$

Thus, the binary representation of a nonzero decimal integer is $1R_{k-1}R_{k-2}\ldots R_1R_0$.

The process is illustrated in the table below for the conversion of decimal 234 to binary form.

	Q_0	Q_1	Q_2	Q_3	Q_4	Q_5	Q_6	
Q_i	117	58	29	14	7	3	1	—
R_i	0	1	0	1	0	1	1	1
	R_0	R_1	R_2	R_3	R_4	R_5	R_6	R_7

Conversion of the fractional part of a number from the decimal to the binary system follows a similar procedure, except that a given decimal fraction F_0 is first *multiplied* by 2 rather than divided by 2. The resulting number will have an integral part B_1 (1 if the original fraction is ≥ 0.5 and 0 otherwise) and a remainder fractional part, F_1. These steps are repeated with F_1, and the process is continued, resulting in the sequence on the left below.

$F_0 = 0.40625$	$2F_{i-1}$	B_i	F_i
$2F_0 = B_1 + F_1$	0.8125	0	0.8125
$2F_1 = B_2 + F_2$	1.625	1	0.625
$2F_2 = B_3 + F_3$	1.25	1	0.25
$\vdots$	0.5	0	0.5
$2F_i = B_{i+1} + F_{i+1}$	1.0	1	0.0

(In the leftmost column, note that there is no B_0.) The process terminates if some fractional remainder F_{i+1} becomes 0. The binary fraction will be $0.B_1B_2B_3...B_{i+1}$. The work is conveniently arranged in a table, on the right above for the decimal fraction 0.40625 converted to binary form.

It is possible that the process will not terminate because the fractional remainder never becomes 0. In such a case, the binary form of the fraction does not have a finite number of digits.

From Octal or Hexadecimal to Binary

Conversion of numbers to and from other bases follows one of the patterns just described. However, a simpler procedure is possible in the special cases of conversion from octal to binary and from hexadecimal to binary. The reason for this is that the bases of the octal and hexadecimal systems can be expressed as 2 raised to some power: 2^3 and 2^4, respectively.

Thus, to convert a binary number to octal form, we note that each octal digit can be represented by three binary digits. Hence, a given binary number is partitioned into sets of three digits starting at the binary point and working to the left for the integral part of the number and to the right for the fractional part. Each set of three binary digits is then replaced by the corresponding octal digit. For example,

$$(10 \ \ 110 \ \ 111 \ \ 011.101 \ \ 1)_2 = (2673.54)_8$$
$$2 \quad 6 \quad 7 \quad 3 \ . \ 5 \ \ 4$$

(The spaces between the three-digit binary sets are not normally there; they have been provided for ease of visualization.) In the reverse operation, converting a given octal number to binary form, each octal digit is replaced by the corresponding set of three binary digits. Thus,

$$(356.07)_8 = (011 \ \ 101 \ \ 110.000 \ \ 111)_2$$

Exercise 3 By analogy with the preceding conversions from binary to octal form and vice versa, describe how to carry out the conversion of a binary number to hexadecimal form. Repeat for the conversion of a hexadecimal number to binary form. Using your scheme, carry out the following conversions:

$$(11001011101.011)_2 = (x)_{16}$$
$$(3C9.E)_{16} = (y)_2$$

Answer[8]

---

[8]$x = 65D.6,$ $\qquad$ $y = 001111001001.1110$ $\qquad\qquad\qquad\qquad$ ◆

Binary Arithmetic

A digital computer spends most of its working life carrying out the operations of binary arithmetic. To understand the operation of computers at the basic level, then, requires an understanding of binary arithmetic. However, digital computers have certain limitations. To understand the operation of the computer also means understanding these limitations and the procedures that have been introduced to get around them. This section will help you do that.

The rules of binary arithmetic are similar to those of decimal arithmetic. The concepts of *carry* and *borrow,* for example, apply to binary arithmetic as well.

Addition

In decimal addition, when the sum of two digits equals or exceeds the base, 10, a carry of 1 is produced. In the case of binary numbers, the sum of two binary digits can, at most, *equal* the base. The only nonzero carry that *can* be produced is 1. The second column in Figure 3 shows the result of adding any combination of two binary digits. The only nonzero carry is created when the digits 1 and 1 are added.

As in decimal addition, two multidigit binary numbers are added by starting with the least significant bits. Any carry resulting is added to the sum of the next significant bits to the left.

An illustration of the addition of two binary numbers follows.[9] The top line represents the carries generated.

	Binary Sum	**Decimal Equivalent**
Carry:	10111.1	
	1011.11	11.75
	+ 1001.01	+ 9.25
	10101.00	21.00

Note that, because a carry is generated in adding the most significant bits, the binary sum has 5 bits in its integral part although the integral part of each operand has only 4 bits. This is called the carry *out* from the msb.

Digits	**Addition**		**Subtraction**		**Multiplication**
$a_1\ a_2$	$a_1 + a_2$		$a_1 - a_2$		$a_1 \cdot a_2$
	Sum	**Carry**	**Difference**	**Borrow**	**Product**
0 0	0	0	0	0	0
0 1	1	0	1	1	0
1 0	1	0	1	0	0
1 1	0	1	0	0	1

Figure 3 Binary arithmetic operations.

---

[9]In all illustrations that follow, you should independently carry out the operations indicated, and thus confirm the given results.

Subtraction

In decimal subtraction, whenever a digit to be subtracted is larger than the corresponding digit in the minuend, a 1 is *borrowed* from the first nonzero digit to its left. The concept of borrowing is applied in binary subtraction as well. Thus, whenever a binary 1 is to be subtracted from a 0 in some position, a digit of 1 must be borrowed from the first nonzero digit to the left of this position. This will leave a 0 in that position and will change all the intervening 0 digits to 1 up to the position in question. The 0 at that position becomes 10 (decimal 2); subtracting the 1 leaves a digit of 1. The process is illustrated by the following example. The top two rows correspond to the digits borrowed into a given position and the changes made in the digits to the right of the position from which a 1 is borrowed.

Binary Subtraction	Details	Decimal Equivalent
	11 .1 : *Borrow*	
	01 0 : *Changed digits*	
10011.01	10011.01	19.25
−01100.11	−01100.11	−12.75
	00110.10	6.50

(You can verify the answer by adding it to the subtrahend; the result should be the minuend. Is it?) The case in which one number is to be subtracted from a smaller number is discussed in the next section.

Multiplication

Multiplication of two binary numbers is carried out by multiplying the multiplicand by each digit of the multiplier and adding these partial products, with the position of each successive partial product shifted one unit to the left. When a digit of the multiplier is 0, all digits of the corresponding partial product are 0; when a digit of the multiplier is 1, the corresponding partial product is the same as the multiplicand. An illustration follows.

Binary Multiplication	Decimal Equivalent
10011	19
× 101	× 5
10011	95
00000	
10011	
1011111	

Division

The steps in carrying out the division of two binary numbers are quite similar to those for decimal division, but with some characteristic differences. At each step, the n-bit divisor goes into the highest n bits of the dividend, that start with a 1, either one time or zero times. In the latter case, the corresponding bit of the partial quotient is 0. Then the next most significant bit of the dividend is appended, and the process continued until a 1 appears in the partial quotient. Then the divisor is subtracted from the dividend at that point and the process continued. An example follows.

$$
\begin{array}{l}
\textit{Dividend} \;\rightarrow\; 100101011 \quad \underline{/\,10111} \leftarrow \textit{Divisor} \\
\qquad\qquad\quad \underline{10111} \qquad\quad 01101 \leftarrow \textit{Quotient} \\
\qquad\qquad\quad 011100 \\
\qquad\qquad\quad \underline{10111} \\
\qquad\qquad\quad\; 010111 \\
\qquad\qquad\quad\; \underline{010111} \\
\qquad\qquad\quad\;\; 00000 \qquad\qquad \leftarrow \textit{Remainder}
\end{array}
$$

Complements: Two's and One's

Repeated additions and multiplications of binary numbers will lead to results in which the number of digits can become indefinitely large. However, the "chips" that form the heart of the computational unit in a digital computer can handle only numbers with a fixed number of digits. This number is referred to as the *word length*. Earlier computers had word lengths of 8. In recent times, chips with 64-bit word lengths are common.

If the length of the number resulting from an arithmetic operation on binary numbers exceeds the word length n of the computer, what is retained is a number consisting of the least significant n bits; the more significant bits will constitute an *overflow*. There is a mathematical mechanism for dealing with this overflow, but we will not take up this topic here.

Numbers on which a digital computer operates can be positive or negative. Negative numbers can arise when a larger number is subtracted from a smaller number. It would be extremely valuable if there were a mechanism that permitted the computer to follow a single procedure whether the operation to be performed is addition or subtraction, and whether the numbers are positive or negative. There must be a unique way of determining from the result of the operation whether the resulting number is positive or negative. That mechanism is the subject of this section.

When a digit, other than 0, is subtracted from the base of the number system in which it is expressed, the result is the *complement with respect to the base*. In the decimal system, this is the complement with respect to 10, or the *ten's complement* (or 10's complement). The ten's complement of 3 is 7, for example. In general, the ten's complement of a decimal integer is obtained by subtracting it from 10 raised to a power equal to the number of digits in the integer. Thus, the ten's complement of 461 is $10^3 - 461 = 539$.

The same concept can be applied to binary numbers. Let N be a binary integer having n digits. The *two's complement* (or 2's complement) of N, labeled N_{2c}, is obtained by subtracting the number from 2^n expressed in binary form, that is, subtracting the binary number from 1 followed by n 0's:

$$
N_{2c} = 2^n - N, \quad \text{all expressed in binary} \tag{4}
$$

To illustrate the two's complement, let $N = 110010$; $n = 6$. Then $N_{2c} = 1000000 - 110010 = 001110$. (Verify this.)

A rather simple way of obtaining the two's complement of a number N can be generalized from the illustration:

- The least significant 1 of the number is retained.
- Any less significant 0's than that 1 are also retained.
- All other digits are replaced by their complements: 0 by 1 and 1 by 0.

The different ways of handling the least significant 1 and the other digits is the result of "borrowing" when carrying out the subtraction. Only the msb in $(2^n)_2$ is a 1; when this is borrowed, all the subsequent 0's up to the position of the least significant 1 of the number become 1. Now, subtracting a 1 from a 1 leaves 0; subtracting a 0 from a 1 leaves 1. Thus, subtracting each digit in N from 1 means replacing each digit by its complement, except for the least significant 1. In this case 1 is subtracted from (binary) 10, leaving a 1.

Even this difference between the least significant 1 and the others would disappear if we subtracted 1 from the result. That is, given a binary integer N with n digits, form the two's complement: $N_{2c} = (2^n)_2 - N$. Now subtract 1:

$$N_{1c} = N_{2c} - 1 = (2^n)_2 - 1 - N \tag{5}$$

This is known as the *one's complement* (or 1's complement), N_{1c}. The one's complement of 1011010 is

$$10000000 - 1 - 1011010 = 1111111 - 1011010 = 0100101$$

Thus, the one's complement of a binary integer is just that binary integer with each 1 bit replaced by a 0 and each 0 bit replaced by a 1. How much simpler can it get![10]

Given a binary integer N, the two's complement of N can be obtained from that integer in two ways: either directly, as previously described, or by first finding the one's complement (trivially interchanging 0's and 1's, as just described) and then adding a 1 to the least significant bit.

Exercise 4   Find the two's complement of the following binary numbers directly, and then confirm each answer by first finding the one's complement: (*a*) 11111, (*b*) 10000, (*c*) 0001, (*d*) 00100.
Answer[11]
(This exercise merely requires carrying out a specific algorithm. However, if all you do is carry out the work, mark the result *answer,* and go on, you will miss important opportunities to observe and learn. Observe in (*b*), for example, a number whose two's complement is identical to it. Can you think of other cases that have that property? Can you detect any other cases in the exercise where a number and its two's complement have some interesting property?)

Like the binary system, the two's complement is another system for representing numbers. Verify that it, too, is a positional number system.

Although it is not evident from the preceding development, the two's complement provides a way of handling negative numbers. The following mixed representation of numbers will be adopted.

---

[10]The same concept in the decimal system is the *nine's complement;* try it out on a decimal number.

[11](*a*) 00001,　　　(*b*) 10000,　　　(*c*) 1111,　　　(*d*) 11100　　　　　　　　◆

A nonnegative number (positive number or zero) will be represented by its binary form; a negative number will be represented by the two's complement of its magnitude.

This mixed form is known as the *two's complement number representation*.

Given a string of binary digits, how does one tell if it represents a positive or a negative number? A clue comes from the fact that this is an either-or, or binary, decision. If we let "positive" be represented by one of the two binary digits and "negative" by the other one, we could let a particular position in a binary string represent the sign of the number. The only position that could possibly be used for this purpose is the msb position, since that is the only position whose digit can be controlled, as will now be discussed.

We can ensure that the msb will always be the same for a positive number if the word length is long enough. In that case, the msb for a positive number must be 0. Suppose the word length is 6. The largest positive number that can be represented in the contemplated system is 011111, or decimal 31. If larger numbers are expected, then, for a positive number, a greater word length is needed to make the msb be 0.

Exercise 5   Determine the largest (positive) decimal integer that can be represented in the two's complement number representation for the currently common word length of 64 bits.   ◆

Exercise 6   Let A be an n-bit positive binary integer. Assume that the word length is at least 2^{n+1}. Show that the msb of the two's complement number representation will be 0 for A and 1 for $-A$.   ◆

Exercise 7   The number of distinct n-bit negative and nonnegative integers that can be represented by a two's complement number representation is called its *range*. Determine the range in terms of n and specify how many integers are negative.   ◆

Addition Of Binary Numbers

The grounds are now prepared to take up the addition or subtraction of two n-bit integers $\pm A$ and $\pm B$, where both A and B are nonnegative. Remember that two's complement number representation is to be used; that is, nonnegative numbers will be represented in binary and negative numbers by the two's complement of the corresponding magnitude. We assume that the word length of the machine that will carry out these operations is long enough so that the msb of both A and B are 0.

We are planning to take the sum or difference of two integers, each of which can be either positive or negative. The most general case can be written $(\pm A) \pm (\pm B)$. There will be a total of eight combinations of the + and − signs. However, since subtraction is equivalent to the addition of the negative of the subtrahend, these cases are not all distinct; thus $A - (-B)$ is the same as $A + B$. So there are really only three distinct cases, and they involve only addition. (Confirm this conclusion by examining all the possibilities.)

The three distinct cases that must be considered are

Case 1: A + B   The sum of two nonnegative integers
Case 2: -A + (-B)   The sum of two negative integers
Case 3: A + (-B)   The sum of a positive and a negative integer

These three cases are discussed in turn.

Case 1 A + B The Sum of Two Positive n-Bit Integers

Even though both A and B may lie within the range, their sum may exceed the range. In this case there will be an overflow. An overflow can be detected by observing the carry at the msb. Since neither of the integers to be added is negative, both most-significant bits are 0. Hence, the carry out *from* the msb must be 0. However, if the sum is not within range, then the carry *into* the msb position (which becomes the msb) will be 1. Thus, the overflow is detected as follows.

If the carry into the msb is the same as the carry out (both 0), then there is no overflow; if the carry out (0) is different from the carry in (1), then there *is* an overflow. In this case, we simply note the fact and disregard it. Two illustrations follow for $n = 6$; the range in this case is $2^{6-1} - 1 = 31$.

	$\underline{0}0111$	*Carry*		$\underline{0}1111$
14	001110		14	001110
+ 11	+ 001011		+ 22	+ 010110
25	011001		36	100100

In the case of $14 + 11 = 25$, the sum is in range; hence, there should be no overflow. This is confirmed by the fact that the (underlined) carry into and the carry out from the msb position are the same: both 0. The sign bit of the sum is 0, indicating a positive number, as it should. In the second case, the sum is not in range, and the presence of an overflow is signaled by the carries at the position of the msb: they are different. The sign bit erroneously shows the sum to be negative. But since the overflow has been detected, this is ignored.

Case 2 -A + (-B) The Sum of Two Negative Numbers

Since both numbers are negative, they will both be represented in two's complement. Thus,

$$(2^n - A) + (2^n - B) = 2^n + [2^n - (A + B)]$$

Since both $-A$ and $-B$ are assumed to be in range, each of the two's complement representations will have a sign bit of 1. Hence, when they are added, the carry out from the msb must be 1. Since the sum of two negative numbers must be negative, the msb of the sum must also be 1. This will happen only if the carry into the msb position is also 1—equal to the carry out from the msb. Again, if the carry into (0) and the carry out from (1) the msb are not the same, an overflow will occur. Note that the first 2^n on the right in the preceding expression represents the carry out from the msb. Two illustrations for $n = 6$ follow.

	Decimal	**Binary**	**Two's Complement**	
			1111	*Carry*
	(-18)	(-010010)	101110	
	$+ (\underline{-11})$	$+ (\underline{-001011})$	$+\underline{110101}$	
Sum	(-29)	(-011101)	100011	

Decimal	Binary	Two's Complement	
		1011	*Carry*
(−18)	(−010010)	101110	
+ (−27)	+ (−011011)	+100101	
(−45)	(−101101)	010011	

In the first case, the sum (−29) is within the range −32 to 31. When the two's complements of the two numbers are added, the carry into and the carry out from the msb are both seen to be 1, implying no overflow. The result of the addition is the two's complement of −29, the correct answer.

In the second case, the sum (−45) is out of range, so we expect an overflow. The overflow is detected by the fact that the carry into the msb position is different from the carry out. Note that the 6-bit result of taking the sum erroneously has 0 as its msb, which normally designates a positive number. But because an overflow has been detected, we correctly interpret this number as negative.

Case 3 A + (−B) *The Sum of a Positive and a Negative Integer*

$$A + 2^n - B = 2^n + (A - B) \quad \text{for } A \geq B \qquad (a)$$
$$= 2^n - (B - A) \quad \text{for } A < B \qquad (b)$$

Exercise 8   Carry out the analysis of this case. From the results of two's complement number representation, determine how it will be possible to tell whether the difference will be in range, whether there will be an overflow, and whether the carry into and out from the msb position are the same for the two cases. Illustrate for $n = 6$ and numerical values $25 - 17$ and $25 - 30$. (The answer is too long to put in a footnote; it follows. Don't look before carrying out the work.)

Answer   If both A and B are in range, then the difference will be in range for both cases. For $A \geq B$, the 2^n in (a) shows that there will be a carry out of 1 from the msb position, which is ignored. For $A < B$, the difference will be negative and in range, so there will be no carry out from the msb position. Since one of the sign bits is 1 and the other 0, and since the carry out from the msb is 0, the carry into the msb must also be 0. This will result in the msb in the difference being 1, meaning the difference is negative. Examples:

	111111	*Carry*			
25	011001		25	011001	
+(−17)	+ 101111		+ (−30)	100010	
8	001000		−5	111011	

In both cases, the carry into the msb position is the same as the carry out (1 and 0, respectively). Hence the result is in range and there is no overflow, as expected. The sign bits show a positive difference in the first case and a negative one in the second case. The last result is the two's complement of −5.   ◆

The reason for considering only addition and treating subtraction as a special case is that computers actually do that. Although it is possible to construct a subtractor in hardware, it is customary for computers to use only adders to perform subtraction by adding the negative of the subtrahend.

4 CODES AND CODE CONVERSION

Information that humans convey to each other is expressed by means of sets of symbols. Each set of symbols constitutes an *alphabet*. The letters A, B, ..., Z constitute the alphabet in which the English language is expressed, the decimal digits constitute the alphabet in which numeric information is expressed, and so on. To form an English word, specific letters are placed end-to-end (*concatenated* or *juxtaposed* in a *queue*).

The English language can be looked upon as a *code*. To each message to be conveyed (e.g., "computer," "book," "peace") a sequence of letters (the digits in the alphabet) are assigned; the result is a *code word*. The code is the collection of all these words, namely, the English language.

Thinking more generally, an arbitrary set of symbols constitutes an alphabet of which these symbols are the digits. Sequences of digits are called words. The assignment of a code word to each message in a set of messages constitutes the code.

The binary number system is an alphabet with only two digits. It can be used to code any desired information. The number of bits to be used in each code word depends on the total number of distinct messages that are to be conveyed in the desired information. If the number of messages is just two, then a code word of just one bit is enough.

Suppose an object can have one of only four colors: red, yellow, green, and blue. A binary code in which each word has two bits can be used. The four possible combinations of two bits are: 00, 01, 10, and 11. Each word can be assigned to one of the colors, but there are no a priori criteria for matching the colors to the words. Thus, red can be assigned to any one of the words; if a particular choice is made for red, then blue can be assigned to any one of the remaining three words; and so on.

Exercise 9 For the color code example just described, determine how many different codes exist.
Answer[12]

Now consider the reverse perspective. Suppose each code word has n binary bits. How many distinct messages can be formed? This number will be the number of distinct sequences that can be formed with n bits, namely, 2^n. For $n = 3$, for example, the $2^3 (= 8)$ 3-bit words that can be formed are 000, 001, 010, 011, 100, 101, 110, 111.

In the subsections that follow, we will study a number of different codes that find application in digital systems in general, and digital computers in particular. In later chapters of this book certain digital circuits that play a role in the coding and decoding process will be investigated.

---

[12] $4! = 4 \times 3 \times 2 \times 1 = 24$

♦

Binary-Coded Decimal

Although almost all digital systems operate with binary signals, some systems carry out arithmetic in the decimal system. Hence, there is a need to express decimal numbers in some binary code. One possible code might be to express each decimal number by its binary equivalent; for example, decimal 213 would be coded as 11010101. This is a very awkward and inefficient system, requiring as many code words as the number of decimal numbers that may be encountered.

A better coding scheme starts with the recognition that any decimal number is made up of a sequence of the 10 decimal digits. If a code word is assigned to each decimal digit, then any decimal number can be coded as a sequence of these words. Coding the 10 decimal digits requires a minimum of 4 bits. (Confirm that 3 bits is not enough.) Just as in the coding of colors previously mentioned, there is a large number of different ways of assigning 4 bits to the decimal digits. One particular choice is to code each decimal digit by its 4-bit binary equivalent. This particular code is called the *binary-coded decimal*, or BCD, system. Since the number of messages that can be expressed with a 4-bit code is $2^4 = 16$, and there are only 10 decimal digits, six of the possible code words are unused in the BCD code.

Figure 4 shows several different codes for the decimal digits that find some application; the BCD code is shown in the second column.

Weighted Codes

In the BCD code just discussed, the position of each binary digit corresponds to a power of 2. We say that the digits are *weighted* by these powers of 2; the weights are 8421, in that order. The sum of the weights of all the 1 bits in a word equals the corresponding decimal digit. (Convince yourself of this by examining the codes for a few decimal digits.)

Instead of the weights corresponding to powers of 2, it is possible to assign different weights to each of the positions of a 4-bit word. Each such assignment of weights will generate a different weighted code for the decimal digits; even negative weights can be assigned. A few possibilities besides the 8421 weighted code (BCD) are shown in the next three columns of Figure 4.

Decimal Digit	BCD 8421	2421	84-2-1	Excess-3	Gray	One-hot	Biquinary 5043210
0	0000	0000	00 0 0	0011	0010	1000000000	0100001
1	0001	0001	01 1 1	0100	0110	0100000000	0100010
2	0010	0010	01 1 0	0101	0111	0010000000	0100100
3	0011	0011	01 0 1	0110	0101	0001000000	0101000
4	0100	0100	01 0 0	0111	0100	0000100000	0110000
5	0101	1011	10 1 1	1000	1100	0000010000	1000001
6	0110	1100	10 1 0	1001	1101	0000001000	1000010
7	0111	1101	10 0 1	1010	1111	0000000100	1000100
8	1000	1110	10 0 0	1011	1110	0000000010	1001000
9	1001	1111	11 1 1	1100	1010	0000000001	1010000

Figure 4 Binary codes for the decimal digits.

Something curious becomes evident from examining the weights in the different weighted codes in Figure 4: the code word for a decimal digit is not necessarily unique. For example, in the 2421 code, 2 different bits have the same weight. The code word for decimal 2 is given as 0010 in the column corresponding to the 2421 code in Table 2. However, the word 1000 also represents decimal 2 in that code.

Exercise 10   Construct a weighted code for the decimal digits having the weights 642-3. Specify any decimal digits that can be represented by two different code words.   ◆

When there is a choice of code words to represent a decimal digit in a particular weighted code, how do we choose? What implication will the choice have? Indeed, what was the criterion for choosing used in Figure 4? Examine the two possibilities for decimal 2 in the 2421 weighted code: 0010 and 1000. Suppose the 0 and 1 bits are interchanged in the first of these; the result is 1101, which corresponds to decimal 7 in the 2421 code. But 7 is the nine's complement of 2. Now examine from this perspective all the other code words for the decimal digits in the 2421 code. You will find that interchanging the 0's and 1's for each decimal digit leads to the nine's complement of that digit. Such a code is called *self-complementing*. That is, generally, a self-complementing code is one in which the code word for the nine's complement of a decimal digit d (namely, $9 - d$) is obtained by replacing each binary bit by its complement.

Exercise 11   Determine whether the BCD code is a self-complementing code. Repeat for the 2421 code if the other possible choices of binary code words for decimals 2 and 7 are made. Repeat for the 642-3 code if the other possible choices of binary code words for decimals 3 and 6 are made.
Answer[13]

Gray Code

Many other codes are possible for the decimal digits, a few of which are listed in Figure 4. Some of these utilize more than four digits; they have redundancies and are consequently inefficient for computation. However, they might find use for other purposes, such as error correcting or controlling output devices in a digital system. Some are mainly of historical interest. For example, the excess-3 code, as its name implies, is obtained by adding 0011 to the BCD code words for the decimal digits. Because it exhibited some computational simplicity, excess-3 code was commonly used in earlier computers. It is not currently popular for computation.

A code with a very useful characteristic is one in which successive code words differ in exactly one bit position.[14] Such codes are said to be *cyclic*. A

---

[13]For BCD, no.   For 2421, yes.   For 642-3, yes.   ◆

[14]This is particularly useful in A-to-D (analog-to-digital) converters. A continuous (analog) function is first approximated by discrete steps. This discrete information is then coded. Ambiguity and possible errors are reduced if adjacent discrete values are given adjacent code words; that is, code words that differ in only one bit.

1 bit	2 bits	3 bits	4 bits
0	0 0	0 00	0 000
1	0 1	0 01	0 001
	1 1	0 11	0 011
	1 0	0 10	0 010
		1 10	0 110
		1 11	0 111
		1 01	0 101
		1 00	0 100
			1 100
			1 101
			1 111
			1 110
			1 010
			1 011
			1 001
			1 000

Figure 5 Gray code generation by reflection.

cyclic code having just 2 bits, for example, is 00, 01, 11, 10. Only one bit is changed in going from one word to the adjacent word, including returning from the last word to the first.

A specific example of a cyclic code is a *Gray code*. The method of generating the $(n + 1)$-bit code from the n-bit code is illustrated in Figure 5. The 3-bit code, for example, is obtained by *reflecting* the 2-bit code about an axis drawn below the latter; then a most significant bit of 0 is appended above the axis and an msb of 1 below the axis. Because of the way it is generated, such a code is also called a *reflected* code.

The full n-bit Gray code has 2^n words. If the decimal digits are to be coded in a 4-bit Gray code, which 10 of the 16 words should we use? One of the many possibilities is similar to the excess-3 code; it omits the first three and the last three words from the last column of Figure 5.

Seven-Segment Code

In many applications of digital systems (clocks, radios, watches, timers, VCRs), displaying the decimal digits is an important function. This can be done with the use of light-emitting diodes (LEDs) or liquid crystals (LCs). Seven of these light sources are arranged in the standard configuration shown in Figure 6; each one can be turned on or off independently. To display one of the decimal digits, a specific subset of the segments must be turned on. For example, to display a 1, segments S_4 and S_5 are illuminated; all others are turned off.

Exercise 12   Let digits 1 and 0 represent the states *on* and *off*, respectively. Construct a table in which the first column lists the decimal digits. The headings of the other columns (seven of them) will be the segment labels S_1 through S_7. The entries in each row of the table will be 0 and 1 bits: 1 if the corresponding segment is *on* and 0 if it is *off*. Thus, each row in the table is a 7-bit code word rep-

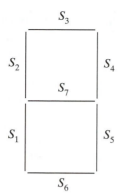

Figure 6 Seven-segment code display.

resenting a decimal digit. To illuminate the digit 1, for example, the corresponding row in the table will be 0 0 0 1 1 0 0. The entire table is the *seven-segment code.*

Alphanumeric Codes

All codes discussed so far are representations of the decimal digits. In many data-processing applications, however, it is necessary to handle lists of members of a class: utility customers, voters organized by election district, insurance policy holders, people on the payroll, and so on.

In such cases, representing the names, addresses, political party designations, salaries, and similar data requires a code for the letters of the alphabet, marks of punctuation, the decimal digits, and other symbols. A binary code that represents both literal and numeric symbols, as well as possible other special symbols, is called an *alphanumeric* code.

Exercise 13   (*a*) Specify the minimum number of bits needed in an alphanumeric code for coding both the English alphabet (capitals only) and the decimal digits. How many additional symbols (e.g., parentheses, brackets, marks of punctuation) could be accommodated by this code? (*b*) Repeat if both capitals and lowercase letters are to be used.   ◆

A 7-bit code can represent $2^7 = 128$ symbols. This is enough for the decimal digits, lower case and uppercase letters of the English alphabet, and 66 other symbols. One of the possible 7-bit codes has been adopted and called the American Standard Code for Information Interchange (ASCII). (The ASCII 8-bit code for the standard symbols is simply the ASCII 7-bit code preceded by an msb of 0. An msb of 1 can be used for any future expansion of the standard symbols. An 8-bit code was adopted because computers generally treat 8 bits together as an entity called a *byte.*) A subset of the ASCII code is shown in Figure 7, both in binary form and in the simpler hexadecimal form.

An example of the use of ASCII code in hexadecimal form follows.

44 41 56 45 20 53 4D 49 54 48 2C 20 32 37 20 4D 41 49 4E 20 53 54 2E

D A V E   S M I T H ,   2 7   M A I N   S T .

Character	8-Bit Binary	Hexadecimal	Character	8-Bit Binary	Hexadecimal
A	0100 0001	41	Blank	0010 0000	20
B	0100 0010	42	$	0010 0100	24
C	0100 0011	43	(	0010 1000	28
D	0100 0100	44	)	0010 1001	29
E	0100 0101	45	*	0010 1010	2A
F	0100 0110	46	+	0010 1011	2B
G	0100 0111	47	,	0010 1100	2C
H	0100 1000	48	-	0010 1101	2D
I	0100 1001	49	.	0010 1110	2E
J	0100 1010	4A	/	0010 1111	2F
K	0100 1011	4B	0	0011 0000	30
L	0100 1100	4C	1	0011 0001	31
M	0100 1101	4D	2	0011 0010	32
N	0100 1110	4E	3	0011 0011	33
O	0100 1111	4F	4	0011 0100	34
P	0101 0000	50	5	0011 0101	35
Q	0101 0001	51	6	0011 0110	36
R	0101 0010	52	7	0011 0111	37
S	0101 0011	53	8	0011 1000	38
T	0101 0100	54	9	0011 1001	39
U	0101 0101	55	=	0011 1101	3D
V	0101 0110	56			
W	0101 0111	57			
X	0101 1000	58			
Y	0101 1001	59			
Z	0101 1010	5A			

Figure 7 ASCII alphanumeric code.

5   ERROR DETECTION AND CORRECTION

The process of generating code words to represent information is called *coding* or *encoding*. To be of any use, this coded information must be *processed* in some ways: it must be transmitted through a transmission channel, stored, retrieved, and so on. In any such processes there is a possibility of changes occurring in a signal due to noise, interference from other signals, malfunctioning of circuit components, or some other cause.

For coded digital signals the only significant error is when one or more bits change value (0 to 1 or 1 to 0). The probability that a single such error will occur is always nonzero, although possibly small. The probability of two or more bits having erroneous values simultaneously is far smaller. To a first order, if the occurrence of all single errors can be detected, we can assume that almost all possible errors have been found.

Error-Detecting Codes

Suppose that some message is to be transmitted. The first step is to encode the message. Suppose that the BCD code is to be used for this purpose and that one of the words transmitted is 0001. Now suppose that an error occurs and one of the bits in this word changes value. The message received will then be one of the following four possibilities: 1001, 0101, 0011, or 0000. Each of these is a valid BCD code word. We were supposed to receive the decimal digit 1, but we erroneously received the digit 9, 5, 3, or 0. But what is even more serious, we are not even aware that there has been an error. When an error occurs, it would be tremendously useful if some action took place as a consequence—perhaps setting off an alarm or, more useful, causing the message to be repeated automatically.

Now suppose that instead of BCD, a different code is used, and it has the property that the occurrence of an error in any one bit converts a valid code word to an invalid one, that is, to gibberish. If an invalid code word is received, we will definitely know that an error has occurred. Such a code is called an error-detecting code (single error).

An arbitrary 4-bit code for the decimal digits is not an error-detecting code. However, such a code can be made into an error-detecting code by appending a fifth bit, called a *parity* bit. The parity of a code word refers to the number of 1-bits in the word; the parity (number of 1's) is either even or odd. Suppose that the parity of every word in a given code is even. A single error will cause the parity of the word to become odd; either a 0 is changed to a 1, thus increasing the number of 1-bits, or a 1 is changed to a 0, thus decreasing the number of 1-bits. Observe, parenthetically, that *any odd number* of errors will give the same result. (A similar result would apply if the original parity of every word in a given code were to be odd.)

If the parity of words in the data received in a given case is observed and if a word is found to have a parity different from what it is supposed to have, then an error must have occurred. In the BCD code the parity of words is not uniform; the parity is odd for some words and even for others. But a fifth (parity) bit can be appended to each word so as to make the parity of all the new 5-bit words the same, either even or odd. The case for odd parity, with the parity bit in the lsb position, is shown in the last column of Figure 8. Such a code is an *error-detecting code*. The middle column in the table gives another error-detecting code. As its name implies, the 2-out-of-5 code gives all possible ways of assigning two 1-bits in a 5-bit word; all words in this code have even parity.

Whether or not a specific code is an error-detecting code can be decided in terms of the number of bits that must be changed to transform a valid code word into an invalid one. Let us define the *distance* between two code words as the number of bits in one word that must be complemented in order to transform it into the second word. For example, the distance between 0011 and 0010 is 1; the distance between 0011 and 0100 is 3. In the BCD code the distance between any two words ranges from 1 to 4. The *minimum distance* of a code is the smallest number of bits by which any two words of the code vary.

Exercise 14   By examining the BCD code and the two codes in Figure 8, specify the minimum distance of each one.
Answer[15]

---

[15]BCD, 1;   odd-parity BCD, 2;   2-out-of-5 code, 2.   ◆

Decimal Digit	2-out-of-5 Code	Odd-Parity BCD
0	00011	00001
1	00101	00010
2	00110	00100
3	01001	00111
4	01010	01000
5	01100	01011
6	10001	01101
7	10010	01110
8	10100	10000
9	11000	10011

Figure 8 Error-detecting codes.

Exercise 15 Demonstrate that a code cannot be an error-detecting code if its minimum distance is less than 2.   ◆

Suppose that the minimum distance of a certain code is m. Suppose also that errors occur and that, in a particular word, up to $m - 1$ bits are complemented. The received word will be an invalid word and the error can be detected. Thus, the larger the minimum distance in a code, the larger the number of simultaneous errors that can be detected.

Error-Correcting Codes

The previous section showed that by proper design of a code having a minimum distance of m, we could ensure that the occurrence of up to $m - 1$ errors can be detected. That's no mean feat! Once it has been established that an error has occurred, what would be even better is for the error to be corrected and the correct word to be reconstructed. If a code has the property that the correct code word can always be deduced from the erroneous word, we say it is an *error-correcting code.*

An illustration of a simple error-correcting code is shown in Figure 9. Verify that the minimum distance in this code is 3. If two simultaneous errors occur, the resulting word is still a distance of 1 away from being a valid code word. That means that the erroneous code word cannot be interpreted as a valid code word, so a double error can be detected. Now suppose that a single error occurs in one of the code words; say that the correct transmitted word 00011 (message M_4) is received as any one of the following erroneous words: 10011, 01011, 00111, 00001, or 00010. Comparing these with any other words in the code shows that each received word is a distance of at least 2 away from any of them. (Do it!) Hence, if one of these erroneous words is received and we assume that only a single error has occurred, it could not have come from any word other than 00011 (M_4). The latter, correct, word is reconstructed by complementing the one bit we know to be in error. (The appropriate bit is detected by comparing the erroneous code word with the valid code word.)

Message	Code
M_1	11111
M_2	00100
M_3	11000
M_4	00011

Figure 9 An error-correcting code.

From the preceding discussion, it is clear that using a minimum-distance-3 code permits *correcting* any single error or *detecting* any double error. The same concept can obviously be applied to codes with larger minimum distances.

Hamming Codes[16]

The previous section merely introduced the concept of error-correcting codes. Now we need to specify such a code that is useful and practical. There are many possibilities; the class of codes to be described here is called *Hamming codes*.[17] To start the development, suppose that messages are coded in BCD or another 4-bit code. Several *parity-checking* bits are to be appended to the code words but dispersed among the *message* bits. The position of each bit in the resultant word is designated by counting to the right from the most significant bit, whose position is 1.

How many parity-checking bits should be used, and how should their positions in the resultant word be distributed? The first question can be answered by observing that there must be enough parity-checking bits to make the minimum distance at least 3 so that the code can be error correcting. In the preceding section it was observed that appending a single parity bit to a BCD code yields a minimum-distance-2 code. To convert a 4-bit code to an error-correcting code requires a minimum of 3 additional bits. If n-bit codes are used (with $n > 4$), then more than 3 parity-checking bits might be needed.

Exercise 16 By determining the number of possible six-digit code words needed for single errors to be unique, prove that at least 3 parity-checking bits are needed to convert a 4-bit code to an error-correcting code.
Answer[18]

Now we turn to the second question: how the parity-checking bits should be distributed in the word. A 7-bit code word consisting of 4 message bits and 3

---

[16]The material in this section is sometimes described as "advanced." It is certainly not beyond your capacity, but you may skip it if you wish.

[17]These are named after Richard W. Hamming (1915–1997), a 30-year employee of Bell Telephone Laboratories, who first proposed them in 1947. After retiring from Bell, he had another 21-year career at the Naval Postgraduate School. He died on the last day of 1997.

[18]With only 2 additional bits, each code word would have 6 bits. The 10 decimal digits would require 10 (correct) words. Single errors could occur in each bit of a word; so for each valid code word there would be 6 distinct invalid code words, making a required total of $10 + 60 = 70$ different six-digit code words. But for six digits there are only $2^6 = 64$ combinations. *Conclusion:* Using only 2 parity-checking bits is insufficient to make an error-correcting code from a 4-bit code of the decimal digits; at least three parity-checking bits are needed. ◆

parity-checking bits can be written as follows:

$$d_1 d_2 d_3 d_4 d_5 d_6 d_7$$

where d stands for digit. The idea is to perform three parity checks by selecting, in three different ways, 4 bits out of the 7. The results of these three checks will be a 3-bit word $c_1 c_2 c_3$. Each bit c_i is to be 1 if an error is detected by the corresponding check and 0 if not. The 3-bit word is called a *syndrome;* its decimal equivalent is to be the position of the digit in the 7-bit code word in which the error has occurred. Of course, the syndrome 000 means that no error has occurred.

$$\text{Syndrome: } c_1 c_2 c_3$$

The positions of the 3 parity-checking bits in the 7-bit word are to be chosen so that the corresponding syndrome has a single nonzero bit: 001, 010, or 100 (decimal 1, 2, or 4). Each of these bits will therefore be independent of the others. With the parity-checking bits in the 1, 2, and 4 positions in the preceding 7-bit word, the distribution of the message and parity-checking bits is as follows:

$$\text{Positions: } 1 \quad 2 \quad 3 \quad 4 \quad 5 \quad 6 \quad 7$$
$$\text{Bits: } p_1 \, p_2 \, m_1 \, p_3 \, m_2 \, m_3 \, m_4 \tag{6}$$

Suppose that c_3 is 1; this means that a parity check indicates an error. Then, no matter what the values of c_1 and c_2, the syndrome must be odd (1, 3, 5, or 7). The digits in positions 3, 5, and 7 are message digits. Suppose that p_1 is selected in each code word so that the digits in positions 1, 3, 5, and 7 (p_1, m_1, m_2, m_4) have *even* parity. If a single error occurs in any of these positions, the parity check will yield a syndrome in which c_3 is 1.

Similarly, suppose that c_2 is 1; with c_1 and c_3 taking on their four possible combinations, the possible values of the syndrome are 2, 3, 6, and 7. (Confirm these values.) Position 2 is occupied by parity-checking bit p_2 while the other three positions correspond to message digits. As in the preceding case, we select p_2 in each code word so that p_2, m_1, m_3 and m_4 together have even parity. If an error occurs in any one of these bits, then the parity check will yield $c_2 = 1$.

In the final case, $c_1 = 1$ can occur for syndromes 4, 5, 6, and 7. Hence, p_3 is chosen in each code word so that p_3, m_2, m_3, and m_4 together have even parity. This time an error in any one of these bits will result in a syndrome with $c_1 = 1$. The preceding discussion is summarized in the tables in Figure 10; the parity-checking bits are selected so that even parity exists among the sets of 4 bits shown in the right column of Figure 10a.

The Hamming code for BCD constructed in this manner is given in Figure 11. By comparing the code words, it is easy to see that its minimum distance is 3.

Parity-Check Bit	Even Parity in Positions:
p_1	1, 3, 5, 7
p_2	2, 3, 6, 7
p_3	4, 5, 6, 7

Message Digits	Parity	Value of p_i
m_1, m_2, m_4	Even	$p_1 = 0$
m_1, m_3, m_4	Odd	$p_2 = 1$
m_2, m_3, m_4	Even	$p_3 = 0$

(a)   (b)

Figure 10 Determination of parity bits in example.

| Decimal | Position: | 1 | 2 | 3 | 4 | 5 | 6 | 7 |
Digit	Digit:	p_1	p_2	m_1	p_3	m_2	m_3	m_4
0		0	0	0	0	0	0	0
1		1	1	0	1	0	0	1
2		0	1	0	1	0	1	0
3		1	0	0	0	0	1	1
4		1	0	0	1	1	0	0
5		0	1	0	0	1	0	1
6		1	1	0	0	1	1	0
7		0	0	0	1	1	1	1
8		1	1	1	0	0	0	0
9		0	0	1	1	0	0	1

Figure 11 Hamming code for the decimal digits based on BCD.

Parity Check in Positions:	Received Digits	Parity	c_i
4, 5, 6, 7	0010	Odd	$c_1 = 1$
2, 3, 6, 7	1110	Odd	$c_2 = 1$
1, 3, 5, 7	1100	Even	$c_3 = 0$

Figure 12 Hamming code syndrome for example.

To illustrate the generation of the parity-checking bits, consider the code word for 5: $m_1 m_2 m_3 m_4 = 0101$. The parities of the message digits and, hence, the selections of the p_i, are as shown in Figure 12. The resulting code word is 0100101; this confirms the code word for decimal digit 5 in Figure 11.

The error-correcting feature of the Hamming code works as follows. Suppose that the transmitted word is 1110000 (decimal 8) but the received word is 1110010, with only one bit being in error. The three parity checks yield the result in Figure 12. Thus, an error is detected in position 6. The error is corrected by complementing the digit in position 6 (m_3).

Transmitted word	1110000
Received word	1110010
Syndrome	110

(7)

The subject of error-detecting and error-correcting codes is fascinating. However, we will not pursue the topic any further here.

CHAPTER SUMMARY AND REVIEW[19]

This chapter has introduced basic concepts and the mathematical foundation on which logic design is based. Following is a summary of what the chapter covered.

---

[19]A Chapter Summary and Review section is provided at the end of each chapter. Each reviews all the significant concepts and principles, procedures, and devices discussed in the chapter. As you read each entry in a chapter summary, you should attempt to explain to yourself, as you might to a learner, the relevant ideas, definitions, principles, theorems, devices, algorithms, diagrams, and other features. When you are unsure of something, turn back to the appropriate pages and review the content until you are sure.

- The binary number system
- The octal and hexadecimal number systems
- Converting from one number system to another
- Converting to the decimal system from binary
- Converting from the decimal system to binary
- Converting from octal (or hexadecimal) to binary
- Representation of negative numbers
- The basics of binary arithmetic

  - Addition
  - Subtraction
  - Multiplication
  - Division

- Number codes
- Weighted codes
- Binary-coded decimals
- Gray code
- Seven-segment code
- ASCII alphanumeric code
- Error detection and error-detecting codes
- Error-correcting codes
- Hamming codes
- Parity checking
- A syndrome

PROBLEMS

1 Several pairs of decimal numbers follow.

$$(18, 7) \quad (75, 28) \quad (11.3, 23.5) \quad (31.25, 17.58)$$

  a. Convert each decimal pair to binary.
  b. Find the sum of each binary pair; verify by finding the decimal sum and converting it to binary.
  c. Find the difference of each binary pair; verify by converting the decimal difference to binary.
  d. Find the product of each binary pair; verify by converting the decimal product to binary.

2 Express the following irrational numbers in the binary system, with enough digits in the fractional part so that the accuracy is to no less than four decimal places in the decimal equivalent: (a) pi (π), (b) base of natural logarithms e, (c) $e^{-\pi/2}$, (d) $\sqrt{15}$.

3 The following decimal numbers are given: 347.8, 2989, 625.7.

  a. Convert each number to the binary system.
  b. Convert each number from decimal to the octal system. Verify by converting the binary form to octal.
  c. Convert each number from decimal to the hexadecimal system. Verify by converting the binary form to hexadecimal.

4 Following is a set of binary numbers:

$$10111, \quad 110100.1, \quad 111001.01, \quad 1100101.101$$

    a. Convert each to a decimal number.
    b. Convert each to an octal number.
    c. Convert each to a hexadecimal number.

5 Following is a set of octal numbers:

$$247, \quad 153.4, \quad 374.25, \quad 5120.7$$

    a. Convert each to a binary number.
    b. Convert each to a decimal number directly; verify by converting the binary equivalent to decimal.
    c. Convert each to hexadecimal directly; verify by converting the binary equivalent to hexadecimal.

6 The base in which the numbers in the following operations (on the left-hand side of the equality) are represented is b. The numbers on the right-hand side of each equality are decimal numbers. Find all possible values of base b limited to the number systems described in this chapter.

    a. $\sqrt{121} = 8$
    b. $14.2 \times 30 = 510$
    c. $346 + 543 = 1111$

7 The conversions between number systems outlined in section 3 are conveniently formulated so that decimal arithmetic is used. Describe how one would convert directly from any base b_1 to another base b_2 without using decimal numbers. (The method should not include converting base b_1 to decimal.)

8 Perform the following arithmetic in binary.

    a. $1001 + 0110/0011$
    b. $(0110 - 0101) \times 1100$
    c. $(0010 + 0001)^2$

9 Perform the following binary multiplication operations. Use as many bits as necessary to represent the result.

    a. 100010×001010
    b. 001100×011001
    c. 000100×010101

10 Perform the following binary division operations. Represent each result as a quotient and remainder.

    a. $001100/0101$
    b. $010010/0010$
    c. $110011/0100$

11   **a.** Write the one's complements of the following binary numbers.

$$0011001 \quad\quad 1110011 \quad\quad 111111$$
$$1001011 \quad\quad 1010101 \quad\quad 000001$$

    b. Write the two's complements of the same numbers.

12 Express the following decimal numbers in the two's complement number representation.

$$36, \quad 101, \quad -49, \quad -75$$

13 Specify the range of decimal numbers that can be represented in the two's complement number representation using a word length of 8 bits.

14 Using the two's complement number representation limited to an 8-bit word length, carry out the following arithmetic operations. (The given numbers are decimal.) In each case, using only the result of the arithmetic operation (not your knowledge of the decimal equivalent), specify whether the resulting integer is positive or negative and describe how you decided.

a.   76 + 12   b.   83 + 60   c.   75 – 203
d.   – 28 – 64   e.   37 + 80   f.   253 – 182

15 Find the decimal equivalent of the following two's complement numbers.

a.   11111111   b.   10000001   c.   01010101
d.   10011100   e.   01110000   f.   10101010

16 Under what condition can an overflow occur if a positive two's complement number is added to a negative two's complement number?

17 Subtract the binary number 0110 from itself by adding its one's complement. Is the result what you expected?

18 Let $V_{2c}(a_{n-1}a_{n-2}...a_1a_0)$ represent the value of the binary string $a_{n-1}a_{n-2}...a_1a_0$ in two's complement number representation.

    a. Prove the following property: If the sign bit a_{n-1} is replicated any number of times, the value does not change. That is,

$$V_{2c}(a_{n-1}a_{n-1}...a_{n-1}a_{n-2}a_{n-3}...a_1a_0) = V_{2c}(a_{n-1}a_{n-2}...a_1a_0)$$

    This is called the *sign extension* property.

    b. The following strings are in two's complement representation. Using the sign extension property, show how you would add each pair of strings; verify your answers.

       101   and   0101001,   0101   and   0101001,   101   and   1101001

    c. Prove the following property:

$$V_{2c}(a_{n-1}a_{n-2}...a_1a_0) = 2^0(0 - a_0) + 2^1(a_0 - a_1) + 2^2(a_1 - a_2) + \cdots + 2^{n-1}(a_{n-2} - a_{n-1})$$

19 *Booth's algorithm* accepts two numbers in two's complement representation and computes their product, also in two's complement. It proceeds as follows:

(*a*) Append a 0 to the multiplier at its least-significant end. Then break the result up into overlapping 2-bit groups. Thus, if the multiplier is $a_{n-1}a_{n-2}...a_1a_0$ in ascending order, the groups are

$$(a_0, 0), (a_1, a_0), (a_2, a_1),..., (a_{n-1}, a_{n-2})$$

(*b*) Initialize the "partial product" (pp) to zero.
(*c*) If a 2-bit group is $(0, 1)$, then increment pp by the multiplicand; if a 2-bit group is $(1, 0)$, then decrement pp by the multiplicand; if a 2-bit group is $(0, 0)$ or $(1, 1)$, then keep pp unchanged.
(*d*) Double the multiplicand.
(*e*) Repeat steps *c* and *d* until all 2-bit groups, in ascending order, are exhausted.

Based on the result of Problem 18c, prove that the result of Booth's algorithm represents the product of the original multiplier and multiplicand.

20 Prove the overflow detection criterion for addition in the two's complement number representation—that is, that the result is out of range if and only if the carry into the sign bit is different from the carry out of the sign bit.

21 Add the following binary numbers, assuming they are

   a. One's complement numbers
   b. Two's complement numbers

Identify cases where an arithmetic overflow occurs.

0011	1001	1000	0101	0111
+0011	+0101	+1010	+1100	+0010

22 Is the addition of fixed-precision, signed binary numbers associative? In other words, for 8-bit two's complement numbers, does $(A + B) + C = A + (B + C)$? (*Hint:* What happens if there is overflow?)

23 Represent each decimal number on the left in each of the codes on the right.

Number	Code
43	Binary
257	BCD
5823	643-2
	Excess-3
	Gray

24 Find a self-complementing code for the decimal digits using each of the following weights.

   a. 443-2   **b.** 3321   **c.** 731-2   **d.** 87-4-2

25 Four messages are encoded in the following code words.

$$M_1: 01101, \qquad M_2: 10011, \qquad M_3: 00110, \qquad M_4: 11000$$

   a. Determine the minimum distance of this code.
   b. The following words are received and it is known that only a 1-bit error has occurred. Specify the correct message in each case.

00010	11101	01001
11010	00011	10010
00111	11010	10110

26 Coded words are transmitted in Hamming code. The following words are received.

0101000	0011101	1100100	1100110
1110011	1111001	1101001	1000010

   a. If any of the words is correct, specify the corresponding decimal digit.
   b. If any of the words has a single error, specify the bit that is in error and specify the correct decimal digit.
   c. Specify any of the received words that has a double error.
   d. Describe how a triple error can be detected.

27 **a.** Suppose that a code has a minimum distance of n, where n is odd. If an extra parity bit is added to each code word to establish even parity over the code word, describe how it will affect the minimum distance of the code.

    b. Repeat part a, except that n is even.

    c. Repeat part a, except that the added parity bit yields odd parity.

    d. Repeat part a, except that n is even and the added parity bit establishes odd parity.

28 **a.** Using the results of Problem 26, suggest and fully justify a method for constructing a single-error-correcting *and* double-error-detecting (SEC-DED) code from a single-error-detecting Hamming code.

    b. Illustrate this method by constructing a SEC-DED code in which each code word contains a 3-bit message.

29 Prove or disprove the claim that, for an m-error-detecting code, it should be possible to specify a decoding algorithm such that if the received word contains exactly j errors, for every j from 1 through m, the algorithm signals that the received word is in error.

30 Prove or disprove the claim that it is possible to generate, for any m-error-detecting code, a decoding algorithm signaling that a received word contains exactly j errors whenever it does, for every j between 1 through m, although it may not be able to identify the error positions.

31 Prove or disprove the claim that, if m or fewer errors have occurred in an m-error-correcting code, it is possible to generate a decoding algorithm that specifies how many errors have occurred and in what positions.

32 Prove or disprove the claim that, for an m-error-correcting code, if a decoding algorithm specifies that there are exactly j errors in a received word, with j from 1 through m, then j errors actually have occurred.

33 Given a code word with a minimum distance of 4, describe the circumstances when you would want to use the code for triple-error detection and when for single-error correction *and* double-error detection (SEC-DED).

34 Prove or disprove each of the following claims.

    a. For a code of minimum distance d, it is not possible to detect *any* d-bit error.

    b. For a code of minimum distance d, it is not possible to detect *every* d-bit error.

35 For a single-error-correcting Hamming code, suppose that the number of message bits, m, and the number of parity bits, k, are such that $2^k \geq m + k + 1$. Specify whether the error position indication obtained from parity checks on the received word would refer to a nonexistent position, that is, a position greater than $m + k$. Explain why such an indication is impossible or what such an indication would mean physically.

36 Suppose the parity-checking bits in a Hamming code for BCD are to be selected so that the parity of the three words to be checked is odd instead of even. Construct a Hamming code table similar to Figure 10b in the text to detect and correct a single error in the transmitted code.

37 **a.** Suppose the minimum distance of a code is an odd number n. An extra parity bit is to be added to each code word to establish even parity for the overall code word. What effect will this have on the minimum distance of the code? Justify your answer.

    b. Repeat part a, except that n is even.

    c. Repeat part a, except that the added parity bit establishes odd parity.

    d. Repeat part a, except that n is even and the added parity bit establishes odd parity

38 **a.** Based on the result of Problem 37, suggest a method for constructing a single-error-correcting *and* double-error-detecting (SEC-DED) code from a single-error-correcting Hamming code. Justify your method.

    b. Illustrate your method by constructing a SEC-DED code in which each code word contains a 3-bit message.

    c. A decoder for the SEC-DED code constructed in part *b* is to be designed for use on the receiving end. The decoder receives the incoming word and emits one of these three outputs:

- MSG: Contains the correct 3-bit message when no error or a correctable error has occurred.
- CE: Signals a correctable error and that MSG contains the corrected message.
- UE: Signals an uncorrectable error. MSG is to be disregarded in this case.

    d. Discuss what happens in the decoder in part *c* if the received word contains a triple error.

    e. Suppose that the code constructed in part *b* is to be used for triple error detection rather than single error correction and double error detection. What would be an appropriate set of outputs for the receiving-end decoder?

39 Prove or disprove each of the following.

    a. For an *m*-error-detecting code, it should be possible to give a decoding algorithm such that, for every *j* from 1 through *m*, if the received word contains *j* errors, the algorithm signals that the received word is in error.

    b. For an *m*-error-correcting code, it should be possible to give a decoding algorithm such that, for every *j* from 1 through *m*, if the received word contains exactly *j* errors, the algorithm signals that the received word has exactly *j* errors—though it may not be possible to identify the error positions.

    c. For an *m*-error-correcting code, there is a decoding algorithm such that, if a number of errors up to *m* have occurred, the algorithm indicates exactly how many and in what positions.

40 Suppose that, for a single-error-correcting Hamming code, the number of message bits *m* and the number of parity bits *k* are such that $2^k > m + k + 1$. State if the error position indication, obtained from the parity checks on the received word, refer to a nonexistent position (i.e., greater than $m + k$). Explain why such an indication is impossible or what such an indication would mean physically.

41   **a.** Suppose that the minimum distance of a code is an odd number *n*. An extra parity bit is to be added to each code word to establish even parity for the overall code word. What effect will this have on the minimum distance of the code? Justify your answer.

    b. Repeat part *a*, except that *n* is even.

    c. Repeat part *a*, except that the added parity bit establishes odd parity.

    d. Repeat part *a*, except that *n* is even and the added parity bit establishes odd parity.

Chapter 2

Switching Algebra and Logic Gates

The word *algebra* in the title of this chapter should alert you that more mathematics is coming. No doubt, some of you are itching to get on with digital design rather than tackling more math. However, as your experience in engineering and science has taught you, mathematics is a basic requirement for all fields in these areas. Just as thinking requires knowledge of a language in which concepts can be formulated, so any field of engineering or science requires knowledge of certain mathematical topics in terms of which concepts in the field can be expressed and understood.

The mathematical basis for digital systems is *Boolean algebra*.[1] This chapter starts with a brief exposition of Boolean algebra that lays the groundwork for introducing the building blocks of digital circuits later in the chapter.

1 BOOLEAN ALGEBRA

Boolean algebra, like any other axiomatic mathematical structure or algebraic system, can be characterized by specifying a number of fundamental things:

1. The *domain* of the algebra, that is, the set of *elements* over which the algebra is defined
2. A set of *operations* to be performed on the elements
3. A set of *postulates,* or *axioms,* accepted as premises without proof
4. A set of consequences called *theorems, laws,* or *rules,* which are deduced from the postulates

As in any area of mathematics, it is possible to start from different sets of postulates and still arrive at the same mathematical structure. What is proved as a theorem from one set of postulates can be taken as a postulate in another set, and what was a postulate in the first set can be proved as a theorem from

---

[1]This designation comes from its originator, the Briton George Boole, who published a work titled *An Investigation of the Laws of Thought* in 1854. This treatise was a fundamental and systematic exposition of logic. The book languished in obscurity for many decades.

Table 1 Huntington's Postulates

---

1. **Closure.** There exists a domain B having at least two distinct elements and two binary operators (+) and (•) such that:
   a. If x and y are elements, then $x + y$ is an element.
      The operation performed by (+) is called *logical addition*.
   b. If x and y are elements, then $x \cdot y$ is an element.
      The operation performed by (•) is called *logical multiplication*.
2. **Identity elements.** Let x be an element in domain B.
   a. There exists an element 0 in B, called the *identity element with respect to* (+), having the property $x + 0 = x$.
   b. There exists an element 1 in B, called the *identity element with respect to* (•), having the property that $x \cdot 1 = x$.
3. **Commutative law**
   a. Commutative law with respect to addition: $x + y = y + x$.
   b. Commutative law with respect to multiplication: $x \cdot y = y \cdot x$.
4. **Distributive law**
   a. Multiplication is distributive over addition: $x \cdot (y + z) = (x \cdot y) + (x \cdot z)$
   b. Addition is distributive over multiplication: $x + (y \cdot z) = (x + y) \cdot (x + z)$
5. **Complementation.** If x is an element in domain B, then there exists another element x', the *complement* of x, satisfying the properties:
   a. $x + x' = 1$
   b. $x \cdot x' = 0$
   The complement x' performs the *complementation* operation on x.

---

another set of postulates. So how do we decide on postulates? Clearly, one requirement for a set of postulates is *consistency*. It would not do for the consequences of one postulate to contradict those of another. Another requirement often stated is *independence*. However, independence involves the customary objective of ending up with a *minimal* set of postulates that still permits the derivation of all of the theorems. So long as a mathematical rule is consistent with the others, it can be added as a postulate without harm. If it is dependent on the previous postulates, however, the added rule is derivable from them and so need not be taken as a postulate.

The postulates we shall adopt here are referred to as *Huntington's postulates* and are given in Table 1.[2] Study them carefully. Note that Boolean algebra is like ordinary algebra in some respects but unlike it in others. For example, the distributive law of addition (Postulate 4b) is not valid for ordinary algebra, nor is the complement operation (Postulate 5). On the other hand, the subtraction and division operations of ordinary algebra do not exist in Boolean algebra.

The set of elements in Boolean algebra is called its *domain* and is labeled B. An *m-ary operation* in B is a rule that assigns to each ordered set of m elements a unique element from B. Thus, a *binary* operation involves an ordered *pair* of elements, and a *unary* operation involves just one element. In Boolean algebra two binary operations (logical addition and logical multiplication) and

---

[2]They were formulated by the British mathematician E. V. Huntington, who made an attempt to systematize the work of George Boole exactly 50 years after the publication of Boole's treatise.

one unary operation (complementation) are defined. Many Boolean algebras with different sets of elements can exist. The terminology *Boolean algebra* is a generic way of referring to them all.

Duality Principle

An examination of Huntington's postulates reveals a certain symmetry: the postulates come in pairs. One of the postulates in each pair can be obtained from the other one

- By interchanging the two binary operators, and
- By interchanging the two identity elements when they appear explicitly.

Thus, one of the commutative laws can be obtained from the other by interchanging the operators $(+)$ and $(\cdot)$. The same is true of the two distributive laws. Consequently, whatever results can be deduced from the postulates should remain valid if

- The operators $(+)$ and $(\cdot)$ are interchanged, and
- The identity elements 0 and 1 are interchanged.

This property of Boolean algebra is referred to as the *duality principle*. Whenever some result (theorem) is deduced from the postulates, the duality principle can be invoked as proof of the dual theorem.

Fundamental Theorems

We will now establish a number of consequences (theorems, rules, or laws) that follow from Huntington's postulates and from the duality principle. The proofs will be carried out step by step, with explicit justification for each step given by referring to the appropriate postulate or previously proved theorem. Two of the general methods of proof used in mathematics are

- Proof by contradiction
- Proof by the principle of (mathematical) induction

A proof by contradiction proceeds by assuming that the opposite of the desired result is true, and then deducing from this assumption a result that contradicts an already-known truth. This means that the opposite of the desired result is not true; therefore, the desired result itself must be true.

A proof by the principle of induction proceeds as follows. A proposition $P(i)$ is claimed to be true for all integers i. To prove the claim, it is necessary to do two things:

- Prove that the claim is true for some small integer, say $i = 1$.
- Assume it to be true for an arbitrary integer k and then show that it must, therefore, be true for the next integer, $k + 1$.

The latter step means that since the result is true for $i = 1$, it must be true for the next integer $i = 2$ $(1 + 1)$; then it must be true for $i = 3$ $(2 + 1)$; and so on for all other integers.

Another general method of proof, which is discussed in the next subsection, is especially valid for a Boolean algebra with only two elements.

Note that the symbol for logical multiplication ($\cdot$) is often omitted for simplicity, and $x{\cdot}y$ is written as xy. However, whenever there might be confusion, the operator symbol should be explicitly shown. Confusion can arise, for example, if the name of a logical variable itself consists of multiple characters. Thus, a variable might be called OUT2, designating output number 2, rather than the logical product of O and U and T and 2. In this chapter the variables are given simple names; hence, we will often omit the logical product symbol. Later we will show it explicitly whenever necessary to avoid confusion. Another possible notation when variable names consist of more than one symbol is to enclose the variable names in parentheses. Thus, the parentheses in (OUT2)(OUT3) permit the omission of the logical multiplication symbol. Now on to the theorems.

Theorem 1 Null Law

1a. $x + 1 = 1$
1b. $x{\cdot}0 = 0$

Note that each of these laws follows from the other one by duality; hence, only one needs explicit proof. Let's prove the second one.

$$
\begin{array}{ll}
x{\cdot}0 = 0 + (x{\cdot}0) & \text{Postulate 2a} \\
\phantom{x{\cdot}0} = (x{\cdot}x') + (x{\cdot}0) & \text{Postulate 5b} \\
\phantom{x{\cdot}0} = x{\cdot}(x' + 0) & \text{Postulate 4a} \\
\phantom{x{\cdot}0} = x{\cdot}x' & \text{Postulate 2a} \\
\phantom{x{\cdot}0} = 0 & \text{Postulate 5b}
\end{array}
$$

Theorem 1a follows by duality.   ■

Theorem 2 Involution $(x')' = x$

In words, this states that the complement of the complement of an element is that element itself. This follows from the observation that the complement of an element is unique. The details of the proof are left for you (see Problem 1d).   ■

Theorem 3 Idempotency

3a. $x + x = x$
3b. $x{\cdot}x = x$

To prove Theorem 3a,

$$
\begin{array}{ll}
x + x = (x + x){\cdot}1 & \text{Postulate 2b} \\
 = (x + x){\cdot}(x + x') & \text{Postulate 5a} \\
 = x + x{\cdot}x' & \text{Postulate 4b} \\
 = x + 0 & \text{Postulate 5b} \\
 = x & \text{Postulate 2a}
\end{array}
$$

Theorem 3b is true by duality.   ■

Theorem 4 Absorption

4a. $x + xy = x$
4b. $x(x + y) = x$

To prove Theorem 4a,

$$
\begin{aligned}
x + x{\bullet}y &= x{\bullet}1 + xy & &\text{Postulate 2b}\\
&= x{\bullet}(1 + y) & &\text{Postulate 4a}\\
&= x{\bullet}1 & &\text{Postulate 3a and Theorem 1a}\\
&= x & &\text{Postulate 2b}
\end{aligned}
$$

Theorem 4b is true by duality.   ∎

Theorem 5 Simplification

5a. $x + x'y = x + y$
5b. $x(x' + y) = xy$

To prove Theorem 5b,

$$
\begin{aligned}
x(x' + y) &= xx' + xy & &\text{Postulate 4a}\\
&= 0 + xy & &\text{Postulate 5b}\\
&= xy & &\text{Postulate 2a}
\end{aligned}
$$

Theorem 5a is true by duality.   ∎

Theorem 6 Associative Law

6a. $x + (y + z) = (x + y) + z = x + y + z$
6b. $x(yz) = (xy)z = xyz$

To prove Theorem 6a requires some ingenuity. First, form the logical product of the two sides of the first equality:

$$A = [x + (y + z)]{\bullet}[(x + y) + z]$$

Then expand this product using the distributive law, first treating the quantity in the first brackets as a unit to start, and going on from there; and then treating the quantity in the second brackets as a unit to start, and going on from there. The result is $A = x + (y + z)$ in the first case and $A = (x + y) + z$ in the second case. (Work out the details.) The result follows by transitivity (if two quantities are each equal to a third quantity, they must be equal to each other). Since the result is the same no matter how the individual variables are grouped by parentheses, the parentheses are not needed and can be removed.

Theorem 6b follows by duality.   ∎

Theorem 7 Consensus

7a. $xy + x'z + yz = xy + x'z$
7b. $(x + y)(x' + z)(y + z) = (x + y)(x' + z)$

To prove Theorem 7a,

$$
\begin{aligned}
xy + x'z + yz &= xy + x'z + yz(x + x') && \text{Postulate 5a} \\
&= xy + x'z + yzx + yzx' && \text{Postulate 4a} \\
&= (xy + xyz) + (x'z + x'zy) && \text{Postulate 3b and Theorem 6a} \\
&= xy + x'z && \text{Theorem 4a}
\end{aligned}
$$

Theorem 7b is true by duality. ∎

Theorem 8 De Morgan's Law

8a. $(x + y)' = x'y'$
8b. $(xy)' = x' + y'$

Prove Theorem 8a by showing that $x'y'$ satisfies both conditions in Postulate 5 of being the complement of $x + y$.

Condition 1

$$
\begin{aligned}
(x + y) + x'y' &= (x + x'y') + y && \text{Postulate 3a and Theorem 6a} \\
&= (x + y') + y && \text{Theorem 5a} \\
&= x + (y' + y) && \text{Theorem 6a} \\
&= x + 1 && \text{Postulate 5a} \\
&= 1 && \text{Theorem 1a}
\end{aligned}
$$

Condition 2

$$
\begin{aligned}
(x + y)(x'y') &= xx'y' + yx'y' && \text{Postulates 3b and 4a} \\
&= 0 \cdot y' + x'(yy') && \text{Postulates 5b and 3b} \\
&= 0 && \text{Postulates 5b and Theorem 1b}
\end{aligned}
$$

Theorem 8b is true by duality. ∎

A number of other important results are left for you to prove in the problem set, but we will use them here as if proved. They include the following:

1. The identity elements 0 and 1 are distinct elements.
2. The identity elements are unique.
3. The inverse of an element is unique.

Exercise 1 Prove that each identity element is the complement of the other one. ◆

Switching Algebra

For the Boolean algebra discussed so far in this book, the domain has not been restricted. That is, no limitation has been placed on the number of elements in the Boolean algebra. From Huntington's postulates, we know that in every Boolean algebra there are two specific elements: the identity elements. Hence, any Boolean algebra has *at least* two elements. In this book, let us henceforth limit ourselves to a *two-element* Boolean algebra.[3]

---

[3]It is possible to prove that the number of elements in any Boolean algebra is some power of 2: 2^n, for $n \geq 1$.

In a 1937 paper, Claude Shannon implemented a two-element Boolean algebra with a circuit of switches.[4] Now a switch is a device that can be placed in either one of two stable positions: *off* or *on*. These positions can just as well be designated 0 and 1 (or the reverse). For this reason, two-element Boolean algebra has been called *switching algebra*. The identity elements themselves are called the *switching constants*. Similarly, any variables that represent the switching constants are called *switching variables*.

This explains some of the common terminology used in this area, but we have already used some terminology whose source is not evident. The terms *logical multiplication* and *logical addition* were introduced in the first of Huntington's postulates. To explain where the adjective *logical* comes from, we will have to digress slightly.

Over the centuries a number of different algebraic systems have been developed in different contexts. The language used in describing each system and the operations carried out in that system made sense in the context in which the algebra was developed. The algebra of *sets* is one of these; another is a system called *propositional logic,* which was developed in the study of philosophy.

It is possible for different algebraic systems, arising from different contexts, to have similar properties. This possibility is the basis for the following definition.

> *Two algebraic systems are said to be* isomorphic *if they can be made identical by changing the names of the elements and the names and symbols used to designate the operations.*

Propositional logic is concerned with simple propositions—whether or not they are true or false, how the simple propositions can be combined into more complex propositions, and how the truth or falsity of the complex propositions can be deduced from the truth or falsity of the simple ones. A simple proposition is a declarative statement that may be either true or false, but not both. It is said to have two possible *truth values:* true (T) or false (F). Examples are

"The earth is flat."   F
"The sum of two positive integers is positive."   T

It is not the intention here to pursue this subject in great detail. However, it turns out that two-valued Boolean algebra is isomorphic with propositional logic. Hence, whatever terminology, operations, and techniques are used in logic can be applied to Boolean algebra, and vice versa.

To illustrate, the elements of Boolean algebra (1 and 0) correspond to the truth (T) or falsity (F) of propositions; T and F could be labeled 1 and 0, respectively, or the opposite. Or the elements of Boolean algebra, 1 and 0, could be called "truth values," although the ideas of truth and falsity have no philosophical meaning in Boolean algebra.

---

[4]To *implement* a mathematical expression means to construct a model of a physical system, or the physical system itself, whose performance matches the result of the mathematical operation. Another verb with the same meaning is "to realize." The physical system, or its model, so obtained is said to be an *implementation* or a *realization.*

x	y	xy
0	0	0
0	1	0
1	0	0
1	1	1

Figure 1 AND truth table.

One proposition is said to be the *negation* of another proposition if it is false whenever the other one is true. ("It is not snowing" is the negation of "It is snowing.") If p is a proposition, then not-p is its negation. This is isomorphic with the complement in Boolean algebra, and the same symbol (prime) can be used to represent it: not-p is written p'. Nobody will be hurt if we use the term *negation* in Boolean algebra to stand for *complement*.

Similar isomorphic relations exist between the operations of Boolean algebra and the connectives that join propositions together. However, further consideration of these will be postponed to the next section.

2   SWITCHING OPERATIONS

A unary operation and two binary operations, with names borrowed from propositional logic, were introduced in Huntington's postulates. For two-element (switching) algebra it is common to rename these operations, again using terms that come from logic.

The AND Operation

Let's first consider logical multiplication (AND) of two variables, xy. The operation will result in different values depending on the values taken on by each of the elements that the variables represent. Thus, if $x = 1$, then from Postulate 2b, $xy = y$; but if $x = 0$, then from Theorem 1, $xy = 0$, independent of y. These results can be displayed in a table (Figure 1) that lists all possible combinations of values of x and y and the corresponding values of xy.

This table is called a *truth table*. Neither the word *truth* nor the name of the operation, AND, makes any sense in terms of Boolean algebra; the terms are borrowed from propositional logic. The operation $x \cdot y$ is like the compound proposition

"The moon is full (x) and the night is young (y)."

This compound proposition is true only if *both* of the simple propositions "the moon is full" and "the night is young" are true; it is false in all other cases. Thus, xy in the truth table is 1 only if both x and y are 1. Study the table thoroughly.

The OR Operation

Besides "and," another way of connecting two propositions is with the connective "or." But this connective introduces some ambiguity. Suppose it is claimed that at 6 o'clock it will be raining or it will be snowing. The compound proposition will

x	y	$x + y$
0	0	0
0	1	1
1	0	1
1	1	1

Figure 2 OR truth table.

be true if it rains, if it snows, or if *it both rains and snows* — that is, if either or both simple propositions are true. The only time it will be false is if it neither rains nor snows.[5] These results, like the truth table for AND, can be summarized in a truth table for the connective OR. (Confirm the entries in Figure 2.)

Let's see how this compares in switching algebra with logical addition, $x + y$. From Huntington's Postulate 2a, if $y = 0$, then $x + y = x$, so $x + y$ will have whatever value x has. But if $y = 1$, then from Theorem 1a, $x + y = x + 1 = 1$, for both values of x. Verify that these values correspond exactly to the values given in the OR truth table.

The NOT Operation

Finally, the complement operation is isomorphic with negation, or NOT, in logic. It is a simple matter to set up a truth table for this operation; we leave it for you to carry out.

Exercise 2 Construct a truth table for the NOT operator. ◆

Commentary

The isomorphism of switching algebra with propositional logic has introduced a new tool, the truth table, that can be used to establish relationships among switching operations. For example, the theorems proved by application of Huntington's postulates together with previously proved theorems can also be proved from truth tables. We illustrate this by applying the truth tables for the AND, OR, and NOT operations to demonstrate the validity of the first form of De Morgan's law:

$$(x + y)' = x'y'$$

The result is shown in Figure 3. The last two columns are the same for all possible combinations of the switching variables; this proves De Morgan's law.

The procedure just used for establishing De Morgan's law illustrates a general method for proving results in switching algebra.

The truth-table method of proving a relationship among switching variables, by verifying that the relationship is true for all possible combinations of values of the variables, is called the method of perfect induction.

---

[5]The ambiguity in the "or" connective is seen in such a statement as "That animal is a cat or a dog." In this case, it is possible for the animal to be a cat or to be a dog, but certainly not both. In such a case, something other than logical addition is needed to describe the connective. This will be discussed in section 5.

x	y	x'	y'	$x+y$	$(x+y)'$	$x'y'$
0	0	1	1	0	1	1
0	1	1	0	1	0	0
1	0	0	1	1	0	0
1	1	0	0	1	0	0

Figure 3 Truth table for De Morgan's law.

This exhaustive approach can become quite cumbersome if the number of variables is large. Some might consider this approach to be intellectually and aesthetically less satisfying than applying the postulates of Boolean algebra and the theorems already proved. Even so, it is valid.

In De Morgan's law, let each side be called z. Then $z = (x + y)'$ and $z = x'y'$. There are two operations, OR and NOT, on the right side of the first expression. Spurred on by textbooks that they might have consulted in the library, some students might be tempted to think that the way to evaluate the right side is just to figure out which of the two operators "takes precedence." Concentrating on the significance of parentheses in algebraic expressions (both Boolean and ordinary) and the meanings of each Boolean operator should suffice to clarify the matter.

A set of parentheses is a tool used to group some variables into a unit; the operations are to be performed on the unit, not on parts of it within the parentheses. Thus $(x + y)'$ means forming the unit $x + y$ and then taking the complement of this unit. The unit can be given a name, say w. So $(x + y)'$ means w'. In terms of w, you can't even formulate the question of which operation takes precedence and should be performed first! You obviously take the OR of x and y, to get w; and then you take NOT w.

Similarly, for $z = x' \cdot y'$, the question, "Which do I do first—NOT and then AND, or AND first and then NOT?" is meaningless. Perhaps it will be simpler if we define $u = x'$ and $v = y'$; then $z = uv$. There is now no question—we take the AND of two things, u and v, which happen to be the NOT of x and the NOT of y, respectively. Thus, $z = (\text{NOT } x) \text{ AND } (\text{NOT } y)$; it couldn't be clearer.

It is common practice, just for simplicity, to omit parentheses around ANDed variables with the understanding that the AND will be performed before other operations on the ANDed unit. Indeed, the same convention—that the times operation is performed before the plus—is used in ordinary algebra as well. So instead of trying to memorize the order in which operations are to be performed in a given expression, concentrate on the meanings of the fundamental AND, OR, and NOT operations. Then you can't go wrong.

3   SWITCHING EXPRESSIONS

Look back at De Morgan's law in Theorem 8. Each side consists of certain switching variables related by means of AND, OR, and NOT operations. Each is an example of a *switching expression*, which we now formally define:

A switching expression is a finite relationship among switching variables (and possibly the switching constants 0 and 1), related by the AND, OR, and NOT operations.

Some simple examples of switching expressions are $xx' + x$, $z(x + y')'$, and $y + 1$. A more complex example of a switching expression is

$$E = (x + yz)(x + y') + (x + y)'$$

where E stands for "expression."

Note that expressions are made up of variables, or their complements, on which various operations are to be performed. For simplicity, we refer to variables or complements of variables as *literals*. The expression E consists of the logical product of two expressions logically added to another term. (When discussing logical sums and products, we will usually drop the adjective *logical* for simplicity. But remember that it is always implied.) The second term in the product is itself a sum of literals, $x + y'$. The first term in the product cannot be described simply as a sum or a product.

A given expression can be put in many equivalent forms by applying Boolean *laws* (that's what we will call the postulates and theorems for short). But, you might ask, what's the point? Why bother to perform a lot of algebra to get a different form? At this time we'll give only a tentative, incomplete answer. Each switching variable in an expression presumably represents a signal; the logical operations are to be implemented (carried out) by means of units of hardware whose overall output is to correspond to the expression in question. If a given expression can be represented in different forms, then different combinations of hardware can be used to give the same overall result. Presumably, some configurations of hardware have advantages over others. More systematic methods for treating different representations of switching expressions will be taken up in Chapter 3; their implementation will be continued in Chapter 4. Here we are only setting the stage.

We return now to expression E, renamed E_1 in what follows. Equivalent expressions can be found by applying specific laws of switching algebra. Applying the distributive law to the product term and De Morgan's law to the last term leads to E_2; then

$$E_1 = (x + yz)(x + y') + (x + y)'$$
$$E_2 = xx + xy' + xyz + y'yz + x'y'$$
$$E_3 = x + x\,(y' + yz) + x'y' \qquad \text{Theorem 3a, Postulates 4a and 5b}$$
$$E_4 = x + x'y' \qquad\qquad\qquad\quad \text{Postulate 4a and Theorem 4a}$$
$$E_5 = x + y' \qquad\qquad\qquad\qquad\ \text{Theorem 5a}$$

A fairly complicated expression has been reduced to a rather simple one. Note that E_2 contains a term yy', which equals the identity element 0. We say the expression is *redundant*. More generally, an expression will be redundant if it contains

- Repeated literals (xx or $x + x$)
- A variable and its complement (xx' or $x + x'$)
- Explicitly shown switching constants (0 or 1)

Redundancies in expressions need never be implemented in hardware; they can be eliminated from expressions in which they show up.

Minterms, Maxterms, and Canonic Forms

Given an expression dependent on n variables, there are two specific and unique forms into which the expression can always be converted. These forms are the subject of this section.

The expression E_1 in the preceding section had mixtures of terms that were products or sums of other terms. Furthermore, although E_1 seemed to be dependent on three variables, one of these variables was redundant. The final, equivalent form was the sum of two terms, each being a single literal.

In the general case, expressions are dependent on n variables. We will consider two nonredundant cases. In one case, an expression consists of nothing but a sum of terms, and each term is made up of a product of literals. Naturally, this would be called a *sum-of-products* (s-of-p) form. The maximum number of literals in a nonredundant product is n. In the second case to be considered, an expression consists of nothing but a product of terms, and each term is made up of a sum of literals; this is the *product-of-sums* (p-of-s) form. Again, the maximum number of literals in a nonredundant sum is n.

Suppose that a product term in a sum-of-products expression, or a sum term in a product-of-sums form, has fewer than the maximum number n of literals. To distinguish such cases from one in which each term is "full," we make the following definition:

> A sum-of-products or product-of-sums expression dependent on n variables is canonic *if it contains no redundant literals and each product or sum has exactly n literals.*[6]

Each product or sum term in a canonic expression has as many literals as the number of variables.

EXAMPLE 1

An example of an expression having three variables, in product-of-sums form, is E_1 below (not the previous E_1). It is converted to sum-of-products form as follows, with each step justified by the listed switching laws.

$E_1 = (x' + y' + z)(x + y + z')(x + y + z)$
$E_2 = (x' + y' + z)[(x + y)(x + y) + (x + y)(z + z') + zz']$ Postulate 4a
$E_3 = (x' + y' + z)(x + y)$ Postulate 1 and Theorem3b
$E_4 = xy' + x'y + xz + yz$ Postulates 4a and 5b, Theorem 7

In going from E_1 to E_4, redundancies were eliminated at each step. The original expression is in product-of-sums form, with the maximum number of literals in each term; hence, it is canonic. The final form, on the other hand, is in sum-of-products form, but it is not canonic; it has only two literals in each term.

Given a noncanonic sum-of-products expression, it is always possible to convert it to canonic form—if there is some reason to do so! The term xy' in expression E_4, for example, has the variable z missing. Hence, multiply the term by $z + z'$, which equals 1 and so does not change the logical value; then expand. The same idea can be used with the other terms. Carrying out these steps on E_4 leads to the following:

$E_5 = xy'(z + z') + x'y(z + z') + xz(y + y') + yz(x + x')$
$E_6 = xy'z + xy'z' + x'yz + x'yz' + xyz$

[6]Some authors use *canonical* instead of *canonic*.

(You should confirm the last line; note that redundancies that are created in the first line by applying Postulate 4a have to be removed.) This is now in canonic sum-of-products form.   ■

In a sum-of-products expression dependent on n variables, in order to distinguish between product terms having n literals (the maximum) and those having fewer than n, the following definition is made:

A canonic nonredundant product of literals is called a minterm.[7]

That is, a minterm is a nonredundant product of as many literals as there are variables in a given expression. Thus, each term in E_6 is a minterm, and so the entire expression is a sum of minterms.

The same idea applies to a product-of-sums expression. To distinguish between product terms having the maximum number of literals and others, the following definition is made:

A canonic nonredundant sum of literals is called a maxterm.

Each factor in expression E_1 in the preceding development is a maxterm; the entire expression is a product of maxterms. If a product-of-sums expression is not initially canonic, it is possible to convert it to canonic form in a manner similar to the one used for the sum-of-products case, but with the operations interchanged.

Exercise 3   Convert the following expression to a canonic product of maxterms: $E = (x + y')(y' + z')$.
Answer[8]

Generalization of De Morgan's Law

One of the Boolean laws that has found wide application is De Morgan's law. It was first stated as Theorem 8 in terms of two variables. It and its dual form, where the sum and product operations are interchanged, are repeated here:

$$\text{(a) } (x_1 + x_2)' = x_1'x_2' \quad \text{and} \quad \text{(b) } (x_1x_2)' = x_1' + x_2'$$

In words, complementing the sum (product) of two switching variables gives the same result as multiplying (adding) their complements. Suppose we were to increase the number of variables to three; would the result still hold true? That is, we want to perform the following operation: $(A + B + C)'$. Let's rename $A + B$ as D; then what we want is $(D + C)'$. But that's just $D'C'$, by De Morgan's law for two variables; and again by De Morgan's law, $D' = (A + B)' = A'B'$. Hence, the final result is:

$$\text{(a) } (A + B + C)' = A'B'C' \qquad \text{(b) } (ABC)' = A' + B' + C' \tag{1}$$

(The second form follows by duality.)

---

[7]The name doesn't seem to make sense—what is "min" about it? If anything, from the fact that we had to go from terms with two literals to ones with three literals, one would think it should be called "max"! Your annoyance will have to persist until Chapter 3, when the whole thing will be clarified.

[8]$E = (x + y' + z)(x + y' + z')(x' + y' + z')$ ◆

Why stop with three variables? The general case can be written as follows:

$$(x_1 + x_2 + \cdots + x_n)' = x_1'x_2' \cdots x_n' \tag{2}$$

$$(x_1 x_2 x_3 \cdots x_n)' = x_1' + x_2' + \cdots + x_n' \tag{3}$$

In words, (2) says that the complement of the logical sum of any number of switching variables equals the logical product of the complements of those variables. (In different terms, the NOT of a string of ORs is the AND of the NOTs of the variables in that string.) In (3), we interchange the operations AND and OR. Just writing the generalization doesn't make it true; it is left for you to prove.

Exercise 4 Prove one of the generalized De Morgan laws by mathematical induction. That is, assume it is true for k variables, and show that it is true for $k + 1$. Since we know it is true for two variables, then it will be true for three, (as already proved); since it is true for three, then ... and so on. ◆

Something more general can be concluded from De Morgan's law. In (2) and (3), the generalization involves increasing the number of variables in the theorem. On the left sides, the operations whose NOT is being taken are the sum and product. Suppose now that it is these *operations* that are generalized! That is, on the left, we are to take the complement of some expression that depends on n variables, where the $+$ and $\cdot$ operations are performed in various combinations. Will the result be the same if we take the same expression but interchange the sum and product operations while complementing all the variables? The generalization is

$$E'(x_1, x_2, \ldots, x_n, +, \cdot) = E(x_1', x_2', \ldots, x_n', \cdot, +) \tag{4}$$

Note the order of the sum and product operations on each side; it indicates that the two operations are interchanged on the two sides, wherever they appear. The result can be proved (we won't do it here) by mathematical induction on the number of operations. (You can do it, if you want.)

EXAMPLE 2

The following expression is given: $E = xy'z + x'y'z' + x'yz$. To find the complement of E, we interchange the $+$ and $\cdot$ operations and replace each variable by its complement. Thus,

$$E' = (x' + y + z') \cdot (x + y + z) \cdot (x + y' + z')$$

Let us first rewrite this expression in sum-of-products form by carrying out the product operations on the terms within parentheses and using necessary Boolean algebra to simplify. (You carry out the steps before you check your work in what follows.)

$$E' = (x'y + x'z + yx + y + yz + z'x + z'y)(x + y' + z')$$
$$= (x'z + xz' + y)(x + y' + z')$$
$$= xy + yz' + x'y'z + xz'$$

x	y	x'	y'	$x'y'$	E_1: $x + x'y'$	E_2: $x + y'$
0	0	1	1	1	1	1
0	1	1	0	0	0	0
1	0	0	1	0	1	1
1	1	0	0	0	1	1

Figure 4 Truth table for E_1 and E_2

To verify that this expression correctly gives the complement of E, you can take its complement using the same generalization of De Morgan's law and see if the original expression for E results.   ■

Exercise 5   Take the complement of E' in the preceding expression, using (4). Put the result in sum-of-products form and verify that the result is the same E as given in Example 2.   ◆

4 SWITCHING FUNCTIONS

In the preceding section we saw several examples in which a given switching expression was converted to other, equivalent expressions by applying Boolean laws to it. The equivalent expressions have the same logic values for all combinations of the variable values. The concept of "function" plays a very important role in ordinary algebra. So far, such a concept has not been introduced for switching algebra. We will do so now.

The discussion will start by considering two simple expressions:

$$E_1 = x + x'y' \quad \text{and} \quad E_2 = x + y'$$

Each expression depends on two variables. Collectively, the variables can take on $2^2 = 4$ combinations of values. (For n variables, the number of combinations of values is 2^n.) For any combination of variable values, each expression takes on a value that is found by substituting the variable values into it. When this is done for all combinations of variable values, the result is the truth table in Figure 4. (Confirm all the entries.) The last two columns are the same. This is hardly surprising given Theorem 5a. (Look it up.)

This example illustrates the principle that different expressions can lead to the same truth values for all combinations of variable values.

Exercise 6   Using a truth table, confirm that the expression: $E = xy + xy' + y'$ has the same truth values as E_1 and E_2 in Figure 4.   ◆

There must be something more fundamental than equivalent expressions; whatever it is may be identified by its truth values. On this basis, we define a switching function as follows:

A switching function is a specific unique assignment of switching values 0 and 1 for all possible combinations of values taken on by the variables on which the function depends.

x	y	z	f_1	f_2	f_1'	f_2'	$f_1 + f_2'$	$(f_1 f_2)'$
0	0	0	0	1	1	0	0	1
0	0	1	1	0	0	1	1	1
0	1	0	0	1	1	0	0	1
0	1	1	1	1	0	0	1	0
1	0	0	0	1	1	0	0	1
1	0	1	0	1	1	0	0	1
1	1	0	1	1	0	0	1	0
1	1	1	1	1	0	0	1	0

Figure 5 Truth tables for several functions.

For a function of n variables there are 2^n possible combinations of values. For each combination of values, the function can take on one of two values. Hence, the number of distinct assignments of two values to 2^n things is 2 to the 2^n power. Hence,

The number of switching functions of n variables is 2 to the 2^n.

On this basis, there are 16 switching functions of two variables and 256 functions of three variables; the number escalates rapidly for more variables.

Now consider the functions $f_1(x, y, z)$ and $f_2(x, y, z)$ whose truth values are shown in Figure 5. Any switching operation (AND, OR, NOT) can be performed on these functions and the results computed using the truth table. Thus, f_1' can be formed by assigning the value 1 whenever f_1 has the value 0, and vice versa. Other combinations of f_1, f_2, or their complements can be formed directly from the truth table. As examples, $f_1 + f_2'$ and $(f_1 f_2)'$ are also shown.

Switching Operations on Switching Functions

It is clear that functions—and, therefore, expressions that represent functions—can be treated as if they were variables. Thus, switching laws apply equally well to switching expressions as to variables representing the switching elements.

Note carefully that there is a difference in meaning between a switching *function* and a switching *expression*. A function is defined by listing its truth values for all combinations of variable values, that is, by its truth table. An expression, on the other hand, is a combination of literals linked by switching operations. For a given combination of variable values, the expression will take on a truth value.

If the truth values taken on by an expression, for all possible combinations of variable values, are the same as the corresponding truth values of the function, then we say that the expression *represents* the function. As observed earlier, it is possible to write more than one expression to represent a specific function. Thus, what is fundamental is the switching *function*. Of all the expressions that can represent a function, we might seek particular ones that offer an advantage in some way or other. This matter will be pursued further in Chapter 3.

Decimal Code	x	y	z	f	Minterms	Maxterms
0	0	0	0	0	$x'y'z'$	$x + y + z$
1	0	0	1	0	$x'y'z$	$x + y + z'$
2	0	1	0	1	$x'yz'$	$x + y' + z$
3	0	1	1	1	$x'yz$	$x + y' + z'$
4	1	0	0	1	$xy'z'$	$x' + y + z$
5	1	0	1	1	$xy'z$	$x' + y + z'$
6	1	1	0	0	xyz'	$x' + y' + z$
7	1	1	1	1	xyz	$x' + y' + z'$

Figure 6 Truth table for example function.

Number of Terms in Canonic Forms

In treating different switching expressions as equivalent in the previous section, Example 1 gave several different expressions dependent on three variables. We say that all these expressions represent the same function. E_1, in canonic product-of-sums form, has three terms as factors in the product—three maxterms. On the other hand, E_6 is in sum-of-products form and has five terms in the sum—five minterms. The sum of the number of minterms and maxterms equals 8, which is 2^3, the number of possible combinations of 3 bits.

This is an interesting observation. Let's pursue it by constructing a truth table of the function represented by expressions E_1 and E_6. First we write the minterms and the maxterms in the order xyz: 000 to 111.

$$E_6 = x'yz' + x'yz + xy'z' + xy'z + xyz$$
$$E_1 = (x + y + z)(x + y + z')(x' + y' + z)$$

The table is shown in Figure 6. Something very interesting can be observed by examining the expressions for the minterms and the maxterms. Suppose we take the complement of the minterm corresponding to 000. By De Morgan's law, it is exactly the first maxterm. Confirm that this is indeed true for each row in the table. From this we conclude that to find the canonic product-of-sums form is *to apply De Morgan's law to each minterm that is not present.*

The result in this example is a general one. Given the canonic sum-of-products form of a switching function, the way to obtain a canonic product-of-sums expression is as follows:

- Apply De Morgan's law to the complement of each minterm that is *absent* in the sum-of-products expression.
- Then form the product of the resulting maxterms.

Conversely, to obtain a sum-of-products expression from a given product-of-sums expression,

- Apply De Morgan's law to each sum term absent from the product;
- Then form the sum of the resulting minterms.

(You can confirm this from the same truth table.)

Exercise 7   Given the product-of-sums expression in E_1 in this section, use the preceding approach to find the corresponding sum-of-products form.   ◆

This approach avoids the necessity of carrying out switching algebra to convert from one form to another; instead, given one of the two forms, you can find the other one by mentally carrying out some steps merely by inspection.

Shannon's Expansion Theorem

Previous examples have shown that more than one expression can represent the same switching function. Two specific expressions noted are the sum-of-products and product-of-sums forms. The question arises as to whether a given function can *always* be expressed in these specific standard forms. An answer to this question, which is the subject of this section, was provided by Claude Shannon.[9]

Sum-of-Products Form

Suppose that $f(x_1, x_2, ..., x_n)$ is any switching function of n variables. Shannon showed that one way of expressing this function is

$$f(x_1, x_2, ..., x_n) = x_1 f(1, x_2, ..., x_n) + x_1' f(0, x_2, ..., x_n) \qquad (5)$$

On the right side, the function is the sum of two terms, one of them relevant when x_1 takes on the value 1 and the other when x_1 takes on the value 0. The first term is x_1 times what remains of f when x_1 takes on the value 1; the second term is x_1' times what remains of f when x_1 takes on the value 0. The proof of this expression is easy; it follows by perfect induction. That is, if the result is true for all possible values of variable x_1 (there are only two values—1 and 0) it must be true, period. (Confirm that it is true for both values of x_1.)

You can surely see how to proceed: Repeat the process on each of the remaining functions, this time using another variable, say x_2. Continue the process until all variables are exhausted. The result is simply a sum of terms, each of which consists of the product of n literals multiplied by whatever the function becomes when each variable is replaced by either 0 or 1. But the latter is simply the value of the function when the variables collectively take on some specific combination of values; this value is either 0 or 1. So the end result is a sum of all possible nonredundant products of the $2n$ literals (the n variables and their complements), some of which are multiplied by 1 and the remaining ones by 0. The latter are simply absent in the final result. According to Shannon, this will equal the original function. The proof at each step is by perfect induction.

Shannon's expansion theorem in the general case is

$$\begin{aligned} f = {} & a_0 x_1' x_2' \cdots x_n' + a_1 x_1' x_2' \cdots x_{n-1}' x_n + a_2 x_1' x_2' \cdots x_{n-1} x_n' + \cdots \\ & + a_{2n-2} x_1 x_2 \cdots x_n' + a_{2n-1} x_1 x_2 \cdots x_n \end{aligned} \qquad (6)$$

---

[9]Claude E. Shannon, "A Symbolic Analysis of Relay and Switching Circuits," *Trans AIEE*, 57 (1938), 713–723.

Each a_i is a constant in which the subscript is the decimal equivalent of the multiplier of a_i viewed as a binary number. Thus, for three variables, a_5 (binary 101) is the coefficient of $x_1 x_2' x_3$.

Shannon's expansion theorem in (6) is a major result. It shows that

Any switching function of n variables can be expressed as a sum of products of n literals, one for each variable.

Exercise 8 Let *f* be a function of two variables, x_1 and x_2. Suppose that *f* takes on the value 1 for the combinations $x_1 x_2$: 00, 10, 11 and the value 0 for the remaining combination. Determine the expression resulting from Shannon's expansion theorem carried out to completion in this case. As an added task, not dependent on Shannon's theorem, simplify the result if possible.
Answer[10]

In carrying the expansion process in (6) through to completion, a subconscious assumption has been made. That assumption is that, at each step until the last, after some (but not all) variables have been replaced by constants, the function that remains is itself not a constant. That is, what remains still depends on the remaining variables. If, instead, this remaining function reduces to a constant, the process will prematurely terminate and the corresponding term will have fewer than *n* literals. We have already discussed in the preceding section how to restore any missing literals. Hence, even in such a case, the generalization is valid.

Shannon's theorem constitutes an "existence proof." That is, it proves by a specific procedure that a certain result is true. Once the proof is established, then for any particular example it isn't necessary to follow the exact procedure used to prove the theorem. Now that we know that a function can always be put into the stated form, we can search for easier ways of doing it.

Product-of-Sums Form

Shannon's expansion theorem in (6) is a sum-of-products form. It's easy to surmise that a similar result holds for the product-of-sums form. It isn't necessary to go through an extensive proof for this form. The result can be obtained from (5) and (6) by duality. That is, 0 and 1 are interchanged, as are the sum and product operations. Doing this, the counterparts of (5) and (6) become

$$f(x_1, x_2, \dots, x_n) = [x_1 + f(0, x_2, \dots, x_n)][x_1' + f(1, x_2, \dots, x_n)] \tag{7}$$

$$f = (b_0 + x_1' + x_2' + \cdots + x_n')(b_1 + x_1' + \cdots + x_{n-1}' + x_n) \cdots$$
$$(b_{n-2} + x_1 + x_2 + \cdots + x_n')(b_{n-1} + x_1 + x_2 + \cdots + x_n) \cdots \tag{8}$$

where the b_i are the constants 0, 1. These equations are the duals of (5) and (6). The last one shows that

Any switching function of n variables can be expressed as a product of sums of n literals, one for each variable.

If the expansion should terminate prematurely, the missing literals can always be restored by the approach illustrated in the last section. The constants

---

[10] $f = x_1' x_2' + x_1 x_2' + x_1 x_2 = x_1 + x_2'$

♦

		NOT		AND	OR	XOR	NAND	NOR	XNOR
x	y	x'	y'	xy	$x + y$	$x'y + xy'$	$x' + y'$	$x'y'$	$xy + x'y'$
0	0	1	1	0	0	0	1	1	1
0	1	1	0	0	1	1	1	0	0
1	0	0	1	0	1	1	1	0	0
1	1	0	0	1	1	0	0	0	1

Figure 7 Truth table for basic operations.

will not show up explicitly in an actual function after redundancies are removed. If one of the constants is 1, for example, the entire corresponding factor will be 1, since $1 + x = 1$, and the corresponding factor will not be present. On the other hand, if any constant is 0, since $0 + x = x$, this constant does not show up but the corresponding factor will be present.

Exercise 9   Use the process of Shannon's theorem step by step to put the function in Exercise 8 in a product-of-sums form. Simplify the result, if possible.   ◆

5   OTHER SWITCHING OPERATIONS

It was noted earlier that propositional logic is isomorphic with switching algebra. As already observed, the language of propositional logic has permeated the way in which operations and ideas are discussed in switching algebra. We will now borrow some other operations that arise naturally in propositional logic and convert them to our use.

Exclusive OR

In discussing the OR operation in Section 2, an example was given of a compound proposition that sounds as if it should be an OR proposition but isn't: "That animal is a cat or a dog." Although the animal could be a cat or a dog, it can't be *both* a cat and a dog. The operation $x + y$ is "inclusive" OR; it includes the case where both x and y are 1. The "or" in "cat or dog" is different; it has a different meaning in logic. It is called an Exclusive OR, XOR for short, and is given the symbol $\oplus$. Thus, $x \oplus y$ is true when x and y have opposite truth values, but it is false when x and y have the same value. The truth values of $x \oplus y$ are shown in Figure 7. (The table includes the truth values of all the basic logic operations, including some yet to come.)

The Exclusive OR is a function of two variables. The most general sum-of-products form for such a function can be written as follows:

$$x \oplus y = a_0 x'y' + a_1 x'y + a_2 xy' + a_3 xy \tag{9}$$

The values of the a_i coefficients can be read from the truth table: 0 for a_0 and a_3, 1 for a_1 and a_2. Thus, (9) reduces to

$$x \oplus y = x'y + xy' \tag{10}$$

Exercise 10   Starting with the sum-of-products form for the XOR in (10), obtain a canonic product-of-sums form.
Answer[11]

NAND, NOR, and XNOR Operations

Besides the NOT operation, we now have three in our repertoire: AND, OR, and XOR. Three additional operations can be created by negating (complementing, or taking the NOT of) these three:

NAND (NOT AND): $(xy)' = x' + y'$ (11)

NOR (NOT OR): $(x + y)' = x'y'$ (12)

XNOR (NOT XOR): $(x \oplus y)' = (x'y + xy')' = xy + x'y'$ (13)

The right side in (13) can be obtained from (10) by negating the truth values of XOR to obtain the values of a_i. (Confirm it.) The truth values for these three are easily obtained from the three operations of which they are the NOTs. All the results are shown in the preceding truth table (Figure 7).

Exercise 11   Discuss the nature of the right sides of (11) and (12) as sum-of-products or product-of-sums forms.   ◆

Exercise 12   Apply switching laws to find a product-of-sums form of the right side of (13).   ◆

Notice from Figure 7 that XNOR is 1 whenever x and y have the same value. In logic, the function of two variables that equals 1 whenever the two variables have the same value is called an *equivalence* relation. The symbol denoting an equivalence relation is a two-headed arrow, $x \Leftrightarrow y$. Thus, XNOR and equivalence have the same truth values and can be used interchangeably: A XNOR $B = A \Leftrightarrow B = xy + x'y'$.

Comparing two signals (inputs) to see if they are the same is an important operation in many digital systems. Thus, an XNOR gate is a one-bit *comparator;* when its output is 1, we know that the two inputs are the same. We will discuss how to use XNOR gates to compare numbers of more than 1 bit in Chapter 3.

Some of you may have some uneasiness with the introduction of XOR and the other operations in this section. Earlier, a switching expression was defined as one that includes the operators AND, OR, and NOT. Would this mean that something including XOR or the other operations discussed in this section cannot be a switching expression? This apparent contradiction is overcome by noting equations (10) to (13). Each of the operations defined by these operators is expressed in terms of AND, OR, and NOT, so the uneasiness should disappear.

6   UNIVERSAL SETS OF OPERATIONS

Switching algebra was introduced in terms of two binary operations (AND and OR) and one unary operation (NOT). Every switching expression is made up

---

[11]Add xx' and yy', and then use the distributive law: $(x + y)(x' + y')$.   ◆

of variables connected by various combinations of these three operators. This observation leads to the following concept:

A set of operations is called universal *if every switching function can be expressed exclusively in terms of operations from this set.*

(Some use the term *functionally complete* in place of *universal.*) It follows that the set of operations {AND, OR, NOT} is universal.

In the preceding section various other operations were introduced. Two of these are NAND and NOR:

$$x \text{ NAND } y: (xy)' = x' + y' \tag{14}$$

$$x \text{ NOR } y: (x + y)' = x'y' \tag{15}$$

We see that the right sides of these expressions are each expressed in terms of only two of the three universal operations (OR and NOT for the first, and AND and NOT for the second). This prompts a general question: Can one of the three operations be removed from the universal set and still leave a universal set? The answer is "yes, indeed" if one of the operations can be expressed in terms of the other two.

Consider the AND operation, xy. By De Morgan's law, it can be written as

$$xy = (x' + y')' \tag{16}$$

The only operations on the right are OR and NOT. Since every AND can be expressed in terms of OR and NOT, the set {AND, OR, NOT} can be expressed in terms of the set {OR, NOT}. Conclusion: The set {OR, NOT} is a universal set.

Exercise 13 By reasoning similar to that just used, show that the set {AND, NOT} is a universal set.   ◆

The conclusion just reached is that any switching function can be expressed in terms of only two switching operations. But why stop there—why not try for just one? Let's explore the NAND operation. Since the set {AND, NOT} is universal, if we can express both those operations in terms of NAND, then, {NAND} will be universal! Here we go:

$$x' = x' + x' = (xx)' \qquad \text{Theorems 3a and 8a} \tag{17}$$

$$xy = ((xy)')' = [(xy)' \, (xy)']' \qquad \text{Theorems 2 and 3b} \tag{18}$$

The right sides are expressed in terms of nothing but NAND. Conclusion:

Any switching function can be expressed exclusively in terms of NAND operations.

Exercise 14 Show by a procedure similar to the one just used that {NOR} is a universal set.   ◆

The conclusion that results from the preceding exercise is

Any switching function can be expressed exclusively in terms of NOR operations.

The practical point of the preceding discussion is that, in the real world, switching operations are carried out by means of physical devices. If all switching

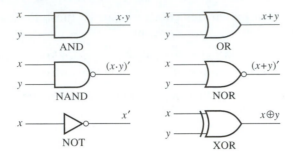

Figure 8 Symbols for logic gates.

functions can be expressed in terms of just one operation, then the physical implementation of any switching function can be carried out with only one type of physical device. This has obvious implications for simplicity, convenience, and cost, which we shall amplify in Chapter 3.

7   LOGIC GATES

Up to this point we have been dealing with rather abstract matters. We have discussed logic operations in an algebra called switching algebra. The variables that are manipulated in switching algebra are abstract switching variables. Nothing has been said so far that relates these switching variables to anything in the real world. Also, no connection has been made between the operations of switching algebra and physical devices.

We are now ready to change all that. For switching algebra to carry out real tasks, physical devices must exist that can carry out the operations of switching algebra as accurately and with as little delay as possible. The switching variables, which take on logic values of 0 and 1, must be identified with real signals characterized by a physical variable, such as voltage.

The generic name given to a physical device that carries out any of the switching operations is *gate*. However, in anticipation of discussing such devices, we will introduce another abstract representation of switching operations, a schematic representation for the real-life devices and for their interconnection in actual switching circuits.

Suppose that each operation in switching algebra is represented by a different schematic symbol, with a distinct terminal for each logic variable. Then any switching expression can be represented by some interconnection of these symbols. Attempts have been made in the past to adopt a set of standard symbols. One set has a uniform shape for all operations (a rectangle), with the operation identified inside.[12] A more common set of symbols for gates uses a distinct shape for each operation, as shown in Figure 8.

The NOT gate is called an *inverter*; each of the others is called by the name of the operation it performs—an AND gate, a NOR gate, and so on. Although each gate (except the inverter) is shown with two input terminals, there may be

---

[12]The Institute of Electrical and Electronic Engineers has published such standards (see IEEE Standard 91-1984). However, they have not been widely adopted.

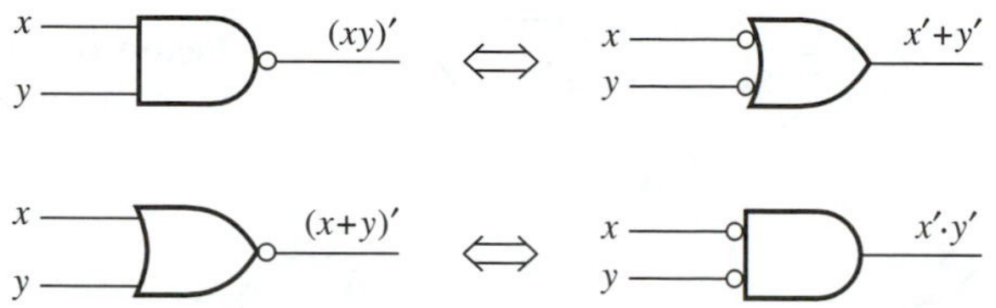

Figure 9 Two equivalent forms of NAND and NOR gates.

more than two inputs. (For the real devices of which these gates are abstractions, there are practical limitations on the number of inputs; these limitations will be discussed shortly.)

There is no problem with having more than two inputs for the two basic Boolean operators AND and OR, because both of these operators satisfy the associative law. The result is the same if we logically add x to $(y + z)$ or add $x + y$ first and then add it to z; unambiguously, the result is $x + y + z$. The same is true of the XOR operation. (Prove this to yourself.) However, the NAND and NOR operations are not associative. The output of a three-input NAND gate is $(xyz)'$; however, this is not the same as the NAND of x and y followed by the NAND with z. That is, $(xyz)' \neq ((xy)'z)'$.

A NAND gate is formed by appending a small circle (called a *bubble*), representing an inverter, to the output line of an AND gate. (The same is true for forming a NOR gate from an OR.) Furthermore, placing a bubble on an input line implies complementing the incoming variable before carrying out the operation performed by the gate.

Alternative Forms of NAND and NOR Gates

Other representations of a NAND gate and a NOR gate can be found by an application of De Morgan's law. For two variables, De Morgan's law is

$$(xy)' = x' + y' \tag{19}$$

$$(x + y)' = x'y' \tag{20}$$

The left sides of (19) and (20) are represented by the NAND and NOR gate symbols in Figure 8. The right side of (19) is an OR operation on inputs that are first inverted. Similarly, the right side of (20) is an AND operation on inverted inputs. These alternative forms of NAND and NOR gates are shown in Figure 9.

As far as ideal logic gates are concerned, the two forms of NAND or NOR gates are equally acceptable. However, when it comes to implementing the logic using real, physical gates, there is a practical difference between the two; we shall discuss this subsequently.

Something else interesting results from taking the inverse of both sides in (19):

$$((xy)')' = xy = (x' + y')'$$

Evidently, an AND gate with inputs x and y can be implemented with a NOR gate whose inputs are the complements of x and y. This result is shown schematically in Figure 10, where the bubbles at the inputs of the NOR gate represent the inverses of the input variables.

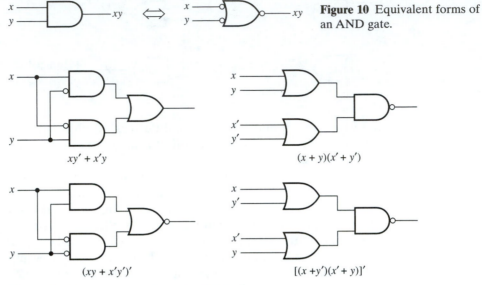

Figure 10 Equivalent forms of an AND gate.

Figure 11 Alternative structures for XOR.

Exercise 15   By a procedure similar to that used above, find an equivalent form of an OR expression and draw the corresponding equivalent gates.   ◆

Exclusive-OR Gates

Several different switching expressions can be written to represent the XOR function; some of these were shown earlier in this chapter. Four equivalent forms are as follows:

$$x \oplus y = xy' + x'y \tag{21}$$
$$= (x + y)(x' + y') \tag{22}$$
$$= (xy + x'y')' \tag{23}$$
$$= [(x + y')(x' + y)]' \tag{24}$$

The first two of these are the canonic s-of-p and p-of-s forms for the XOR function; they were given in (10) and in Exercise 9. Confirm the other two forms.

Not counting complementation, each of the expressions on the right involves three switching operations. Hence, each can be represented by three interconnected gates, as shown in Figure 11. Two of these involve the AND or NAND of two ORs, and two involve the OR or NOR of two ANDs. The number of gates and literals is the same in each case.

Commentary

Some observations can be made on the basis of what has just been discussed. *Logic gates* are schematic representations of switching operations. An interconnection of logic gates is just another way of writing a switching expression. We can say that switching algebra and the interconnections of logic gates are isomorphic systems. When we use schematic symbols called logic gates or draw interconnections of gates to represent switching operations, there need be no implication of anything physical or real.

Suppose, however, that physical devices that perform the operations symbolized by the gates could be constructed, even if they were not perfect. Then we could construct a physical circuit (we might call it a *switching circuit*) that can *implement,* or *realize,* a specific switching expression.[13] The circuit would be an *implementation,* or *realization,* of the expression. This is what makes switching algebra, an abstract mathematical structure, so useful in the physical design of real circuits to carry out desired digital tasks.

As a simple illustration, consider an XOR gate with inputs x and y. We might be interested in comparing the inputs. It is clear from (21) that $x \oplus y = 1$ whenever the values of the two inputs are different. (The opposite would be true for an XNOR gate.) Thus, either gate can play the role of a comparator of 1-bit numbers.

As another illustration, it may not be necessary to explicitly use an inverter to obtain the complement of a switching variable. A certain device that generates a switching variable commonly generates its complement also, so that both are available.[14] Hence, if x and x' both appear in an expression, then in the corresponding circuit, either we can show separate input lines for both of them or we can indicate the complement with a bubble. If you glance back at Figure 11, you will see that each of these approaches is used in half the cases.

8   POSITIVE, NEGATIVE, AND MIXED LOGIC

No matter how electronic gates and circuits are fabricated, the logical values of 0 and 1 are achieved in the real world with values of physical variables—in most cases voltage but sometimes current. The actual voltage values depend on the specific technology; but whatever the two values are, one is higher than the other. Therefore, one of them is designated High (H) and the other Low (L). There is no necessary correlation between the high and low voltage levels and the logic values of 0 and 1. Two schemes are possible, as illustrated in the tables in Figure 12.

Although the 0 and 1 logic values are not numerical values, if we interpret them as such, positive logic (with 1 corresponding to H) appears to be the natural scheme. Negative logic is superfluous; it contributes nothing of value that

Logic Value	Voltage Level
0	L
1	H

(a)

Logic Value	Voltage Level
0	H
1	L

(b)

Figure 12 Correlations between logic values and voltage levels. (*a*) Positive logic. (*b*) Negative logic.

---

[13] See footnote 4 for the meanings of *realize* and *implement.*

[14] As will be discussed in Chapter 5, a device called a *flip-flop* has two outputs, one of which is the complement of the other.

is not achievable with positive logic. Hence, there is no useful reason for adopting it. However, it may be useful to adopt positive logic at some terminals of devices in a circuit and negative logic at other terminals. This possibility is referred to as *mixed logic*. Its greatest utility occurs in dealing with real physical circuits. The next few paragraphs will discuss the common terminology in the use of mixed logic.

Two concepts are utilized in addressing the issues surrounding mixed logic. One is the concept of *activity*. The normal state of affairs, it might be assumed, as *inactivity*. The telephone is normally quiet (inactive) until it rings (becomes active), and conversations can then be carried on. Lights are normally off (inactive) until they are activated by a switch. When a microprocessor is performing a READ operation, it is active. Otherwise it isn't doing anything; it is inactive.[15] Thinking of logic 1 in connection with activity and logic 0 with inactivity is just a habit.

At various terminals in a physical system that is designed to carry out digital operations, voltages can have either high (H) or low (L) values. As just discussed, the two ways in which voltage levels and logic values can be associated are $1 \leftrightarrow H$ and $0 \leftrightarrow L$, or the opposite, $1 \leftrightarrow L$ and $0 \leftrightarrow H$. What complicates life even more is that at some points in a circuit the association can be made in one way and at other points in the opposite way.

Because of the connection of activity with logic 1, if the association of $1 \leftrightarrow H$ is made, the scheme is said to be "active high." If the association $1 \leftrightarrow L$ is made, it is said to be "active low." But this description employs two steps, two levels of association: Activity is first connected with logic 1; then logic 1, in turn, is associated with a high or a low voltage. Schematically, the train of thought is

$$\text{activity} \rightarrow \text{logic 1} \leftrightarrow \text{high voltage} = \text{active high}$$

$$\text{activity} \rightarrow \text{logic 1} \leftrightarrow \text{low voltage} = \text{active low}$$

"Active high" means that logic 1 is associated with a high voltage. More simply, one could say 1-high or 1-H instead of "active high." The adjective *active* just adds distance to the real association to be established between a logic value and a voltage level.

Another concept sometimes used as a synonym for activity is the notion of *asserting*. Thus, "asserted high" and "asserted low" mean the same as "active high" and "active low" which, in the end, mean the association of logic 1 with a high voltage and with a low voltage, respectively. The reasoning goes something like this: To assert is to affirm, to state that something is so; what is asserted might be thought to be "true." In propositional logic, a proposition is either true or false. So when something is asserted, this is equivalent to saying that the proposition is true. The dichotomy true/false in the logic of propositions is associated with switching constants 1 and 0. But there is no one-to-one correspondence between 1-T and 0-F; it is simply customary to associate 1 with "true" and 0 with "false."

---

[15]One might even think of Newton's first law as exemplifying this contention. Normally, things are in stasis; if you want something to change, you have to become active and exert a force.

Anyway, *if* we associate 1 with "true," and *if* assertion means "true," then assertion will be associated with 1:

$$\text{asserted} \Rightarrow \text{true} \Rightarrow \text{logic } 1$$

Hence, saying something is "asserted high" means that 1 corresponds to a high voltage. The use of "asserted" here adds nothing to the identification of 1 with a high voltage; you might as well say "1-high", or 1-H. In the same way, saying that something is "asserted low" means that 1 corresponds to a low voltage. Here the terminology "asserted" adds nothing to the identification of 1 with a low voltage; you might as well say 1-L.

One final comment on terminology. Sometimes the verb "to assert" is used in a way that does not connect it to "high" or "low." It might be said, for example, that a certain signal must be asserted before something else can happen. In such a usage, there is no commitment to asserting high or asserting low—just to asserting. This terminology does not require a commitment to mixed logic or positive logic; it can be applied in either case. Hence, it can be useful terminology.

When physical devices are used to implement a logic diagram, it is possible to use a different correspondence of voltage levels with logic values at different input and output terminals—even at internal points. For the actual implementation phase of a design, it might be useful to consider what is called mixed logic, in which the correspondence 1-H is made at some device terminals and 1-L at others. To achieve this purpose, special conventions and notations are used to convey the information as to which terminals correspond to 1-H and which to 1-L.

In this book, we will stick with positive logic, so we will not deal with this special notation. Indeed, the gate symbols shown in Figures 8–11 are based on positive logic.

9   SOME PRACTICAL MATTERS REGARDING GATES[16]

Up to this point we have adopted the following viewpoint: The specifications of a logic task, followed by the procedures of switching algebra, result in a switching expression. A schematic diagram containing logic gates is then constructed to realize this expression. Ultimately, we expect to construct the physical embodiment of this diagram using real, physical devices, generically referred to as *hardware*.

The way in which physical gates are designed and built depends on the technology of the day. As already mentioned, the first switching circuits utilized

---

[16]This section is largely descriptive. Although a couple of exercises are slipped in, these do not require a great deal of effort and creativity to complete. What is discussed is highly relevant to laboratory work. If you wish, you can be a largely passive reader; but we urge you to become engaged, to take notes, to formulate questions, and to refer to other books and manufacturers' handbooks for specific items.

mechanical devices: switches and relays.[17] The first switching devices that were called "gates" were designed with vacuum tubes. Vacuum tubes were later replaced by semiconductor diodes and, later still, by bipolar transistors and then MOSFETs. Each gate was individually constructed with discrete components.

The advent of integrated circuits permitted, first, an entire gate to be fabricated as a unit, and then several independent gates. Such *small-scale integrated* (SSI) units still find use. Soon it became possible to incorporate in the same integrated circuit an interconnection of many gates that together perform certain specific operations; we will discuss a number of such *medium-scale integrated* (MSI) circuits in Chapter 4. In time, an entire circuit consisting of hundreds of thousands of gates—such as a microprocessor—came to be fabricated as a single unit, on a single "chip." Some characteristics of integrated circuits will be discussed here. Design using specific circuits of this type will be carried out in Chapter 8.

Logic Families

As mentioned briefly above, the design and implementation of logic gates in any given era is carried out using the particular devices available at the time. Each type of device operates optimally with specific values of power-supply voltage. Different designs can result in different high and low voltage levels. A set of logic gates using a single design technology is referred to as a *logic family*. Some that were considered "advanced" just four decades ago (such as resistor-transistor logic, or the RTL family) are now obsolete but can still be found in museums.

Different designs are developed and promoted by different manufacturers and are suited to different requirements. The ECL (emitter-coupled logic) family came out of Motorola, while Texas Instruments is the creator of the TTL (transistor-transistor logic) family. The basic TTL design has been modified and improved over time in different ways to enhance one property of a gate (say speed) at the expense of some other property (say power consumption). In this way, different subfamilies are created within the TTL family. A table listing a number of the TTL subfamilies is given in Figure 13.

While the TTL and ECL logic families utilize bipolar transistors as the switching element, CMOS (complementary metal-oxide semiconductor) technology utilizes the MOSFET transistor. The subfamilies of CMOS are listed in Figure 14. A major question arises as to whether the inputs and outputs of CMOS gates can be interconnected with TTL gates without any special conversion circuits. If they can, we say that CMOS logic families are TTL-compatible. It turns out that with a power supply between 3.3 and 5 volts, CMOS families are indeed compatible with TTL.

---

[17]Although early switches were mechanical devices, the most basic modern electronic devices also act as switches—namely, transistors, both the bipolar junction type and, especially, the MOSFET variety. (See the Appendix for descriptions of MOSFETs and BJTs.) Hence, implementations of switching circuits with switches can be brought up to date utilizing MOSFETs. The logical operations of AND and OR would then be accomplished with series connections and parallel connections of such switches. Since most contemporary suppliers have a vested interest in the currently used technology, it is unlikely that a switch will be made. (Pun intended!) Seriously, though, some are returning to the switch circuits pioneered by Shannon, and some books are beginning to reintroduce such circuits.

Designation	Name	Power	Speed
LS	Low-power Schottky	Low	Slow
ALS	Advanced LS	Low	Moderate
S	Schottky	Medium	Fast
AS	Advanced Schottky	Medium	Fast
L	Low-power	Low	Moderate
F	Fast	High	Very fast

Figure 13 Some subfamilies within the TTL family of gates.

Designation	Name	Power	Speed
HC/HCT	High-speed CMOS	Very low	Moderate
AC/ACT	Advanced CMOS	Low	Fast
C	CMOS	Very low	Moderate
LCX/LVX/LVQ	Low-voltage CMOS	Very low	Moderate
VHC/VHCT	Very-high-speed CMOS	Low	Fast
CD4000	CMOS	Low	Moderate

Figure 14 Some subfamilies within the CMOS family of gates.

ECL circuits, on the other hand, are not compatible with either TTL or CMOS. Special conversion circuits are required to connect ECL gates to TTL or CMOS gates. Except for the CD4000 subfamily, the CMOS families are pin-compatible with the TTL families of Figure 13. That is, two chips with the same functionality have the same pin assignments for inputs, outputs, power, and ground.

How does one decide what subfamily to use in a given case? What is needed is a metric that takes into account both speed and power consumption. The product of these two values might be considered, but that won't work since the goal is to increase one and reduce the other. Instead of speed, we use the delay of a signal in traversing a gate; low delay means high speed. So the product *delay-power* is used as a metric. The lower this product, the better.

In this book, we will not study the operation of the electronic devices with which logic gates are constructed. We will, however, use the TTL and CMOS families of gates for illustrative purposes, but only in a descriptive way.

Input/Output Characteristics of Logic Gates[18]

Physical switching circuits are made up of interconnections of physical logic gates in accordance with a switching expression derived to accomplish a particular digital task. An illustration of a circuit is shown in Figure 15. The output from one gate becomes the input to one or more other gates.

Ideally, there would be no interaction among gates; that is, the operation of gate 4, for example, would have no influence on the proper operation of gate 2.

---

[18]This subsection is not essential to what follows. Even if your previous knowledge does not permit you to assimilate it thoroughly, you should read it anyway, at least to become exposed to the terminology. See Texas Instruments' *Data Book for Design Engineers* for further information.

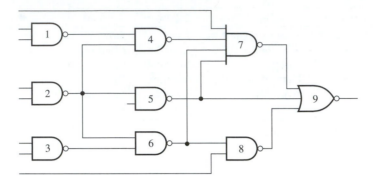

Figure 15 Logic circuit illustrating the loading of gates.

But gate inputs and outputs are connected to electronic devices internal to the gates. When these devices are conducting, currents flow into and out of gate terminals. These currents must necessarily flow through preceding-gate output terminals or succeeding-gate input terminals. Since there are practical limits imposed by the properties of the electronic devices on the level of such currents that can be carried, there are limits on the extent to which gates can be interconnected. Logic designers need not know the details of semiconductor technology and logic-gate implementation, but must understand the input/output characteristics of the technology in use to be able to analyze the interactions between gates.

The interactions between gates are technology specific, so it is necessary to distinguish CMOS and TTL. Let's analyze the input/output characteristics of CMOS first and then TTL. A logic gate output is driven high or low as a function of the input states. From the Appendix we learn that the MOSFET is a voltage-controlled switch. Thus, the inputs to a CMOS logic gate are connected to the gate terminals of MOSFETs. When an input is in one state it closes a MOSFET switch; when in the alternate state it opens a MOSFET switch. The output of a CMOS gate contains one or more MOSFET devices configured to drive the output high for certain input states. If the inputs are such that the output is high, then a set of switches between the output and power supply are closed, connecting the output to the power supply. (The power-supply voltage thus corresponds to logic 1, in positive logic.)

Similarly, a CMOS gate contains one or more MOSFETs to drive the output low when required. These MOSFETs connect the output to ground (corresponding to logic 0) for certain inputs. For our analysis we can assume that one MOSFET is used to pull the output high and one MOSFET is used to pull the output low. The typical input and output MOSFET connections for CMOS technology are shown in Figure 16.

When a MOSFET is switched on, it behaves not as a short circuit, but rather as a source of current. This is an important point. If it did behave as a short circuit, gates would have zero delay and zero power dissipation! We know from the Appendix that the MOSFET has very high input impedance between gate

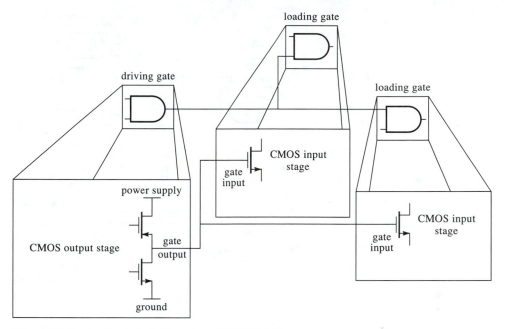

Figure 16 Typical input/output stages of CMOS logic gates.

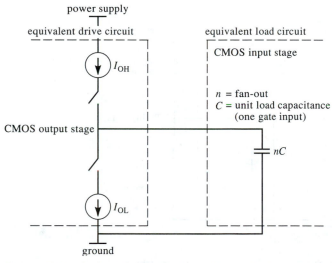

Figure 17 Equivalent circuit for the analysis of logic gate connections in CMOS technology.

and ground. The gate insulator forms the dielectric of a capacitance, and it is this capacitance that creates delay in a gate. A connection from the output of one gate to the inputs of other gates can be modeled as shown in Figure 17.

In a CMOS logic gate the switches shown in Figure 17 are never closed at the same time. Thus, one of the currents flows until the output voltage (which

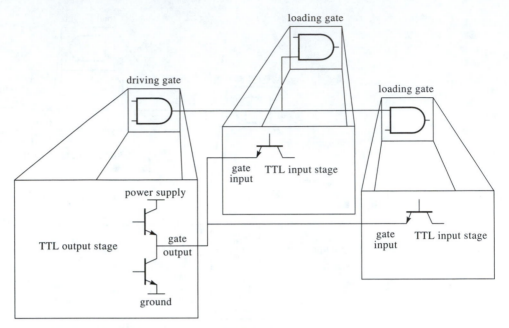

Figure 18 Typical input/output stages of TTL logic gates.

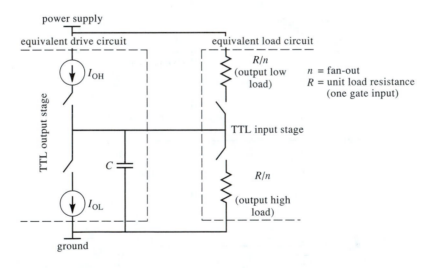

Figure 19 Equivalent circuit for analyzing TTL logic gate connections.

is proportional to the charge on the capacitor) reaches its final value. When the circuit is in steady state, no current is flowing. When the circuit switches state, the capacitance is charged or discharged to the new state. The difference between the input/output interactions of CMOS and TTL is due mainly to the difference in the input stages of the two logic gates. Input/output stages of TTL are shown in Figure 18.

An input of a TTL logic gate is connected to the emitter of a bipolar transistor, so the load seen by a gate output is resistive (as opposed to capacitive for CMOS). The bipolar transistors in the output stage of a TTL gate behave as a current source when switched on, and as an open circuit when switched off. The equivalent circuit for the analysis of gate interaction in TTL is shown in Figure 19.

Fan-out and Fan-in

We can learn some things by making some observations about the structure of Figure 15. The output from gate 1 goes to the input of gate 4 and nowhere else. We say gate 1 *drives* gate 4 and, conversely, that gate 4 *loads* gate 1. The output of gate 2 goes to the inputs of three other gates: gate 2 drives gates 4, 5, and 6, and each of the latter gates loads gate 2. The number of gate inputs driven by, or loading, the output of a single gate is called the *fan-out* of the driving gate. The fan-out of gate 2 in Figure 15, for example, is 3.

In CMOS technology the transistors of a gate output do not have to provide any static current to the inputs of the loading gates. Thus, a CMOS gate can have unlimited fan-out without affecting the logic (voltage level) of the circuit. However, as will be discussed shortly, the fan-out in CMOS technology has a substantial impact on circuit delay.

In TTL circuits a gate output must provide a static current to its fan-out gates. For example, in Figure 15 when the output of gate 2 is low, according to the last paragraph of the preceding subsection, the transistors at the inputs of the driven gates (4, 5, and 6) are conducting. Their currents must all flow through the output transistor of the driving gate (2). This process of returning the current at the input of a gate to ground through the output transistor of the driving gate is called *sinking* the current. For a given family of TTL gates, when the input transistor is conducting, the amount of current is called a *standard load*. With this terminology, gate 2 in Figure 15 must sink three standard loads since its output drives the inputs of three other gates. The proper operation of the output transistor of gate 2 imposes a limit on the number of standard loads that it can sink. Thus, for each TTL circuit design, there is a maximum fan-out (in the mid single digits!), which should not be exceeded.

Continuing the inspection of Figure 15, we notice that most of the gates have two inputs, but gate 9 has three inputs and gate 7 has four inputs. We refer to the number of inputs to a gate as its *fan-in*. Although there is no strong limitation on the fan-in of a gate imposed by its proper operation—as there is on its fan-out—there is, nevertheless, a practical consideration. Gates are manufactured with specific fan-ins, a fixed number of input terminals. If the logic design of a circuit calls for a gate with such a high fan-in that it is not commercially available, the design ought to be changed to utilize gates with available fan-in. There are differences between the fan-in constraints of the CMOS and TTL families, but they are beyond the scope of this book. The fan-in constraint in CMOS is revisited in the subsection on speed and propagation delay.

Buffers

In TTL circuits when the fan-out of a gate is high, the gate will need to sink a lot of current. This may cause the gate to become overloaded and to cease to function properly. In CMOS circuits large fan-out can cause increased delay. To overcome

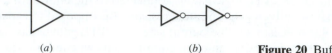

(*a*)   (*b*)   **Figure 20** Buffer.

these difficulties, a (noninverting) *buffer* is introduced at the output of the gate; its function is simply to provide increased drive current that permits continued operation with an increased load. The buffer performs no logical operation except the trivial identity operation. That is, its output is the same as its input. The logic symbol for the buffer is shown in Figure 20*a*; it is an inverter without the bubble. It might be made up of two cascaded inverters, as shown in Figure 20*b*. In this way, both a variable and its complement are available to drive a substantial load.

Power Consumption

Whenever there is current in a physical circuit, some of the associated electrical power will be converted to heat. If the circuit is to continue functioning, this heat must be dissipated so as not to result in excessive heating of the devices in the circuit. The design of gates and integrated circuits—including power-supply requirements, the number and type of transistors, and other factors—will have a lot to do with the power consumption. It is possible to design circuits specifically for low power consumption. You must know by now, however, that you can't get something for nothing. Low power consumption can be achieved at a price—usually at the expense of reduced speed. Review the tables for the TTL family in Figure 13 and the CMOS family in Figure 14; notice the range of power dissipation based on the variety of designs.

Noise Margin

Each logic family has specific nominal voltages corresponding to its high level and its low level. (For the TTL family, the nominal high value is 3.3 V and the nominal low value is 0.5 V.) However, digital circuits operate reliably even when actual voltage levels deviate considerably from their nominal values. Moreover, the high output level of a gate is not necessarily the same as its high input level. What is considered a "high" input or output voltage must extend over a range of values and similarly for "low" values.

The inputs and outputs of the gates under discussion are signals that we *intend* them to have. But there are many ways in which unwanted signals can be generated in a circuit and thus change input and output values from what are intended. We refer to such extraneous signals produced in a system as *noise*. Noise can be generated by a variety of mechanisms within the environment in which a circuit operates, from atmospheric radiation to interference from the 60 Hz power system. Noise can also be generated within the electronic devices in the system itself.

When intended signals are accompanied by noise, the actual signals will be modified versions of the desired ones. Provision must be made in the design of the circuit to enable it to continue to operate in the presence of noise up to some anticipated level. That is, there should be a certain immunity to noise. A measure of the amount of noise that will be tolerated before a circuit malfunctions is the *noise margin, N.* We will not pursue the details of this topic. Suffice it to say that what are considered high voltage levels and low voltage levels in the operation of gates are not fixed values but ranges of values. So long as input and output voltages stay within the range of values specified by the manufacturer, even though contaminated by noise, the gates will operate as if the voltages had their nominal values.

Speed and Propagation Delay

The speed at which logic circuits operate determines how rapidly they complete a task. Limitations on the speed arise from two sources:

- The delay encountered by a signal in going through a single gate
- The number of *levels* in the circuit, that is, the number of gates a signal encounters in going from an input to the output. We call the sequence of gates from an input to an output in a circuit a *logic path.*

Exercise 16   Reexamine Figure 15. Specify the number of levels that each input encounters on its way to the output.
Answer[19]

The delay in a TTL gate stems largely from the fact that transistors within the gate require a nonzero time to switch from a cut off state to a conducting condition and vice versa. This delay, for the most part, is independent of the load seen by a gate. Thus, in TTL circuits it can be assumed that a gate has a fixed delay, and the total delay from a logic input to a logic output can be estimated by accumulating the delays of the gates in the logic path from input to output.

The delay in a CMOS gate comes not only from the time required for transistors to switch between conducting and nonconducting states, but also from the time required to charge and discharge the load capacitance of the fan-out gates. Let's call the delay due to the switching of state of the transistors in a gate the *intrinsic delay* of a gate, and the delay due to load capacitance the *extrinsic delay.* The intrinsic delay is a strong function of the fan-in of a gate; the extrinsic delay is a strong function of its fan-out. Gates with large fan-in have a greater intrinsic delay than gates with small fan-in.

In CMOS it can be advantageous to increase the number of gate levels to keep the fan-in low in order to decrease the delay of a circuit. For example, three realizations of an eight-input AND gate are shown in Figure 21. In CMOS technology (due to the characteristics of the MOSFET) it is likely that the realizations shown in Figures 21*b* and 21*c* have lower delays than the one in Figure 21*a*; this, however, depends on the characteristics of the specific subfamily of CMOS used for the implementation. The extrinsic delay is caused by

---

[19]Six signals encounter the maximum of four levels, one signal encounters three levels, and two signals encounter one level.   ◆

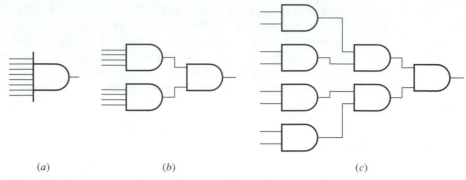

(a)   (b)   (c)

Figure 21 Three alternative realizations of an eight-input AND gate.

the physical constraint imposed by the capacitance: it takes time for the current of the driving gate to charge or discharge the load capacitance to the desired voltage level. The delay in a CMOS gate cannot be estimated accurately by simply counting the number of levels of gates in a logic path. This delay for a given load capacitance can be obtained from manufacturers' data sheets, which contain curves of measured delay as a function of load capacitance.

The transient response shown in Figure 22 illustrates the integrated effect of all the transistors and other components in a gate. The symbols t_{pLH} and t_{pHL} are typically used to denote the low-to-high and high-to-low propagation delay through a gate, respectively. The propagation delay time is the time elapsed between the application of an input signal and the output response. The delay times t_{pLH} and t_{pHL} are not necessarily the same for a specific gate. The delay of a logic path can be determined by the accumulation of t_{pLH} and t_{pHL} for each gate in the path. The symbols t_r and t_f denote, respectively, the rise and fall times of a signal and are defined as the time required for a signal to make a transition from 10 percent to 90 percent of its final value.

10 INTEGRATED CIRCUITS

Historically, both analog and digital circuits, from the simplest device to the most complex, were constructed with discrete components interconnected by conducting wires called *leads*. During the 1960s a method was developed for fabricating all the components that make up a circuit as a single unit on a silicon "chip." The generic name for this process is "integrating" the circuit. Thus, an *integrated circuit* (IC) is a circuit that performs some function; it consists of multiple electrical components (including transistors, resistors, and others), all properly interconnected on a semiconductor chip during the fabrication process.[20]

---

[20]If you visit a historical museum, among the collections you will find such objects as millennia-old bone knives, flint arrowheads, and hammers formed by a stone lashed with bark to a tree branch. Thousands of years old, these objects are now referred to as *primitive*. In the same way, single gates, buffers, and inverters are described as *primitive devices*, even though they were created only a few decades ago. That will tell you something about the rapidity of change in contemporary times.

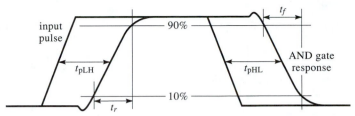

Figure 22 Waveforms of two-input AND gate output in response to a pulse at one input, with the other input held at logic 1.

The complexity of a digital circuit fabricated as an IC ranges from just a few gates to an entire microprocessor on a single chip. The following classifications give a measure of the complexity.

- SSI (*small-scale integration*): ICs with up to a dozen gates per chip, often multiple copies of the same type of gate.
- MSI (*medium-scale integration*): ICs with a few hundred gates per chip, interconnected to perform some specific function.
- LSI (*large-scale integration*): ICs with a few thousand gates per chip.
- VLSI (*very-large-scale integration*): ICs with more than about 10,000 gates per chip.[21]

Each integrated circuit is packaged as a single unit with a number of terminals (pins) available for connecting inputs, outputs, and power supplies. A few typical SSI integrated circuits are illustrated in Figure 23. Consult manufacturers' handbooks for others.

Some Characteristics of ICs

Digital circuits using ICs have a number of highly significant characteristics:

Size and space. The starting point for an integrated circuit is a semiconductor *wafer* about 100 mm in diameter and only 0.15 mm thick. Each wafer is subdivided into many hundreds of rectangular areas called *chips*. A complete circuit is fabricated on each chip. Although each circuit is mounted in a standard-size packet, a considerable size advantage is obtained.

Reliability. Very refined manufacturing and testing techniques are used in the production of ICs. This, together with the fact that all components and their interconnections are fabricated at the same time in the initial process, gives integrated circuits more reliability. For the same reason, since multiple copies of the same type of components (say transistors) are fabricated at the same time, it is much easier to obtain "matched" components, both in terms of parameter values and sensitivity to temperature. Such matched components are often required in a circuit design.

---

[21]Someone in Britain suggested a further category: VLSII, standing for VLSI, indeed!

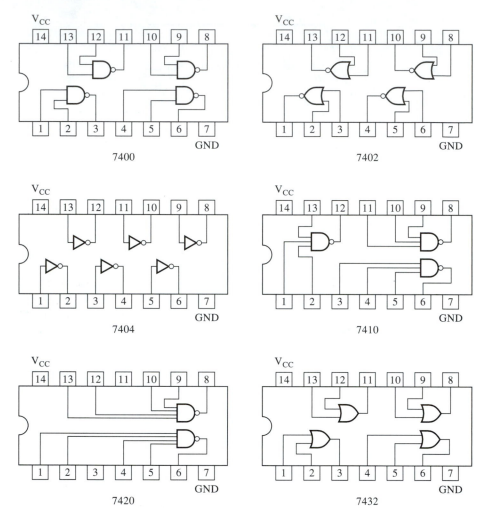

Figure 23 Typical SSI integrated circuits.

Power requirements. The heat generated in electrical circuits must be removed if the circuits are not to overheat. (In early computers, so much heat was generated that as much room for air conditioning equipment was needed to cool the computer as for the huge computer itself.) The problem is greatly reduced if the power consumption of the circuit is low. The much-reduced power requirement of the ICs and their small size thus have a symbiotic relationship. With a lower power demand, circuits can be packed more densely without fear of overheating.

Cost. The process of fabricating a single transistor is not much different from that of fabricating a gate or an SSI integrated circuit; the same chip can be used as the beginning point. The number of details of manufacturing operations to be performed is similar.

The packaging cost for a single transistor is only marginally less than the packaging cost for a more complicated IC. The same considerations apply to two different ICs with different numbers of gates: the cost of an IC with 30–40 gates is only slightly lower than the cost for one with 80–90 gates. Thus, the cost of an IC is not directly proportional to the number of individual components packed in a package; steps in the manufacturing process cost only marginally more if the number of gates in the package is doubled, for example.

Design Economy

When logic circuits were constructed with discrete ("primitive") gates, costs could be reduced by reducing the number of gates required to perform a particular function. Procedures for finding such minimal circuits became very highly developed. Such minimization procedures still play a role, especially in connection with logic design using what are called *programmable logic devices* (PLDs), as we will see in Chapters 4 and 8. We will therefore devote some attention to them in the following chapter.

The cost of a circuit designed with SSI ICs, however, depends not simply on the gate count but also on the IC package count.[22] Thus, adding a gate or two to a design may cost nothing extra if SSI packages with unused gates are already present in the design. Or suppose that a particular task can be achieved with a 25-gate circuit in two SSI packages, and a 40-gate package (with gates having the appropriate number of inputs) is available; it might be less costly to use the single package in the design and "waste" the unused gates rather than using the two packages with fewer gates.

In some cases another part of the circuit to be designed may call for a few gates of the kind left unused in the design just described. These previously "wasted" gates could now be used for this purpose, with no further hardware cost except for the cost of making pin connections, something that would be required anyway. In any case, design economy no longer depends on gate count but on integrated circuit package count and of the number of connections that have to be made in constructing the circuit.

One way in which to economize on package count, for example, is to avoid using inverters (e.g., 74LS04s). Instead, use an available fraction of a NAND, NOR, or XOR package. Consider the following expressions:

$$A' = (A + A + \cdots + A)' = (A + 0 + \cdots + 0)' \tag{25}$$

$$A' = (A \cdot A \cdot \cdots \cdot A)' = (A \cdot 1 \cdot \cdots \cdot 1)' \tag{26}$$

$$A' = A \oplus 1 \tag{27}$$

From (26) we note that the complement of a variable is the output of a NAND gate with any number of inputs all connected to that variable, or one input connected to that variable and all the rest connected to a high voltage (logic 1, assuming positive logic). Thus, if one-quarter of a 74LS00 (or one-third of a 74LS10) is unused in a package whose remainder already forms part of a circuit, this fraction of the package can be used to implement an inverter at no further hardware cost.

---

[22]In practice, multiple chips in a design are mounted on a *printed-circuit board* (PCB). So design economy involves not only the IC count but the PCB count as well. Such practical matters are easily learned when one is engaged in practice and need not concern us here.

Exercise 17 In a similar way, interpret the expressions in (25) and (27) for implementing an inverter. ◆

As we shall see in Chapter 4, many combinational-circuit units that perform some rather complex functions have been available for some time in MSI integrated circuits. There is no longer any need to design such functions by combining individual gates in SSI packages.

Application-Specific ICs

The integrated circuits and chips discussed so far have been general purpose. The SSI chips illustrated in Figure 23, for example, the MSI chips in which the applications to be described in Chapter 4 are implemented, or even the LSI chips that embody larger units can be used in the design of a wide variety of applications. A more recent development is the possibility of designing a chip for one particular application. If only a few of the resulting chips will ever be used, then it would make no economic sense to spend the required engineering design resources. On the other hand, if the number of copies needed is large, it may be economically feasible to design an entire integrated circuit just for this particular application. It should be no surprise that this type of IC is called an *application-specific integrated circuit (ASIC)*.

Two kinds of costs are associated with an ASIC. Once the unit is designed and debugged, there will be production costs. Of course, the actual design and debugging itself will entail a substantial cost, but this will not depend on later production costs; it is a *nonrecurrent engineering (NRE) cost*. Why undertake an enterprise requiring high NRE costs unless there is some cost or performance advantage? Indeed, ASICs do provide such advantages: generally, the system uses use fewer chips, has smaller physical size, and consequently lower power consumption and smaller time delays, and thus smaller delay-power products. We will not pursue this topic any further.

11 WIRED LOGIC

It is sometimes convenient to perform logic functions using special circuit configurations and connections rather than logic gates (for example, simply by connecting two or more wires together). Two types of gates are described in this section that enable the use of such special configurations and connections. The usefulness of these gates is probably not obvious to you at this point, but it will be when we get to Chapter 4. We consider them here because their implementation depends on the transistor circuits described in the previous section.

Tristate (High-Impedance) Logic Gates

Consider the equivalent circuit for the output stage of a CMOS gate as shown in Figure 16. In a conventional CMOS gate the logic inputs to the gate control the state of the switches (the output transistors). For example, in a two-input NAND gate the switch between the output and ground is closed if both inputs are logic 1; otherwise the switch between the output and the power supply is closed. The

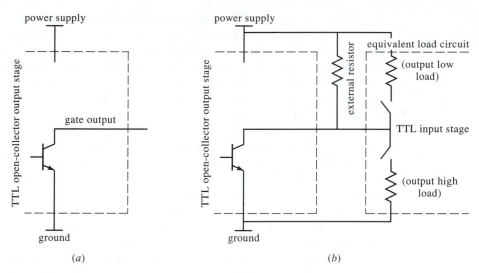

Figure 24 (*a*) The output stage of an open-collector TTL gate. (*b*) The
circuit configuration for the open-collector gate.

two switches are never closed at the same time, since this would result in an output voltage level that does not correspond to either a logic 1 or a logic 0. (In a
conventional gate the output is always driven to one of the valid logic levels.

Is there any advantage to opening both switches at the same time? If both
of the switches are open, then the output is not driven at all; that is, it is float-
ing. In this state the output has a very high impedance, so it is called the *high-
impedance* (or high-*Z*) state. A gate can be given the capability of entering this
third state by adding an additional input that, independent of the logic inputs,
can open both output switches. This additional input is often called an *enable*
input. When it is active the gate behaves as a conventional gate, but when it is
inactive the gate is in the high-impedance state. Such a gate is called a high-
impedance or *tristate* gate. The outputs of two tristate gates can be connected
directly to one another if they are never enabled at the same time. The operation of a tristate TTL logic gate is similar.

Exercise 18   Explain the behavior of two conventional logic gates if their outputs are connected directly to one another. (Does it make a difference whether
the gates are TTL or CMOS?)   ◆

Open-Collector and Open-Drain Logic Gates

Suppose a new gate is constructed from a conventional TTL gate (Figure 18) in
which one of the transistors in the output stage is removed. The new gate would
have the ability to drive its output to one logic state but not the other. Let's remove the transistor that drives the output high. The new circuit is shown in
Figure 24*a*. Is such a circuit useful? Perhaps, but only if there is some way to
drive the output high when necessary. This can be accomplished using an external resistor as shown in Figure 24*b*.

When the inputs of the gate are such that the output is driven low, it is driven low through the TTL gate output; the resistor acts as an additional load in parallel with the fan-out gates. The resistance value must not be so small that it overloads the gate. When the inputs are such that the output should be driven high, the ouput is not driven by the TTL gate; the external resistor provides a path for current to drive the fan-out gates instead.

Notice that when the output is driven high, the external resistor is in series with the equivalent input stages of the fan-out gates; thus, its value should be chosen such that an output voltage corresponding to a valid logic level is achieved. This kind of gate is called an *open-collector* gate because the collector of the output transistor is an open circuit without any internal connection on the chip. A similar gate can be constructed in CMOS and is called an *open-drain* gate because, internal to the chip, the drain terminal of the output transistor is not connected.

The usefulness of this circuit is not obvious. Furthermore, we need more components than we started with (an external resistor). However, if we connect the outputs of two or more such gates, we have a circuit that always provides a valid logic level as long as the external resistor has an appropriate value. If one or more of the gates pull the output low, then the external resistor acts as an additional load. If none of the gates pull the output low, then the external resistor provides a conductive path for current and drives the output high. For this reason the connection is called a *wired-AND*. If one or more of the outputs wired together correspond to a logic 0, then, assuming positive logic, the signal does likewise; otherwise, it corresponds to a logic 1.

CHAPTER SUMMARY AND REVIEW

This chapter introduced Boolean algebra and the results that follow from its use. It also introduced the primitive devices, called gates, on which logic design is based. The following topics were included.

- Huntington's postulates
- Boolean algebra

  - Involution
  - Idempotency
  - Absorption
  - Associative law
  - Consensus

- De Morgan's law
- Switching operations: AND, OR, NOT, NAND, NOR, XOR, XNOR
- Switching expressions

  - Sum-of-products form
  - Product-of-sums form
  - Canonic form

- Minterms and maxterms
- Switching functions
- Shannon's expansion theorem: s-of-p and p-of-s
- Universal operations

- Logic gates
- Equivalent forms of NAND and NOR gates
- Positive, negative, and mixed logic
- Families of logic gates: TTL, CMOS
- Gate properties

  - Fan-out
  - Fan-in
  - Power consumption
  - Propagation delay
  - Speed
  - Noise margin

- Integrated circuits: SSI, MSI, LSI, VLSI, ASIC
- Wired logic

  - Tristate gates
  - Open-collector gates
  - Open-drain gates

PROBLEMS

1 Prove the following theorems of Boolean algebra.

  a. The identity elements are distinct; that is, they are not the same.
  b. The identity elements are unique; that is, there are no others.
  c. The inverse of an element is unique; that is, there aren't two different inverses.
  d. The complement of an element is unique. From this, prove $(x')' = x$.

2 Use the distributive law and any other laws of switching algebra to place each of the following expressions in simplest sum-of-products form. (These are useful results to remember.)

  a. $f_1 = (A + B)(A + C)$
  b. $f_2 = (A + B)(A' + C)$

3 Complete the details of the proof of the associative law (Theorem 6) outlined in the text.

4 Prove the consensus theorem by *perfect induction*, that is, by showing it to be true for all possible values of the variables.

5 Prove that a Boolean algebra of three distinct elements, say $\{0, 1, 2\}$, does not exist. (If all of Huntington's postulates were satisfied, then it *would* exist.)

6 The following set of four elements is under consideration as the elements of a Boolean algebra: $\{0, 1, a, b\}$. The identity elements that satisfy Postulate 2 are to be 0 and 1. Fill in the tables for (+) and (•) in Figure P6 such that the algebraic system defined by these tables will be a Boolean algebra. Find the complement of each element, and confirm that all of Huntington's postulates are satisfied.

7 The operations of a four-element algebraic system are given in Figure P7. Determine whether this system is a Boolean algebra. If so, which are the identity elements?

8 In a natural-food restaurant, fruit is offered for dessert but only in certain combinations. One choice is either peaches or apples or both. Another choice is either cherries and apples or neither. A third choice is peaches, but if you choose peaches, then you must also take bananas. Define Boolean variables for all of the fruits and write a logical expression that specifies the fruit available for dessert. Then simplify the expression.

9 Write a logical expression that represents the following proposition: The collector current in a bipolar transistor is proportional to the base-emitter voltage v_{BE}, provided the transistor is neither saturated nor cut off.

(+)	0	a	b	1
0				
a				
b				
1				

(a)

(•)	0	a	b	1
0				
a				
b				
1				

(b)

Figure P6

x	x'
a	c
b	d
c	a
d	b

(a)

(+)	a	b	c	d
a	a	a	a	a
b	a	b	b	a
c	a	b	c	d
d	a	a	d	d

(b)

(•)	a	b	c	d
a	a	b	c	d
b	b	b	c	c
c	c	c	c	c
d	d	c	c	d

(c)

Figure P7

10 Construct a truth table for each function represented by the following expressions.

a. $E = (A' + B)(A' + B' + C)$

b. $E = (((A' + B)' + C)' + A)'$

c. $E = xz' + yz + xy'$

11 Carry out the proof of Shannon's expansion theorem given as (6) in the text.

12 Use one or both of the distributive laws (repeatedly if necessary) to place each of the expressions below in product-of-sums form.

a. $f_1 = x + wyz'$

b. $f_2 = AC + B'D'$

c. $f_3 = xy' + x'y$

d. $f_4 = xy' + wuv + xz$

e. $f_5 = BC' + AD'E$

f. $f_6 = ABC' + ACD' + AB'D + BC'D$

13 Construct a truth table for each function represented by the following expressions.

a. $E = xy + (x + z')(x + y' + z') + xy'z$

b. $E = (w + x'z' + y)(y' + z') + wxy'z$

c. $E = (AB'C + BC'D) + AB'D + BCD'$

14 Apply De Morgan's law (repeatedly if necessary, but without any simplifying manipulations) to find the complement of each of the following expressions.

a. $f = AB(C + D') + A'C'(BD' + B'D)$

b. $f = [AC' + (B' + D)(A + C')][BC + A'D(CE' + A)]$

15 Verify the following expression using the rules of Boolean algebra.

$$x'y + y'z + yz' = x'z + y'z + yz'$$

16 Using the rules of Boolean algebra, simplify the expressions that follow to the fewest total number of literals.

a. $f = AB' + ABC + AC'D$

b. $f = wyz + xy + xz' + yz$

c. $f = B + AD + BC + [B + A(C + D)]'$

17 Use switching algebra to simplify the following expressions as much as possible.

    a. $xyz' + xy'z' + x'y$
    b. $(wx')'(w + y)(x'y'z')'$
    c. $x'(y + wy'z') + x'y'(w'z' + z)$
    d. $(w + x)(w' + x + yz')(w + y')$

18   **a.** Carry out appropriate switching operations on the following expressions to arrive at noncanonic sum-of-products forms. Then restore missing variables to convert them to canonic form.
    b. In each case, using the distributive law and other Boolean laws, convert the non-canonic sum-of-products form to product-of-sums form.
    c. In each case, restore the missing variables to the forms determined in part *a* to convert them to canonic form.
    d. In each case, construct the truth table and confirm the canonic sum-of-products form.
    e. In each case, by applying De Morgan's law to the complements of the minterms absent from the canonic sum-of-products form, find the canonic product-of-sums form.

$$E_1 = (x + y')(y + z')$$
$$E_2 = (x'y + x'z)'$$
$$E_3 = (x + yz')(y + xz')$$
$$E_4 = (B + D')(A + D')(B' + D')$$
$$E_5 = (AB)'(B + C'D)(A + D')$$

19 Use switching algebra to prove the following relationships.

    a. $x' \oplus y = x \oplus y'$
    b. $x' \oplus y' = x \oplus y$
    c. $x'y \oplus xy' = x \oplus y$

20 Use appropriate laws of switching algebra to convert the expression $(x \oplus y)(x \oplus z)$ to the form $A \oplus B$; specify A and B in terms of x, y, z.

21   **a.** Express the Exclusive-OR function, $x \oplus y$, in terms of only NAND functions.
    b. Express $x \oplus y$ in terms of only NOR functions.

22 Prove that the Exclusive-OR operation is associative. That is, $(x \oplus y) \oplus z = x \oplus (y \oplus z)$.

23 Show that {NOR} is a universal set of operations.

24 The *implication* function $f = (x \Rightarrow y)$ is the statement "If x is true, then y is true." A switching expression for the implication function is $f = x' + y$. This function can be implemented by a two-input OR gate in which one of the input lines has a bubble. It might be called an *implication* gate.

    a. Construct a truth table for the implication function. Are any of the rows puzzling?
    b. Show that the implication function constitutes a universal set; that is, any switching expression can be represented in terms of implication functions only.
    c. Write the following expression in terms of implication functions only: $f = AC + BC'$.

25 Each of the following expressions includes all three of the Boolean operations. Using appropriate Boolean theorems, convert each expression to one that includes

    a. Only AND and NOT operations
    b. Only OR and NOT operations

    c. Only NAND operations
    d. Only NOR operations

$$E_1 = AB' + AC + B'C$$
$$E_2 = (x' + y')(y' + z)(y + z')$$
$$E_3 = xy' + (y + z')(x + z)$$

26   **a.** Use the result of Figure 2 in the text to convert a two-level AND-OR circuit consisting of two double-input AND gates followed by an OR gate into an all-NAND circuit.

    b. Use the result of Exercise 15 in the text to convert a two-level OR-AND circuit consisting of two double-input OR gates followed by an AND gate into an all-NOR circuit.

27 Show that the OR operation in the first form of Shannon's theorem can be replaced by the Exclusive-OR.

28 Determine the relationship between the Exclusive-OR function and its dual.

29 The objective of this problem is to convert one kind of two-input gate to a different kind of gate but with inverted inputs. Let the inputs be A and B. Use an appropriate Boolean theorem to carry out this objective for the following gate types:

    AND: AB
    OR: $A + B$
    NAND: $(AB)'$
    NOR: $(A + B)'$

30 An OR gate has two inputs A and B. B is obtained from A after A goes through three consecutive inverters. Assume that each inverter has a 1 μs delay. Suppose that A has been 1 for some time and then drops to 0. Draw the waveform of the resulting output.

31 A standard load for an LS-TTL gate is 3 mA of current; such a gate can drive three standard loads of the same kind of gate. A standard load for an S-TTL gate is 7mA. How many S-TTL gates can be driven by an LS-TTL gate?

32 Assuming the logic-high and logic-low voltage levels of two CMOS and TTL subfamilies are compatible, what are the consequences of driving a TTL gate with a CMOS gate and vice versa?

33 The intrinsic delay of a gate is 0.5 ns and the extrinsic delay is 0.2 ns per load. What is the fan-out limitation of this gate, given that its delay cannot exceed 1.2 ns?

34 Can the outputs of conventional gates be connected to the outputs of open-collector gates? If so, show such a circuit and describe its operation. If not, explain why they cannot be connected together.

35 What is the consequence of connecting many tristate outputs together? How many tristate outputs can be connected without affecting the proper functionality of a circuit?

36 What is the consequence of connecting many open-collector outputs together? How many open-collector outputs can be connected without affecting the proper functionality of a circuit?

37 Can the outputs of open-collector and tristate gates be connected together? Explain why or why not.

Chapter 3

Representation and Implementation of Logic Functions

The preceding chapter dealt with switching algebra. There we found that entities called switching variables are related by a number of switching operations, with different combinations of these operations called switching, or logic, expressions. We found that a switching function can be represented by one or more expressions.

Suppose that we interpret the variables as signals. Then a switching expression $E(x_1, x_2, \ldots, x_n)$ can be interpreted as operations performed on the input signals x_i, giving an output signal in accordance with the expression. As briefly described in the preceding chapter, the ultimate goal is to implement switching expressions by means of physical circuits and hardware. Since a switching function can be represented by many different expressions, it can be implemented by many different embodiments of hardware. Some hardware arrangements may be more desirable than others in some way; they may be simpler, more convenient, cheaper, faster, or less power demanding. We will examine such matters in detail starting with the next chapter.

This chapter will explore alternative methods of expressing logic functions: geometrical, algebraic, and tabular methods. We will investigate approaches to converting expressions to simple forms after first deciding what "simple" means. We will also discuss the implementation of logic functions by means of primitive gates in small-scale integrated circuits. More extensive implementations in MSI and LSI circuits will be discussed in later chapters.

1 MINTERM AND MAXTERM LISTS

As defined in the preceding chapter, a switching function of n variables is a specific assignment of binary values to each of the 2^n combinations of values of the n variables. A switching function is thus completely identified by its truth table. But a truth table can have a large number of rows and columns—16 rows and 5

Decimal Code	x y z	f	Minterm	Maxterm
0	0 0 0	0	$x'y'z'$	$x + y + z$
1	0 0 1	1	$x'y'z$	$x + y + z'$
2	0 1 0	0	$x'yz'$	$x + y' + z$
3	0 1 1	1	$x'yz$	$x + y' + z'$
4	1 0 0	1	$xy'z'$	$x' + y + z$
5	1 0 1	0	$xy'z$	$x' + y + z'$
6	1 1 0	0	xyz'	$x' + y' + z$
7	1 1 1	0	xyz	$x' + y' + z'$

Figure 1 Truth table for a given function f.

columns for four variables, for example. What we need is a simpler procedure for specifying a switching function.

The discussion will be initiated with the following example of a switching function represented by a mixed-form expression:

$$f(x, y, z) = (x' + y'z')(z + xy')$$

This can be converted to noncanonic sum-of-products form by applying the distributive law and other Boolean rules; it can then be converted to canonic form. Carry out this suggestion; only then check your result with what follows.

$$f = x'z + xy'z' = x'z(y + y') + xy'z' = x'y'z + x'yz + xy'z'$$

Each term in the final form is a minterm. Each minterm is 1 for a specific combination of values of the variables. Hence, the function has the value 1 for three different combinations of variable values: 001, 011, and 100, as shown in the truth table of Figure 1. (Temporarily ignore the last two columns; they are not part of the truth table but are added for future utility.)

Minterm Lists and Sum-of-Products Form

It is clear that the truth table contains a lot of redundant information. To identify a function, it isn't necessary to specify both those places where the function is 1 *and* those where it is 0; it is enough to list only the specific combinations of values for which the function is 1, or only the values for which the function is 0. Simpler yet is to list their decimal equivalents. Thus, the function corresponding to the preceding truth table can be identified simply as:

$$f(x, y, z) = \Sigma(1, 3, 4)$$

Such a representation of a function is called a *minterm list*. The summation symbol means that the minterms identified by their decimal values are to be added. When a switching expression is given in any form, one way of determining the minterm list is to first construct the truth table; the minterm list can then be written by inspection. Alternatively, the given expression can be first converted to canonic sum-of-products form; the minterm number corresponding to each minterm is then obtained by converting to decimal form the binary number representing that minterm. Thus, for four variables, a minterm $ABC'D'$ takes on the value 1 for the combination 1100, or decimal 12.

The converse is, of course, also possible. Given a minterm list, the minterm corresponding to each minterm number can be generated. The sum of all the minterms represents the canonic sum-of-products form. Take, for example, the following minterm list:

$$f(A, B, C) = \Sigma(2, 3, 5, 7)$$

Minterm number 5 has the value 1 for the combination 101, which corresponds to minterm $AB'C$, and similarly for the other minterms.

For ease of reference, a minterm can be designated m_i, where i is the minterm number. (We use a lowercase m because it is a *min*term, get it?) For a function of three variables, for example, $A'BC' = m_2$. In this notation, the canonic sum-of-products form of the function whose minterm list is given in the preceding expression can be written as

$$f(A, B, C) = m_2 + m_3 + m_5 + m_7$$

Exercise 1   A truth table was given for a function f_1 in Figure 5 of Chapter 2. Give three representations of this function: by the minterm list, by a summation of appropriate m_i, and by summing the actual minterms. (The answers are given in footnotes; don't peek until you work them out.)
Answer[1]

Maxterm Lists and Product-of-Sums Form

Because of the principle of duality, what was done in the previous subsection in terms of sums of products, minterms, and minterm lists can also be carried out in terms of products-of-sums, maxterms, and what are called *maxterm lists*. A switching function can certainly be identified by specifying its 1's. It can also be identified by specifying its 0's, that is, the combinations of variable values for which the value of the function is 0. In the truth table of Figure 1, for example, the function is 0 for five combinations of variable values, corresponding to the decimal equivalents: 0, 2, 5, 6, 7. (Don't just read this; verify!) The maxterms corresponding to these rows of the table will go to zero for the specified variable values. Hence, the *product* of these five maxterms, and no others, will be 0. Symbolically, therefore, the function can be identified as

$$f(x, y, z) = \Pi(0, 2, 5, 6, 7)$$

This representation of a function is a *maxterm list*. The product symbol Π (pi) means that the corresponding maxterms are to be multiplied. With the expressions for the maxterms given in the last column in the truth table, the canonic product-of-sums form of the function can be written as follows:

$$f = (x + y + z)(x + y' + z)(x' + y + z')(x' + y' + z)(x' + y' + z')$$

Just as a minterm is represented symbolically by m_i, a maxterm can be represented symbolically as M_i. (Capital M is used because it's a *Max*term,

---

[1] $f_1 = \Sigma(1, 3, 6, 7) = m_1 + m_3 + m_6 + m_7 = x'y'z + x'yz + xyz' + xyz$

see?) In this notation, the preceding canonic product of sums can be written as

$$f = M_0 M_2 M_5 M_6 M_7$$

Exercise 2   For the same truth table used in Exercise 1, give three representations of the function: one in terms of maxterm lists, one symbolically using a product of appropriate M_i, and the last by multiplying the actual maxterms. **Answer**[2]

2   LOGIC MAPS

The preceding section showed how simple symbolic expressions—namely, minterm lists and maxterm lists representing canonic sum-of-product and product-of-sums forms—can be written for any switching function. But canonic forms are very wasteful of literals. Most of the time, minterms can be combined by the elimination of redundant literals to yield expressions with fewer literals. We shall now describe a geometrical way of representing a switching expression, in preparation for a method of attack on the problem of simplifying such expressions.

Logical Adjacency and Geometrical Adjacency

Given a canonic sum-of-products expression, it would be useful to have a foolproof method of arriving at an expression, equivalent to the canonic one, that has the fewest number of terms. If two or more terms in an expression could be combined into a single term, that would certainly lead to fewer terms. Consider two minterms: $wxyz$ and $wxy'z$. Their sum is $wxyz + wxy'z = wxz(y + y') = wxz$, a simpler expression than the two minterms. We need a name for describing how two such minterms are related. Toward this end, let's make the following definition:

> *One combination of input-variable values is said to be logically adjacent to another if the two combinations differ in one bit position only.*

In a three-variable expression, for example, the input combinations (truth values) 110 and 111 are logically adjacent since they differ in one bit position only—the last one. The objective now is to create a geometrical (sort of tabular) representation of a function. We will find that such a representation is useful in arriving at a minimal sum-of-products or product-of-sums form.

Let's start with a function of two variables: x and y. The skeleton of a geometrical representation is shown in Figure 2a. To complete the skeleton, we need to specify the order of the x and y values. There are four possibilities; let us arbitrarily assign the x and y values as shown in Figure 2b. (Think of the other possibilities.) For this assignment of x and y values, the possible xy combinations are shown within the cells in Figure 2c. We observe that combina-

---

[2] $f_1 = \Pi(0, 2, 4, 5) = M_0 M_2 M_4 M_5 = (x + y + z)(x + y' + z)(x'+ y + z)(x' + y + z')$ ◆

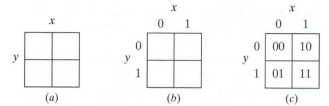

Figure 2 Geometrical representation of a function of two variables. (*a*) Map structure. (*b*) *x* and *y* values. (*c*) Cell coordinates.

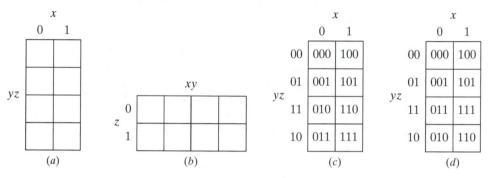

Figure 3 Geometrical structure for the three-variable case.

tions of *x* and *y* values in each row and each column in this figure are logically adjacent, as defined earlier. But note that the two cells in each row and each column are also physically, let's say *geometrically, adjacent* to each other.

Now suppose the order of values of *x* or *y* or both are reversed in this two-variable case (10 instead of 01 for *x* or *y* or both). The same result is true, namely, that the cells in each row and each column are both geometrically and logically adjacent. (Confirm this claim.)

Not much in the way of generalization can result from this simplest, two-variable case, so let us pursue this matter beyond this case. Figures 3*a* and 3*b* illustrate the possible arrangement of the variables when there are three of them—one variable on the horizontal and two variables on the vertical, or vice versa. Which of these arrangements one uses is a matter of taste, and who can account for taste? In this book we will universally use the form in Figure 3*a*, although some books use the one in Figure 3*b*. While the order of assigning the variable values for the single-variable (*x*) axis is not significant (as we noted in the two-variable case), this order does make a difference in the two vertical variables in this case, as we will note in what follows.

Another variation in the order of assigning the variables to the horizontal and vertical axes is also possible. We have assigned the first coordinate to columns and the last two to the rows. The opposite is also possible and is used in some books: assigning the first two coordinates to the rows and the last one to the column.

Suppose that the two-variable (*yz*) values in Figure 3*a* are assigned in BCD code, as shown in Figure 3*c*. The two cells in each row obviously are geometrically adjacent. By inspection, we see that they are also *logically* adjacent.

Next let's look at the geometrically adjacent cells in each column; are they also logically adjacent? (Study Figure 3c and come to a conclusion; only then look at the answer in the footnote.)[3]

This unfortunate circumstance is a result of our having chosen BCD code for ordering the yz values. In going from the second to the third row, the values of both y and z changed in each column. By studying the entries in the cells, we note that if the yz values in the third and fourth rows are interchanged (10 with 11), then the geometrically adjacent cells in each row and column will be logically adjacent also! The result of doing this is shown in Figure 3d. Name the code in which the yz values are now expressed.

Answer[4]

It is common practice to identify the cells not by their binary representation, as done in this figure, but by their decimal equivalent. The numerals are written in small type in a corner of each cell because other entries will be made in the cells to describe specific functions. These numerals are exactly the minterm numbers. Perhaps you should do this for yourself until it becomes second nature to deal with the binary equivalents.

Another interesting feature of the three-variable case in Figure 3d can be observed. Note in this figure that in each column, the topmost cell is logically adjacent to the bottommost cell, but the two do not seem to be geometrically adjacent. This anomaly can be explained. Think of reproducing this map on a square piece of paper (using up the entire sheet) and wrapping this paper vertically into a cylinder, with the map on the outside. (Use scotch tape, if you want, to hold the cylinder together so as to make your observations easier.) What *was* the bottom row will now be geometrically adjacent to what *was* the top row. Now, for each value of x (that is, in each column), the geometrically adjacent cells will also be logically adjacent! The upshot of this argument is summarized as follows:

The order of variable combinations is chosen so that any two geometrically adjacent cells are also logically adjacent.

Take, for example, cell 011 (decimal 3) in Figure 3d. The logically adjacent cells are obtained by replacing, one at a time, each bit value by its complement; they are

111 (decimal 7)
001 (decimal 1)
010 (decimal 2)

From the figure, verify that these cells are indeed also geometrically adjacent to 011.

The geometrical structure in which any two geometrically adjacent cells are also logically adjacent is called a *logical* (or *logic*) *map*.[5] We are now ready to

---

[3]In each column, two geometrically adjacent cells *are also* logically adjacent except for those in the second and third rows.

[4]Gray code.

[5]It is very often called a *Karnaugh map* (abbreviated K-map), after Maurice Karnaugh, who first proposed it in 1953. E.W. Veitch had proposed a slightly different form a year earlier. We have chosen to give it a name based on function rather than one based on history, but we might slip once in a while and call it a K-map.

Decimal Code	x y z	f
0	0 0 0	0
1	0 0 1	1
2	0 1 0	0
3	0 1 1	1
4	1 0 0	1
5	1 0 1	0
6	1 1 0	0
7	1 1 1	0

(a)

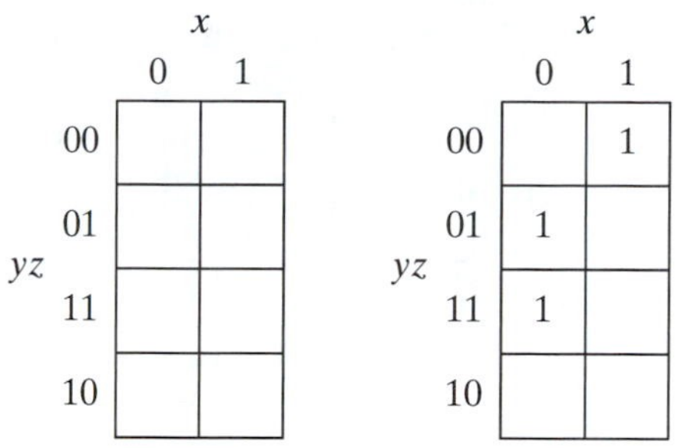

Figure 4 (*a*) Truth table, (*b*) geometrical structure, and (*c*) representation of $f = \Sigma(1, 3, 4)$.

pursue the goal of finding a method for simplifying logical expressions, arriving at minimal sum-of-products or product-of-sums forms.

The truth table of a function of three variables was given in Figure 1. It is shown again in Figure 4a, and the logical map structure for three variables is shown in Figure 4b, using the Gray code for the vertical variables. The logical map of the function itself is in Figure 4c; there are 1's in only three cells corresponding to the minterms. That is, each minterm of a function corresponds to one cell of the map for which the function is 1. A single cell is the smallest unit that makes up the map.

Now we enter in each cell the value of the function given in the truth table. Is it really necessary to enter both the 1 and the 0 values? If only the 1's, for example, are entered, those cells in which there is no entry must carry a 0. In the logical map in Figure 4c, only the 1's are entered. Each minterm of the function represented by this map corresponds to one cell of the map that carries a 1. A single cell is the smallest unit that makes up the map; it is the minimum entity that can be shown in the map. *Aha!* That's the reason for the designation *min*term! There is a similar explanation for the designation *max*term.

Exercise 3 Using a three-variable logic map, give an explanation that makes the designation "maxterm" plausible for a cell containing a 0 of a function. ◆

The logic map tool is very handy. Our next step should be to extend it to functions of more variables. Let's first go on to four variables; in this case we would expect a map to have two variables on each axis. With variables w, x, y, z, we would expect a 2 by 2 logic map (using, say, wx to identify the columns and yz to identify the rows). Since we found the Gray code to be a winner before, it seems reasonable to assign both the wx and the yz values in accordance with Gray code; if this doesn't serve for some reason, we can always try something else.

		wx		
	00	01	11	10
00	0	4	12	8
01	1	5	13	9
11	3	7	15	11
10	2	6	14	10

yz

(a)

		wx		
	00	01	11	10
00				
01	1			1
11	1	1	1	1
10			1	

yz

(b)

Figure 5 Four-variable logical map of $f = \Sigma(1, 3, 7, 9, 11, 14, 15)$.

EXAMPLE 1

The minterm list of a function of four variables follows. The objective is to construct a logical map for this function.

$$f(w, x, y, z) = \Sigma(1, 3, 7, 9, 11, 14, 15)$$

The structure of a four-variable map with the Gray code used to assign values on both axes is shown in Figure 5a. The map for the given function is constructed by inserting the value 1 in each cell corresponding to the minterm number of the function. It is shown in Figure 5b. Confirm the entries in this map. ◆

Let's consider drawing the map in Figure 5a on a sheet of paper and rolling it from top to bottom into a cylinder, with the map on the outside. (You will benefit by actually doing this.) The cells in the bottom row are now adjacent to the corresponding ones in the top row, both geometrically and logically. Now unwrap that cylinder and create another one, this time wrapping it vertically, placing the right boundary geometrically adjacent to the left boundary. Confirm that the cells in the left column are also logically adjacent to the corresponding cells in the right column. That is, the two cells at the end of each row in the flat map are logically adjacent, as are those at the top and bottom of each column. This is what gives the logical map its value.

After you form one of the cylinders described in the preceding paragraph, a further twist will give even more insight. Imagine twisting the cylinder into a torus, where the cross section on one end is geometrically adjacent to that on the other. (You won't be completely successful in doing this unless your sheet is elastic.) The four cells at the corners of the flat map are seen to be geometrically adjacent in pairs, as well as being logically adjacent in pairs. More on this development will be discussed in a subsequent section.

Note that any internal cell in a four-variable map has a common side with four other cells, the ones to which it is geometrically adjacent. Using the cylinder and torus constructions described in the preceding paragraph, demonstrate to yourself that the same is true of the cells around the perimeter of the map. The generalization is that any cell anywhere in the three-variable map is adjacent to three other cells.

	v = 0 wx				v = 1 wx			
	00	01	11	10	10	11	01	00
00	0	4	12	8	16	20	28	24
01	1	5	13	9	17	21	29	25
11	3	7	15	11	19	23	31	27
10	2	6	14	10	18	22	30	26

(yz labels the rows)

Figure 6 Five-variable map structure.

The structure of the five-variable map is given in Figure 6. It consists of two four-variable maps side by side, with the first variable, v, taking the value 0 on the left half and the value 1 on the right half of the map. Notice the order of the wx values for $v = 1$ compared with the ones for $v = 0$. As an exercise, confirm (by rolling the map into cylinders again) that the cells in the leftmost column for $v = 0$ are logically adjacent to the corresponding cells in the rightmost column for $v = 1$, as are those in the top row and the bottom row. Note again that the code in which the five variables are ordered is the Gray code.[6]

Exercise 4 Consider the cell in a five-variable logic map identified by the minterm 01011. Locate this cell on the five-variable map. Specify the minterm numbers of the five cells to which this one is logically adjacent and locate them on the map.
***Answer*[7]**

A six-variable map would consist of two five-variable maps located one below the other. For the top part, the first variable would take on one value, say 0, and the bottom part would represent the value 1. (Draw such a map for yourself and keep it handy for later use.) This scheme can be repeated conceptually for higher-order maps. But since it is difficult to visualize the interrelationships of the cells, the practical utility is clearly diminished for more than six variables.

Cubes of Order k

The ideas just discussed have some far-reaching consequences, which will now be taken up. The subject will be initiated by the function described in Example 1 and specified by the map in Figure 5b, repeated in Figure 7a. Minterms m_3 and m_7 are logically adjacent. The sum of these two minterms is

$$w'x'yz + w'xyz = w'yz(x + x') = w'yz$$

The three common literals in the minterms can be factored by the distributive law, as shown, leaving a factor $x + x' = 1$ by the definition of the complement.

---

[6]Maurice Karnaugh had to *conceive* the idea that if the variable values were listed in the order of the Gray code in an n-variable map, then each cell would be logically adjacent to n others.

[7]Obtain them by replacing, one at a time, each bit by its complement. ◆

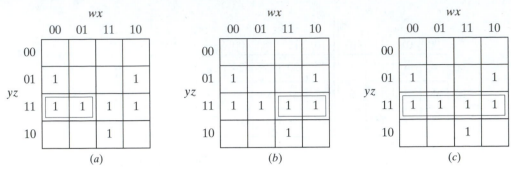

Figure 7 Groupings of minterms on a map.

In this way, the sum of the two minterms is simplified to a single term consisting of the three common literals. This grouping of the two minterms is shown enclosed within a rectangle in Figure 7a.[8]

A similar grouping of minterms m_{11} and m_{15} can be carried out to yield $wx'yz + wxyz = wyz$. This is shown in the map in Figure 7b by enclosing the two minterms. The results of the two preceding groupings of minterms, $wy'z$ and wyz, have the property that all literals but one are common; the odd one appears complemented in one term and not in the other. The pattern should be evident. Adding the two terms should result in eliminating this odd literal as well:

$$w'yz + wyz = yz$$

The result is that a grouping of four minterms has been reduced to a term with only two literals. This grouping of four minterms in a row is shown enclosed in Figure 7c.

Exercise 5

a. The preceding example showed four minterms that were adjacent in two pairs, which were then combined into a 2-cube. There exist another two pairings of these same four minterms that yield the same 2-cube. Find these two pairs.

b. The map in Figure 7 contains four other minterms, in addition to the ones just described, that are adjacent in pairs, forming four different 3-cubes. Appropriate pairs of these 3-cubes together form a 2-cube. Encircle two appropriate pairs (in rectangles) and write a simplified expression for their sum. Repeat with the other two pairs and confirm that the final results are the same.

c. There is another pair of adjacent minterms. Encircle them on the map and write a simplified expression for the sum of the two minterms.

Answer[9]

---

[8]In other books it might be enclosed by means of a circle or an ellipse. For brevity, we might often say "encircle" when we mean "enclose by a rectangle."

[9](a) 0111 and 1111 together form xyz; 0011 and 1011 form $x'yz$; they combine into the same yz.
(b) $\{m_1, m_3\}$ and $\{m_9, m_{11}\}$; $(w'x'y'z + w'x'yz) + (wx'y'z + wx'yz) = x'z$. Adjacency of pairs is shown by drawing an arc joining the two pairs in the table. Another possibility is joining $\{m_1, m_9\}$ and $\{m_3, m_{11}\}$ and then joining the two 3-cubes. Confirm that this grouping yields the same final result.
(c) $\{m_{14}, m_{15}\}$; $wxyz + wxyz' = wxy$. (The map is shown within the text.)   ◆

Part of Exercise 5 answer

What has just been introduced by means of an example can be generalized. A function of n variables will have the value 1 in some cells of its logical map and the value 0 in others. To distinguish between these cells, let's refer to them as 1-cells and 0-cells, respectively. Based on the preceding example, we make the following definition:

> *A set of 2^k 1-cells, each of which is adjacent to k others in the set, is called a* cube *of order k, or a k-cube for short. The k-cube is said to cover each of the 2^k cells.*

Now go back to Figure 7c and apply this new language. The set $\{14, 15\}$ is a 1-cube (it has $2 = 2^1$ cells, so $k = 1$). The sets $\{3, 7, 11, 15\}$ and $\{1, 3, 9, 11\}$ are 2-cubes. (Verify the claim.) Note that within 2-cube $\{3, 7, 11, 15\}$ there are four 1-cubes: $\{3, 11\}$, $\{3, 7\}$, $\{7, 15\}$, and $\{11, 15\}$. However, since these 1-cubes are entirely covered by the larger 2-cube, the ultimate expression need not include them.

Exercise 6   Write the sum of the switching expressions corresponding to the 2-cube $\{3, 7, 11, 15\}$ and the 1-cubes covered by it: $\{3, 11\}$, $\{3, 7\}$, $\{7, 15\}$, and $\{11, 15\}$. Specify the switching law that permits simplification of this expression, and then simplify it as much as possible.
Answer [10]

EXAMPLE 2

A function is specified by the following minterm list:

$$F = \Sigma(2, 4, 6, 9, 10, 11, 12, 13, 15)$$

  a. Construct the logical map.
  b. List all the possible k-cubes.
  c. List all k-cubes that are not covered by any higher-order k-cubes.
  d. List a minimal set of k-cubes that covers all the 1-cells at least once. Repeat for as many minimal sets as you can find.

Answer

  a. The four-variable map is shown in Figure 8.
  b. Start with the lowest minterm number; construct all 1-cubes formed with higher minterm numbers. Repeat with each succeeding minterm number

---

[10]The absorption law.   ◆

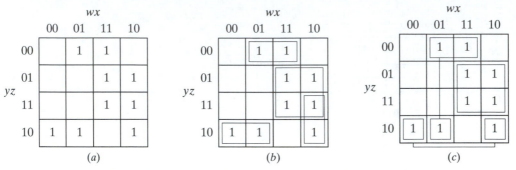

Figure 8 Map of $f = \Sigma(2, 4, 6, 9, 10, 11, 12, 13, 15)$ and its k-cubes.

until all are exhausted (assuming you aren't!). Construct 2-cubes in the same fashion, and then delete all 1-cubes covered by the latter. Higher-order k-cubes are constructed similarly. Result: {2, 6}, {2, 10}, {4, 6}, {4, 12}, {10, 11}, {12, 13}, {9, 11, 13, 15}; plus 1-cubes {9, 11}, {9, 13}, {11, 15}, {13, 15}, which are all covered by the 2-cube. This is obviously an algorithmic approach for which a program could be (and has been) written.

c. All but the last four in b. Seven terms would be needed in a sum-of-products expression if all were used, but not all will be needed.

d. Besides the 2-cube, which covers all the odd minterms, 1-cubes are needed to cover the five remaining minterms, all even: 2, 4, 6, 10, 12. Two 1-cubes alone cannot cover all five minterms, so at least three 1-cubes are needed. Two of these three must each cover two distinct even minterms. Besides the 2-cube, there are four different combinations of three 1-cubes that do the job. Two of the sets of k-cubes that cover all the minterms are shown circled in parts a and c of the figure. (You find the other two.) Each of these four possibilities will have the same number of terms and the same number of literals.[11] Confirm all this and write the four expressions.   ◆

3   MINIMAL REALIZATIONS OF SWITCHING FUNCTIONS

The preceding example shows a case where, with the use of a logic map, four different switching expressions can be obtained that represent the same function, where each of the expressions has the same number of terms, and the same number of literals per term. It is impossible to reduce these expressions further, either by deleting a term or by eliminating a literal from a term, without changing the logical value of the function.

Irreducible and Minimal Expressions

On the basis of the preceding paragraph, we make the following definition:

A sum-of-products expression is irreducible *if no term, or no literal from any term, can be deleted without changing the logical value of the expression.*

---

[11]Thinking in terms of eventual implementation in discrete gates, if each minterm costs the same, each AND operation costs the same, and each OR costs the same, we could say each of these implementations will cost the same. However, few, if any, digital circuits are implemented in discrete gates with current technology. More on this later.

Thus, the expressions obtained in the preceding example are irreducible.

Another idea arises when we compare various expressions representing a function. We define it as follows:

A sum-of-products expression equivalent to a function is minimal *if it has the fewest number of terms of any equivalent such expression; if there is more than one expression having the fewest number of terms, the one having the fewest literals is minimal.*

In Example 2, all four expressions have the same number of terms and the same number of literals per term. No other, equivalent, expression has fewer terms (or fewer literals per term); hence, they are all minimal. This demonstrates that the existence of a minimal expression representing a function does not necessarily imply that it is unique. Other expressions exist, equivalent to those considered in the example (e.g., the sum of all the k-cubes). However, no others are minimal.

Is it possible for an expression to be minimal but not irreducible? Look over the two definitions carefully. Suppose an expression is minimal but it can be reduced by eliminating a term or a literal in a term. That's a contradiction, since the ability to reduce it means that it is not minimal. The conclusion is that *if an expression is minimal, it is necessarily irreducible.*

How about the other way around? Is it possible for an expression to be irreducible but not minimal? The definition of an irreducible expression says absolutely nothing about the number of terms in that expression. Hence, there is no reason why an irreducible expression *must be* minimal; it may or it may not be!

Exercise 7   Return to Example 2 and write the expression that includes the 2-cube and the following 1-cubes: {2, 6}, {4, 6}, {10, 11}, and {12, 13}. Present arguments showing that this expression is irreducible but not minimal. ***Answer***[12]

Prime Implicants

A number of new concepts were introduced in the preceding section: *minimal* and *irreducible* expressions, and the *covering* of one k-cube by another. These concepts will now be formalized and extended to switching functions in general. The concept of covering, for example, can be extended as follows:

Switching function f_1 covers switching function f_2 if, whenever $f_2 = 1$, $f_1 = 1$.

Suppose, as an example, that a function $f = xy'z + wyz' + wx'y + xz'$ and that a function g is equal to the last two terms: $g = wx'y + xz'$. Whenever $g = 1$, then $f = xy'z + wyz' + 1 = 1$. Hence, f covers g.

In a sum-of-products expression, it is evident from this illustration that whenever any one of the terms is 1, the expression itself is 1. That means that a

---

[12]The expression containing the corresponding five terms is not minimal, since the minimal expressions previously found had four terms. That the expression is irreducible, however, is easily established by noting that each of the five cubes includes a minterm that is not included in any other cube. (Make sure you confirm that.)   ◆

sum of products covers every product term in the expression. For this special case of a function covering a product of literals, we make the following definition:

If a function f covers a product of literals, then the product of literals implies f or is an implicant *of f.*

In the immediately preceding illustration for example, the first term, $xy'z$, is a product of literals. Whenever $xy'z = 1$, the entire function f becomes 1. Thus, f covers $xy'z$ and, in the new terminology, $xy'z$ is an implicant of f. Now a product of literals can be a minterm or it can represent a k-cube covering 2^k minterms. *The same language of covering applies to k-cubes as to functions in general.* Since a k-cube covers each of the 2^k 1-cells that make up the k-cube, each cell is an implicant of the k-cube.

As an illustration, return to Example 2 in the preceding section. There the 2-cube {9, 11, 13, 15} covers each of the 1-cubes {9, 11}, {9, 13}, {11, 15}, {13, 15}. Each 1-cube, in turn, covers each of its two minterms. An expression that represents the function is

$$f = wz + xy'z' + w'yz' + wx'y + wxyz$$

Each of the product terms on the right is an implicant of f. But one of these (with two literals) represents a 2-cube, each of those terms with three literals represents a 1-cube, and the last one is a minterm. Suppose the literal x is deleted from $wxyz$, leaving wyz. From the map in Figure 8, this corresponds to a 1-cube covered by the 2-cube {9, 11, 13, 15} represented by wz. Both the 1-cube and the 2-cube are implicants of f. There ought to be some way to distinguish differences of this kind among implicants. The distinction is made in the following definition:

An implicant of a function f is a prime implicant *p if deletion of any literal from p results in a product of literals that is not an implicant of f.*

Thus, in the preceding illustration, $wxyz$ is an implicant, but it is not a prime implicant, because removing one of the literals x or y, or both, will still result in an implicant of f.

There is a distinction between implicants and k-cubes, although there is also a relationship. An implicant (including a prime implicant) is a product of literals; a k-cube, on the other hand, is a set of 2^k 1-cells in a map, each of which is adjacent to k others. Each 1-cell corresponds to a minterm; the sum of the 2^k minterms corresponding to a k-cube constitutes the implicant. If a k-cube is not covered by a higher-order cube, the corresponding implicant is a prime implicant. Thus, to find all the prime implicants of a function, we locate on the logical map of the function all those k-cubes that are not covered by higher-order cubes.

The value of the concept of prime implicant results from the following theorem:

If a sum-of-products expression representing a switching function f is irreducible, then it is a sum of prime implicants.

Exercise 8   Prove the preceding theorem by contradiction. That is, assume that a product is an implicant (every term in a sum *must be* an implicant) but not a prime implicant, and arrive at a contradiction.   ◆

	wx 00	01	11	10		Implicants	k-cubes
00		1	1			{9, 11, 13, 15}	wz
01			1	1		{2, 6}	w'yz'
11			1	1		{2, 10}	x'yz'
10	1	1		1		{4, 6}	w'xz'
						{4, 12}	xy'z'
						{10, 11}	wx'y
						{12, 13}	wxy'

(yz on left axis)

Figure 9 Example function with *k*-cubes and corresponding implicants.

Minimal Sum-of-Products Expressions[13]

The process of finding a minimal sum-of-products expression that is equivalent to a switching function *f* is now clear:

1. Find all the prime implicants.
2. Select the smallest possible subset of prime implicants, ensuring that all minterms are covered.

The only remaining task is to illustrate the process by way of examples.

EXAMPLE 3

The map of the function shown in Figure 8 is repeated in Figure 9, together with a list of all the *k*-cubes not covered by any higher-order ones, and their corresponding implicants. The example will be used to illustrate how to find a minimal sum-of-products expression representing a given function.

Each *k*-cube listed in the figure corresponds to a prime implicant. (Verify this claim.) Since the function has four variables, the 2-cube corresponds to a prime implicant of 4 − 2 = 2 literals, and all the 1-cubes correspond to prime implicants having 4 − 1 = 3 literals. Scanning the listed *k*-cubes shows that all 1-cells appear in more than one cube except cells 9 and 15, which appear in only one. (The example will be continued shortly.) ∎

Although arising from an example, this distinction is the basis of a general definition:

A prime implicant is an essential *prime implicant if it covers at least one minterm not covered by any other prime implicant. Any minterm that is covered by only one prime implicant is called a* distinguished *minterm.*

[13]In earlier times, the interest in finding minimal expressions representing a function had to do with reducing cost by reducing the number of primitive gates or the number of inputs per gate. Although this motivation is less compelling today, it is still significant, especially in what are called *programmable logic* devices, which will be discussed in Chapters 4 and 8.

$$
\begin{array}{c}
\qquad\qquad x \\
\qquad\quad 0 \quad 1
\end{array}
$$

	x = 0	x = 1
yz = 00	1	1
yz = 01		1
yz = 11	1	1
yz = 10	1	

$\{0, 4\}$	$y'z'$
$\{4, 5\}$	xy'
$\{5, 7\}$	xz
$\{7, 3\}$	yz
$\{3, 2\}$	$x'y$
$\{2, 0\}$	$x'z'$

Figure 10 Cyclic map.

The importance of an essential prime implicant is that it *must* be included in any minimal expression representing a function; otherwise the minterm it distinguishes will not be covered.

The process of seeking a minimal expression, hence, proceeds as follows:

1. Identify all essential prime implicants. If this set covers all minterms, the task is complete.
2. Seek the fewest number of nonessential prime implicants that cover the minterms not covered by the essential prime implicants.
3. If there is a choice in step 2, choose the nonessential prime implicants with the fewest literals.

EXAMPLE 3 (*continued*)

Now return to the example. The prime implicant wz corresponding to the 2-cube is essential. It covers four minterms (leaving only five to be covered by other prime implicants), all of which cover two minterms each. Hence, at least three more prime implicants will be needed (two more could cover only four minterms), giving a minimal expression of four prime implicants. In Example 2, using another approach, we already found four minimal expressions containing four of what we can now identify as prime implicants. ∎

EXAMPLE 4

The following function has an interesting structure of prime implicants: $f(A, B, C) = \Sigma(0, 2, 3, 4, 5, 7)$. The map, together with the list of 1-cubes and prime implicants, is shown in Figure 10.

There are six 1-cubes: $\{0, 4\}$, $\{4, 5\}$, $\{5, 7\}$, $\{7, 3\}$, $\{3, 2\}$, and $\{2, 0\}$, all representing prime implicants. Collectively, they have a certain pattern, emphasized by the order of writing the 1-cubes and their constituent minterms. Such a map is said to be *cyclic*. Each minterm appears in exactly two prime implicants. Thus, all minterms can be covered by two sets of three prime implicants each, with no common minterms:

$$f = y'z' + xz + x'y = x'z' + xy' + yz$$

(Encircle the corresponding minterms and note the pattern.) ∎

Exercise 9   Construct the map and find a minimal sum-of-products expression equivalent to the following function:

$$f(A, B, C, D) = \Sigma(0, 2, 3, 4, 8, 9, 10, 14)$$

Comment on anything unexpected you observe.
Answer[14]

Minimal Product-of-Sums Expressions

Because of the duality principle, everything that has been done in terms of sum-of-products expressions can be repeated to arrive at product-of-sums expressions. Note that the idea of adjacency applies to 0-cells as well as to 1-cells. The needed changes in definitions of such concepts as *k*-cubes, covering, irreducible expressions, implicants, and prime implicants are obvious, with obvious changes in terminology. Thus,

>*A set of 2^k 0-cells, each of which is adjacent to k others in the set, is called a k-cube; it covers each of the 2^k 0-cells.*

>*If a function f covers a sum of literals, then the sum of literals* implies *f, or is an* implicant *of f. The implicant is a* prime implicant *if deletion of any literal results in a sum of literals that is not an implicant of f.*

>*If a product-of-sums expression equivalent to a function f is irreducible, then it must be a product of prime implicants. A prime implicant is* essential *if it covers at least one maxterm not covered by other prime implicants. A maxterm that is covered by only one prime implicant is* distinguished.[15]

From all of this, it follows that a minimal product-of-sums expression must contain every essential prime implicant. It could also contain other prime implicants that are not essential.

The process of determining a minimal p-of-s expression from a maxterm list may differ somewhat in detail from the process of finding a minimal s-of-p expression from a minterm list, but all the steps are dual.

EXAMPLE 5

For the function whose minterm list was given in Example 2 and whose logic map was shown in Figure 8*a*,

  a. Write the expression that represents the function as a maxterm list.
  b. Specify a method for obtaining a list of all *k*-cubes that are not covered by any higher-order *k*-cube; then obtain the list.

---

[14]There are five prime implicants, one corresponding to a 2-cube and four corresponding to 1-cubes: {0, 2, 8, 10}, {0, 4}, {2, 3}, {8, 9}, {10, 14}. All four of the 1-cubes are essential, so they *must* be included in the minimal expression. Together they cover all the minterms, so the remaining prime implicant is not needed, even though it has fewer literals. The minimal expression is $f = A'C'D' + A'B'C + AB'C + ACD'$.   ◆

[15]To distinguish between the s-of-p and p-of-s cases, some authors use *implicate* and *prime implicate* for the p-of-s case instead of *implicant* and *prime implicant*. We will use the same term in both cases since the likelihood of confusion is minimal.

    c. Write the sum factors for each prime implicant.

    d. Write a minimal product-of-sums expression for the function. Is there more than one?

Answer

    a. All the squares on the map that are not 1 must be 0. Hence, the list is

$$\prod (0, 1, 3, 5, 7, 8, 14) = M_0 M_1 M_3 M_5 M_7 M_8 M_{14}$$

    b. One method is by the appropriate encirclings of the zeros on the map; the result is $\{1, 3, 5, 7\}$, $\{0, 8\}$, $\{14\}$, $\{0, 1\}$.

    c. The sum factors are $(w + z')$, $(x + y + z)$, $(w' + x' + y' + z)$, and $(w' + x' + y')$.

    d. The first three factors in c are essential prime implicants, and together they cover all the maxterms. Hence, there will be only one such expression:

$$f = (w + z')(x + y + z)\,(w' + x' + y' + z)$$ ■

Two-Level Implementations

The preceding development in this chapter has concentrated on obtaining different expressions to represent a given switching function. It should be possible to *implement* any of these expressions using primitive logic gates, and that's what we are now ready to do.

AND-OR Implementation

One expression that can represent a switching function is a sum of products of switching variables. A logical product such as xyz', for example, is the AND of three variables: x, y, and z'. It can therefore be implemented by an AND gate with three inputs. Every term in an s-of-p expression is in exactly the same form, differing perhaps only in the number of variables. Hence, every term can be implemented by an AND gate, each possibly with a different number of inputs.

    The entire s-of-p expression is the logical sum of such product terms. Since the logical sum is implemented by an OR gate, then the entire expression can be implemented by an OR gate whose inputs are the outputs of the AND gates that implement each logical product.

    Let's use the minimal expression obtained in Exercise 9 as an illustration. It consists of the logical sum of four terms, each one of which is the logical product of three variables. Hence, the expression can be implemented by four three-input AND gates whose outputs (four in all) are inputs to an OR gate. Complemented variables are obtained through inverters. The result is shown in Figure 11.

    Some of the variables in the expression are complemented and some are not. As already mentioned in section 7 of Chapter 2, very often both a switching variable and its complement are available.[16] Hence, nothing special needs to be done to obtain the complement, although in Figure 11 it is assumed that

---

[16]Chapter 5 will show how this comes about.

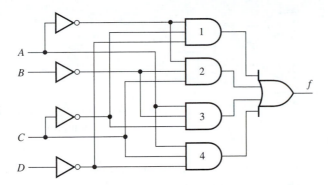

Figure 11 Two-level implementation of the function of Exercise 9.

only the variable is available as an input. (Assume that both the variables and their complements are available. Show the resulting implementation of the function.)

Not counting the inverters, which may not be present anyway, Figure 11 is a two-level circuit: all external signals must go through two gates from the initial input to the output. For TTL implementations, such circuits provide for less delay than any other circuit that might realize the same function.

Of course, the *canonic* sum-of-products expression written from the minterm list has exactly the same form as the *minimal* sum of products found in Exercise 9. Therefore, a circuit that implements it will also be a two-level circuit. (Specify the number of gates this implementation will have and the fan-in of each gate for the preceding case. Compare. Assume that variables and their complements are both available, and sketch the circuit.)

In both of the preceding implementations of the same function, the fan-ins of all AND gates are the same. This is not necessarily true of all s-of-p expressions.

Exercise 10

    a. Using the map of Figure 8*b*, write the minimal s-of-p expression resulting from the circled *k*-cubes.

    b. Draw a two-level AND-OR circuit implementing this expression, and compare the fan-ins of the AND gates.   ◆

NAND Implementation

Primitive gates of the kind used in the preceding examples are universally implemented with small-scale integrated circuits, as described in Chapter 2. Specific SSI circuits typically include several copies of the same kind of gate, all having the same fan-in. (Look back at Figure 23 in Chapter 2.) Using such circuits, two-level AND-OR implementations require the use of two different SSI types, even when all the AND gates have the same fan-in. Physical implementation would be simpler if only one kind of gate were used in logic implementations. The solution to this problem follows from two observations: one comes from Figure 9 in Chapter 2, which shows two equivalent forms of a NAND gate (look it up), and the second comes from noting that two consecutive inversions of a signal result in that same signal.

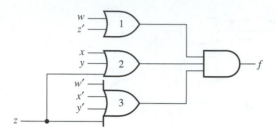

Figure 12 Product-of-sums implementation.

In the circuit of Figure 11, suppose that all the AND gates were replaced by NAND gates. The result would be unacceptable since the output of each AND gate has been inverted. No problem, however: if another inversion were introduced at each input to the OR gate, the two consecutive inversions in each line from AND gate output to OR gate input would result in no change in output. Furthermore, the OR gate with inverted inputs is another form for a NAND gate. Hence, NAND gates can replace all gates in the AND-OR circuit. (Go through the steps and draw the resulting NAND circuit.) In this particular example, the motivation of using only one kind of SSI chip has not been fully achieved; all gates are NAND, all right, but the output gate does not have the same fan-in as the others.

OR-AND Implementation

Another form in which a switching function can be expressed is the product-of-sums form. To implement such an expression, each logical sum is implemented by an OR gate, and the final logical product is implemented by an AND gate. The OR gates may or may not have the same fan-in.

As an illustration, we will use the p-of-s expression found in Example 5, repeated here:

$$f = (w + z')(x + y + z)(w' + x' + y' + z)$$

This is the logical product of three factors, each of which constitutes an input to an AND gate. These factors are themselves the outputs of OR gates whose inputs are the literals within the parentheses. We will assume that the variables and their complements are both available. The resulting implementation in primitive gates is shown in Figure 12.

Compare the structure of this circuit with that of Figure 11. They are both two-level implementations and hence have approximately the same delay. In the present case, the three OR gates do not have the same fan-in; in fact, they are all different.

It is also true, just as it was in the s-of-p case, that the implementation of a canonic p-of-s expression is a two-level circuit. Assuming a canonic p-of-s form for the preceding function, how many first-level gates will there be in an OR-AND implementation, and what will be their fan-in? What will be the fan-in of the second-level AND gate?

Exercise 11   Use the fact that two consecutive inversions of a variable result in the variable itself, together with Figure 9 in Chapter 2, to convert

Figure 12 into an all-NOR circuit. Generalize your result to all two-level OR-AND realizations. ◆

4 IMPLEMENTATION OF LOGIC EXPRESSIONS

Earlier parts of this chapter have presented different ways of *representing* a switching function, some of which seem to be "simpler" in some ways. The only implementations of logic functions discussed so far were two-level AND-OR and OR-AND circuits. Since several different expressions can be written for a given switching function, it is likely that different implementations can be obtained. Some of these may have more desirable features than others. We will consider such matters here.

One factor in the cost of an implementation is the number of gates. However, in an integrated circuit, not only do the gates take up part of the IC area, but so also do the connections between a gate output and the input to the next level gate. Furthermore, in a two-level implementation, the output gate has as many input terminals as the number of first-level gates. This number, the fan-in, can be high. In some IC technologies, the performance of a gate with high fan-in is correspondingly degraded. Hence, an increase in the number of levels is sometimes appropriate.

Exercise 12   The following six different expressions can be written for a certain switching function.

$$f(w, x, y, z) = (y + z)(wx + w'x') \qquad (a)$$
$$= y(wx + w'x') + z(wx + w'x') \qquad (b)$$
$$= wx(y + z) + w'x'(y + z) \qquad (c)$$
$$= wxy + w'x'y + wxz + w'x'z \qquad (d)$$
$$= m_1 + m_2 + m_3 + m_{13} + m_{14} + m_{15} \qquad (e)$$
$$= (y + z)(w' + x)(w + x') \qquad (f)$$

a. Confirm that each of these expressions represents the same function.
b. Assume that each expression can be implemented by AND, OR, and NOT gates. Assume also that both variables and their complements are available from an external source.[17] Find an implementation of each expression.
c. Assume that all gates have the same delay, t_p. Construct a table whose rows correspond to the different implementations and whose columns give the number of gates, the number of internal connections, the greatest delay from input to output, and the largest fan-in of any gate.

Each of the expressions given here consists of the operations AND, OR, and NOT only. Hence, you should have no difficulty in determining an implementation for each expression. Carry out the implementation of each one yourself, before you confirm it by referring to the ones given in Figure 13. Even

---
[17]Recall that devices called *flip-flops*, to be studied in Chapter 5, always put out both a variable and its complement.

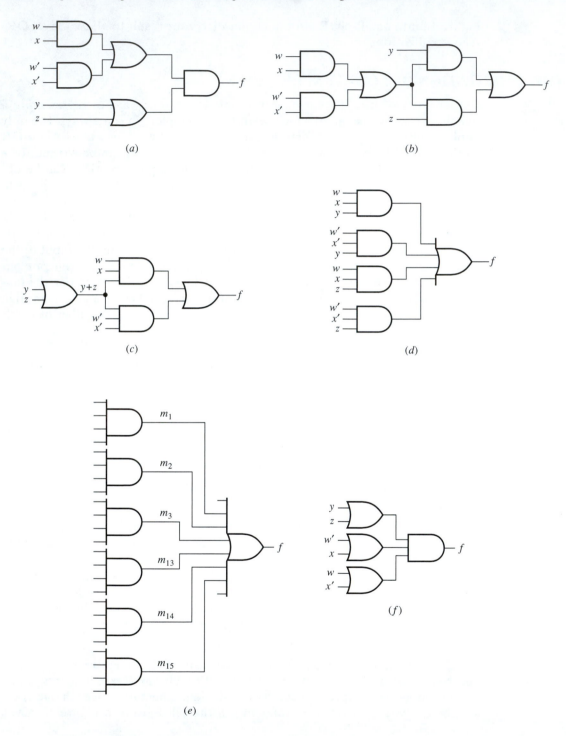

Figure 13 Different circuits implementing the same function.

though we previously carried out only two-level implementations of switching functions, we have seen how a gate represents each switching operation. The first expression, for example, is the AND of two expressions. Hence, the output comes from a two-input AND gate. Each of the two inputs, in turn, is the output of a two-input OR gate. Thus, the implementation is carried out backward from the output. Carry on!   ◆

Earlier in this chapter we spent a considerable amount of time obtaining canonic or minimal s-of-p or p-of-s expressions for a given function. These are implemented in two-level circuits. As the preceding exercise shows, it is possible to obtain multilevel implementations that have fewer internal connections and lower fan-ins. As the expressions for circuit diagrams (*a*) through (*c*) show, it's a matter of finding appropriate factors of the logic expression.

Exercise 13

    a. For practice in factoring, consider the expression in Figure 13*b*. Factor a common expression from each of the terms to obtain an equivalent expression.

    b. Implement this expression; compare number of gates, the number of internal connections, and the propagation delay through the longest path with those you constructed in Exercise 12.   ◆

Analysis

For a given truth table, minterm list, logic map, or logic expression, we have described how to implement a circuit satisfying the specified information. But how can we know that no mistakes have been made and that the circuit obtained actually has the specified outputs?

In physical circuits of all types (not just logic circuits), a process is carried out that involves making measurements (of voltage, say) at appropriate points in the circuit to verify that the measured values are what they are supposed to be theoretically. This process could be called *verification*.

But why physically implement the circuit first, before verification? Once a paper (or software-generated) logic circuit has been obtained, we can *analyze* the circuit to verify that the logic values at any point are indeed the values required by the design specifications. Compared with the process of design, logic circuit analysis is rather simple. A name can be assigned to any gate output. Logic expressions for these outputs can be written in terms of the inputs to those gates. Each input of a gate is either a primary input or the output of another gate.

If the process is carried out for the outputs of all gates, eventually only primary inputs are left in the expressions for any gate output, including those from which the circuit outputs are taken. These expressions are then compared with the given information. That's it; no big deal. In the case of a two-level implementation of a single-output circuit, the process is trivial. In other cases, and with circuits having more than one output, there may be more to it.

Consider as an example the circuit in Figure 13*b,* and let the output of the OR gate on the left be called *u*. The inputs to this OR gate are the outputs of the two AND gates to its left, which are wx and $w'x'$, respectively. Hence, this OR gate output is $u = wx + w'x'$. The circuit output is the output of the OR gate on the right; it is $f = yu + zu$. Substituting the expression for u, this confirms the expression from which the circuit was implemented.

Features of Gate Circuits

In the preceding chapter the following features of logic circuits were discussed: fan-in, fan-out, speed (or the opposite property, propagation delay), and levels. The implementations in Figure 13 differ in these respects. Note that the sum-of-products and product-of-sums implementations, whether canonic or reduced, are two-level circuits and thus have the least delay in TTL technology. However, canonic implementations are the most profligate in terms of the number of gates.[18] From their structure, s-of-p implementations are described as AND-OR circuits while p-of-s implementations are OR-AND circuits. The only gates in each of the implementations in Figure 13 are AND and OR.

As discussed in relation to the implementation of Figure 11, it would be of great value if only one kind of gate were used in each implementation; then SSI packages with only one kind of gate would be needed. In Chapter 2 we found that NAND gates are universal, as are NOR gates. Hence, the function implemented by the circuits in Figure 13 must be realizable by a circuit consisting of only NANDs or only NORs. A way of doing this is to convert each of the circuits in Figure 13 to equivalent forms that utilize only NANDs or only NORs. The key to this is Figure 9 in Chapter 2 (extended to any number of variables), with the further knowledge that a variable remains unchanged after two inversions.

Take as an illustration the s-of-p circuit in Figure 13d. Place an inversion bubble at the output of each AND gate and balance it with another inversion bubble at the corresponding inputs to the OR gate; this means that there is no change in variables on any line. Using the equivalences shown in Figure 9 in Chapter 2, all gates are replaced by NAND gates, with no modifications in any variables.

Exercise 14   Confirm this result by applying involution and De Morgan's theorem to the s-of-p expression.   ◆

Some slight modification to this procedure is needed for each of the multilevel circuits in Figure 13. In Figure 13c, for example, there are no AND gates to use with the input OR gate in order to carry out a double inversion. If a bubble is placed at each input of that OR gate, it cannot be balanced by bubbles at the output of AND gates, but it *can* be balanced by inverting the corresponding input variables. Thus, all gates can be replaced by NAND gates, but the y and z inputs must be complemented. No big deal, since complemented inputs are also available.

---

[18]What is profligate in number of gates may not be the most expensive in terms of SSI chips.

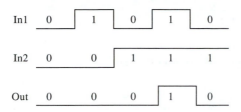

Figure 14 Timing diagram for an AND gate.

Exercise 15   Figures 13*a* and 13*f* differ from the others by having an AND gate at the output. Show that we can obtain an all-NAND equivalent, but at a price. What price?
Answer[19]

Exercise 16   Of all the circuits implementing the same logic function in Figure 13, the one that no other one beats in terms of number of gates, number of levels, number of connections, *and* maximum fan-in is the p-of-s circuit in Figure 13*f*. After converting all of the circuits to all-NAND circuits, compare Figures 13*c* and 13*f* in terms of number of gates, levels, connections, and maximum fan-in. Any comments? (Remember that input variables and their complements are both available.)   ◆

5   TIMING DIAGRAMS

The representations of digital logic circuits described so far characterize only static behavior: for given inputs, the system has specific outputs in the steady state, such as those described in a truth table. In operation, inputs are frequently changing, and outputs make transitions from one level to another in response to input changes. As a result of unavoidable gate delays, however, output transitions are delayed in time relative to input changes. This section briefly considers the effects of such delays on circuit outputs.

Sketching their waveforms as a function of time is a good way to represent input and output transitions. Such sketches, known as *timing diagrams*, are another means of describing the operation of digital systems. A timing diagram that shows the behavior of a two-input AND gate is illustrated in Figure 14. In a timing diagram the transitions between logic levels are typically drawn as vertical lines, though in reality the lines have a finite slope, as illustrated in Chapter 2, Figure 22. If the timing diagram were drawn with estimates of the real gate delays, then it would show the dynamic behavior of the system, as opposed to the static behavior conveyed by a truth table.

If signals propagated through gates without delay, logic value transitions would occur instantaneously. However, signals are inevitably delayed in traveling through gates, and different gates may introduce different amounts of delay. Thus, signals experience different amounts of delay when traversing different paths in a circuit. As a result, while the eventual steady-state logic value at a

---

[19]Increased delay resulting from an added NAND at the output in accordance with (26) in Chapter 2. ◆

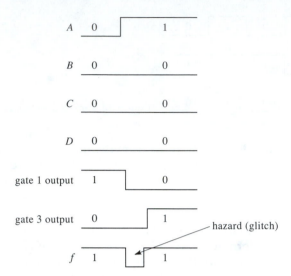

Figure 15 Timing hazard in the circuit in Figure 11.

certain point in a circuit may be the expected value, the output may assume momentary erroneous values in arriving at this steady-state value.

This process can be illustrated by reference to the circuit shown earlier in Figure 11. Suppose the circuit is in steady state with inputs $ABCD = 0000$, and suppose that gate 3 has a greater delay than gate 1. Now suppose that input A switches from 0 to 1. The output transition f can be drawn as shown in Figure 15. The output temporarily switches to 0 before reaching the proper steady-state value of 1. This temporary transition is called a *hazard* (or, colloquially, a *glitch*). Timing diagrams are the only representations of digital systems that have the ability to display such hazards. (The time scale is exaggerated.)

In combinational circuits, hazards are more of a nuisance than a threat. However, they are still undesirable for several reasons. The worst consequence occurs if the output is observed or sampled at the time of the glitch; then an erroneous value is sampled that can result in system failure. Glitches can also cause switching noise and power dissipation that may be undesirable if sensitive circuits are physically nearby (perhaps on the same chip). As we will see later (Chapter 7), there are many circumstances where hazards have no adverse effect on system operation or performance. In such cases we need not concern ourselves with them.

Is it possible to design circuits to be hazard free? To avoid the hazard demonstrated in Figure 15, for example, the circuit in Figure 11 could be designed so that gate 2 has greater delay than gate 1. But this cannot be guaranteed for every copy of the circuit that is fabricated (perhaps millions of copies). Because of the inability to precisely control the processes in the manufacture of integrated circuits, delays of "identical" gates have a certain statistical variation. The precise value of delay in a specific gate cannot be predicted; all that is possible is to specify a minimum and maximum value between which the delay of a gate on a chip will lie. Design of this circuit to guarantee that the delay of gate 2 is larger than that of gate 1 may thus be impossible or it might require substantial increase in the delay of the system.

The proper method for guaranteeing the absence of the hazard is to include an extra gate that holds the output at 0 while input A switches from 0 to 1. If the circuit implementation includes an AND gate implementing the product term $B'C'D'$, then regardless of the relative delays of gates 1 and 3 and the variation in their delays, the output will not have a glitch for this input transition. (See Chapter 7 for a more complete discussion.)

Exercise 17   For the input transition in $ABCD$ from 0000 to 0010, draw a timing diagram for the circuit in Figure 11. Is there a potential hazard? Identify a product term that should be included in the circuit to eliminate the hazard when the inputs switch from 0000 to 0010.
Answer[20]

This brief example of hazards is used to emphasize the importance of timing-diagram representations in digital system design. Hazards are treated in greater detail in Chapter 7 in connection with asynchronous sequential circuits.

6 INCOMPLETELY SPECIFIED FUNCTIONS

In all the discussion preceding this point, it has been assumed that, for every combination of input values, any switching function has a definite specified value: either 1 or 0. Such a function can be called *completely specified.*

Don't-Cares

There are occasions, however, when particular input combinations are known never to occur. In such cases, what values should be assigned to the *output?* The answer is, *we don't care.* The value can be 0 or 1, whichever makes life more pleasant.

A clear case involves a 4-bit code representing the decimal digits. (Consult Chapter 1 for a discussion of codes.) Since four variables result in 16 combinations of values to represent the 10 decimal digits, 6 of the possible combinations do not correspond to decimal digits. If the 4 bits are inputs to a switching circuit, then 6 of the 16 possible input combinations will never occur. Therefore, what the output might be for these particular input combinations is immaterial. Isn't it reasonable, then, to call the outputs *don't-cares?* Since the output in such cases is not completely specified, the function is said to be *incompletely specified* (are you surprised?). In order to show a don't-care output on a map, we need a special symbol. In this book we will use the "times" symbol, $\times$.[21]

We also need some way to include don't-cares when a function is specified by listing minterm numbers. The convention is to append a don't-care list to the minterm list, as follows:

$$f = \Sigma(\text{minterm numbers}) + \Sigma d(\text{don't-care numbers})$$

(The d here is an impetus for some authors to add m before the minterm list. We won't.)

---

[20]The product term $A'B'D'$ should be included to eliminate the hazard.   ◆

[21]There is no standard symbol that is universally used; other books use a hyphen (-) or d.

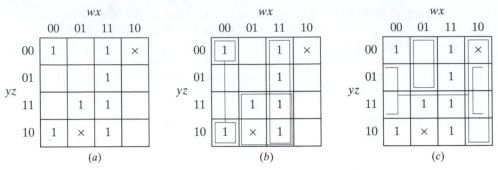

Figure 16 Maps of $f = \Sigma(0, 2, 7, 12, 13, 14, 15) + \Sigma d(6, 8)$.

EXAMPLE 6

An incompletely specified function follows:

$$f(w, x, y, z) = \Sigma(0, 2, 7, 12, 13, 14, 15) + \Sigma d(6, 8)$$

Our goal is to find minimal sum-of-products and product-of-sums expressions that represent this function. Maps with k-cubes of 1's encircled in one case and k-cubes of 0's in the other are shown in Figure 16.

The 2-cube {12, 13, 14, 15} in Figure 16b is essential, distinguished by minterm 13. Minterm 7 can form a 1-cube with minterm 15, but the resulting prime implicant BCD will have three literals. However, the don't-care can be utilized to form a 2-cube {6, 7, 14, 15} yielding a prime implicant BC with just two literals. Only one of the don't-cares is utilized to form higher-order cubes in each map. Nothing is gained in utilizing don't-care 8 in Figure 16b or don't-care 6 in Figure 16c. The number of terms and number of literals in both forms are exactly the same. Minimal expressions obtained from the maps are as follows:

$$f = wx + xy + w'x'z' \text{ and } f = (w' + x)(x + z')(w + x' + y)$$

(Don't just read all this; verify it independently.) ∎

Exercise 18   A combinational circuit with four input lines is to be designed so that the output f becomes 1 whenever the input combination $x_3x_2x_1x_0$ represents a BCD number that equals a power of 2.

    a. Construct a logic map satisfying the conditions of the design.
    b. Using the map, find a minimal s-of-p expression that represents this function.
    c. Draw a logic circuit that implements this minimal s-of-p expression.
    d. Using the map, find a minimal p-of-s expression that represents this function.
    e. Draw a logic circuit that implements this minimal p-of-s expression.
    f. Note the complexities of each circuit.

Answer[22]

---

[22]**a.** The minterm list, including don't-cares, is $f = \Sigma(1, 2, 4, 8) + \Sigma d(10, 11, 12, 13, 14, 15)$. Your map should include 1's in cells 1, 2, 4, 8 and don't-cares in cells 10 to 15.

b. None of the minterms combine with others to form cubes. Use don't-cares to form 1-cubes with

7 COMPARATORS

So far, in the "implementation" phase of combinational circuit design, the problem statements have had a "theoretical" feel to them. This section will address an important computational task: comparison of the magnitudes of two binary numbers having the same length. Dealing with this for the general case is algebraically very complex. For the trivial case of two 1-bit numbers, an Exclusive-OR gate will have an output of 0 when the 2 bits are the same and 1 when they are different. (An XNOR output would be the opposite, of course.) In either case, if they are different, we won't know which one is greater. We will build up the general case by first considering the case of two 2-bit numbers.

2-Bit Comparators

Let $X = x_1x_0$ and $Y = y_1y_0$ be two 2-bit numbers. The objective is to compare these numbers and determine their relative magnitudes. Let's define the outputs:[23]

$$G = (X > Y)$$
$$E = (X = Y)$$
$$L = (X < Y)$$

Similar notation will be defined for the individual bits (e.g., E_i means $x_i = y_i$). A logic function of two variables that becomes 1 when the variables are equal was defined in Chapter 2 as the equivalence relationship XNOR, the complement of the Exclusive-OR. Thus,

$$E_i = (x_i \Leftrightarrow y_i) = (x_i \oplus y_i)' = (x_i y_i' + x_i' y_i)' = x_i y_i + x_i' y_i', \quad i = 1, 0 \qquad (1)$$

$E_i = 1$ only when both corresponding bits are equal: $x_i = y_i$. Then E is the AND of all the E_i; in the present case, $E = E_1 \cdot E_0$, and $E = 1$ when $x_1 = y_1$ and $x_0 = y_0$. In a circuit implementation, once the E_i have been obtained, an AND gate with the E_i as inputs will give E.

Exercise 19

a. Construct a logic diagram whose inputs are x_1, x_0 and y_1, y_0 and whose output is E. Use the expression on the right in (1) and also $E = E_1 \cdot E_0$, and assume that all bits are available in parallel. Although it might be a temptation to look at the diagram provided in Figure 17, consult the figure only to confirm what you yourself have worked out.

---

minterms 2, 4, and 8. There are no higher-order cubes. Minterm 1 (20) does not form any k-cube with other minterms or with don't-cares. So the minimal s-of-p expression is f = x2'x1x0' + x2x1'x0' + x3x1'x0' + x3'x2'x1'x0.

c. On the first level are three 3-input and one 4-input AND gates, with a four-input OR gate at the second level.

d. There are six maxterms, corresponding to the decimal numbers not in the minterm list. Maxterm 0000 combines with no other maxterms or don't-cares. Maxterm 1101 combines with three don't-cares to form a 2-cube. Maxterm 0111 combines separately with each of the other three maxterms and two don't-cares to form three other 2-cubes. The p-of-s expression is $f = (x_3' + x_2' + x_1' + x_0')(x_3 + x_0)(x_2 + x_0)(x_1 + x_0)(x_2 + x_1)$.

e. On the first level are four 2-input and one 4-input OR gates, with a 5-input AND gate at the second level.

[23]G stands for "greater than," and so forth.

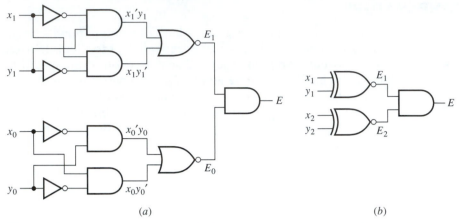

Figure 17 Implementation of a circuit showing the equality of two 2-bit numbers.

b. In Figure 17b, suppose the XNOR (equivalence) gates are replaced by XOR gates. What other compensating change would have to be made?
Answer[24]

Similarly, the condition $G = 1$ is determined by comparing corresponding bits in the two words, starting from the most significant bit. If the most significant bits are the same, the next bits are compared (and so on, conceptually, for words of greater length until the 2 corresponding bits are unequal—"conceptually" because that's not the procedure we will ultimately use for numbers having more than 2 bits).

- If the x-bit is greater than the y-bit, then $G = 1$, regardless of the remaining bits. In our case, if $x_1 > y_1$ (i.e., $x_1 = 1$ and $y_1 = 0$), then $G = 1$, regardless of x_0 and y_0.
- If $x_1 < y_1$, then G cannot be 1; in this case $L = 1$, regardless of x_0 and y_0.
- If $x_1 = y_1$, then we examine the next less significant bit; in this case G can be 1 only if $x_0 > y_0$. (A completely parallel argument can be made for L; carry it out explicitly.)

Exercise 20 Construct a four-variable logic map for G using $X = x_1x_0$ and $Y = y_1y_0$ as the variables. $G = 1$ in the squares for which $x_1x_0 > y_1y_0$. Write a non-minimal expression for G using the 2-cube and the two 0-cubes, factoring any common factors in the latter two.
Answer[25]

---

[24]Inputs to the AND gate would have to be complemented, so $E = E_1'E_0' = (E_1 + E_0)'$. Replace the AND gate by a NOR gate. ◆

[25]$G = x_1y_1' + x_0y_0'(x_1y_1 + x_1'y_1') = x_1y_1' + x_0y_0'E_1$, where $E_1 = x_1y_1 + x_1'y_1'$ is the output of an XNOR gate as shown in Figure 17b. In accordance with (13) in Chapter 2 and the subsequent discussion it also equals $(x_1 \Leftrightarrow y_1) = (x_1 \oplus y_1)' = (x_1y_1' + x_1'y_1)'$. If $E_1 = 0$, that means the first bits of the two numbers are not equal. The first term then shows that the only way G can be 1 is for $x_1 = 1$ and $y_1 = 0$, independent of the second bits. If $E_1 = 1$, the first bits of the two numbers must be the same; hence, the first term in G is 0 and G reduces to x_0y_0'. The only way that G can be 1 is for $x_0 = 1$ and $y_0 = 0$, thus confirming that $X > Y$. ◆

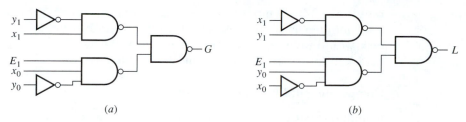

(a)   *(b)*

Figure 18 Partial circuits for (*a*) *G* and (*b*) *L*.

Exercise 21   Repeat Exercise 20, this time for *L*.
Answer[26]

     As a final step, note that the E_1 signal is available in Figure 17 as the output of a NOR gate (or an XNOR gate in Figure 17b); the AND gate outputs $x_1'y_1$ and the others in the expressions for *G* and *L* are also available in that figure. Using these as inputs, a diagram for the implementation of the *G* and *L* outputs is shown in Figure 18. (Confirm that the AND-OR circuits implied by the expressions in footnotes 25 and 26 can be implemented by all-NAND circuits, as shown.) The two diagrams can be combined into a single circuit with the four *x* and *y* inputs and the three outputs *E*, *G*, and *L*. Carry out this step yourself.

Generalization

What has been done so far might be considered trivial since only 2-bit numbers have been compared. For greater word lengths, different approaches are possible depending on whether the number of bits is even or odd. Consider first the even case of 4 bits.

4-Bit Comparators

Four-bit comparators can be obtained by using the results from two 2-bit comparators, one for the 2 higher-order bits and one for the 2 lower-order bits. If the two pairs of higher-order bits are not the same (one pair is greater than the other), then the lower-order bits need not be compared; the decision will be based completely on the two higher-order bits. (Try this out for some examples, e.g., $10x_1x_0$ and $11y_1y_0$. Demonstrate to yourself that the *x*'s and *y*'s in the third and fourth positions don't matter; the second number will always be greater than the first, since the first 2 bits are greater.) So a 2-bit comparator is needed to compare the 2 higher-order bits. If $G = 1$ or $L = 1$ for this comparator, that settles the question; the pair of lower-order bits are irrelevant.

     Only if the higher-order pairs of bits for each 2-bit comparator are the same will it be necessary to compare the 2 lower-order bits. But that case also

---

[26]$L = x_1'y_1 + x_1'x_0'y_1'y_0 + x_1x_0'y_1y_0 = x_1'y_1 + x_0'y_0E_1$. Carry out an analysis as in footnote 25 to confirm that this expression gives the correct result.   ◆

requires a 2-bit comparator; the result from this comparator will decide which 4-bit number is greater. A significant realization problem remains. The whole task would involve either only one or at most two 2-bit comparators. You will be asked to explore the details in the problem set.

Comparators of Even Numbers of Bits

Comparators of any even number of bits can be obtained in a similar fashion. The higher-order pairs of bits are treated first; if $G = 1$ or $L = 1$, that decides the matter without any need to check the lower-level bits. The next highest-level pairs are treated in the same way. Only if $E = 1$ at any pair of bits will it be necessary to check the next lower pairs. The only time it is necessary to check on the lowest-level pairs is when all the preceding bits in the two numbers are the same. The details will be left for you to carry out as a problem.

Comparators of Odd Numbers of Bits

Comparators of two numbers $A = Xa_0$ and $B = Yb_0$ having an odd number of bits can be obtained by putting the lsb of the two numbers aside and first constructing a comparator of the preceding even number of bits, X and Y. Remember that this comparator will have three outputs. If output G represents $X > Y$, then, irrespective of the least significant bits, $A > B$. On the other hand, if L represents $X < Y$, then, irrespective of the least significant bits, $A < B$. Only if $X = Y$ (meaning $E = 1$) do the least significant bits enter the picture. So additional circuitry will be needed only if $X = Y$. Then a procedure similar to what was carried out for two single-bit numbers will be needed, where the two numbers to be compared are a_0 and b_0. The details will be left for you to work out in the problem set.

8   PRIME IMPLICANT DETERMINATION: TABULAR METHOD

The determination of minimal sum-of-products or product-of-sums expressions for a given switching function can become difficult for functions of more than about five or six variables using the methods described earlier in this chapter. What is needed is a systematic approach that is specified clearly so that it can be codified in an algorithm; in that case, it can be programmed and performed by machine. Such a minimization procedure, usually called the *Quine-McCluskey algorithm*, will now be described.[27] Although a computer program

---

[27]Eliminating a few gates from a design may yield only marginal benefits in cost reduction, since designs of small circuits are normally carried out with SSI circuits in which multiple primitive gates of the same type and same fan-in are packaged together. Nevertheless, there is some intellectual benefit in comprehending the concepts involved: it leads to a deeper understanding of the structure of logic circuits. Furthermore, "large" circuits are normally implemented by prepackaged "programmable logic devices" (PLDs) to be discussed in the next chapter. The internal structure of these devices is two-level AND-OR, with a fixed number of ANDs. If a nonminimized function has too many ANDs, even an available PLD can't be used to implement it. Hence, minimization is an essential tool even for functions with a large number of variables. Moreover, the concepts developed in the context of minimizing a switching expression can be utilized in other contexts also—for example, minimizing an experiment for the diagnosis of faults in logic circuits, an advanced area not treated in this book.

to carry out the algorithm will not be discussed here, such programs are available.[28] Two steps are involved in the minimization process:

- Finding all prime implicants
- Finding a minimal set of these that covers the function

Each process is algorithmic.

The basic idea in the method to be described is that two logically adjacent minterms that are canonic products of literals (0-cubes on a map) can be combined into a product in which one of the literals is missing (a 1-cube on the map). Furthermore, two 1-cubes that are adjacent—that is, two products of literals that are the same except that the ith literal appears as x_i in one product and as x_i' in the other—can be combined into a 2-cube. This process can be continued until there are no more adjacent k-cubes. The set of *all* prime implicants is represented by the set of all k-cubes that are not entirely covered by a higher-order k-cube.

Suppose that a four-variable map includes the following minterms, as well as others: {4, 5, 12, 13}. We would tend to encircle all four minterms into a 2-cube, without necessarily noticing that the minterms form *four* different 1-cubes, all covered by the 2-cube to be sure. But if we seek a sure-fire tabular method where higher-order cubes are formed from the next-lower-order ones, we can't leave the k-cubes to a haphazard, chance identification. We must systematically consider *all* possible adjacencies.

Representations of Adjacent k-Cubes

As you know, the binary representations of two adjacent cells differ in exactly one bit position. Consider the function: $f = \Sigma(2, 3, 4, 5, 7, 8, 10, 11, 12, 13)$; with the set of minterms {4, 5, 12, 13}, the following k-cubes can be formed:

1-cube {4, 5}	**1-cube {12, 13}**	**2-cube {4, 5, 12, 13}**	
4 0100	12 1100	{4, 5}	(010-)
(010-)	(110-)		(-10-)
5 0101	13 1101	{12, 13}	(110-)

Each bit position in a binary number corresponds to a power of 2. When two adjacent k-cubes form a $(k+1)$-cube, the position in which the binary values differ is replaced by -, as shown above.

The following question arises: If the binary representations of two adjacent minterms differ in exactly one position, how do their decimal representations differ? Since each position corresponds to a power of 2, and since the bit in this position is 1 in one minterm and 0 in the other, the difference in the two decimal representations is exactly this power of 2. That is,

If two minterms are adjacent, their decimal minterm numbers differ by a power of 2.

---

[28]For a description of a Pascal language program for carrying out the Quine-McClusky algorithm, see pp. 236–243 of Wakerly, *Digital Design*. (See the Bibliography.) The process described in this section is algorithmic and repetitive; it might be considered tedious. Since a computer program for the algorithm discussed exists, you might wish to skip the section. However, it is intellectually satisfying and might be worth at least a less-than-full immersion.

Index (No. of 1's)	0-Cubes (Minterms)	1-Cubes	2-Cubes
1	2✓	2, 3 (1)✓	2, 3, 10, 11 (1, 8) P_1
	4✓	2, 10 (8)✓	4, 5, 12, 13 (1, 8) P_2
	8✓	4, 5 (1)✓	
		4, 12 (8)✓	
		8, 10 (2) P_3	
2	3✓	8, 12 (4) P_4	
	5✓		
	10✓	3, 7 (4) P_5	
	12✓	3, 11 (8)✓	
		5, 7 (2) P_6	
3	7✓	5, 13 (8)✓	
	11✓	10, 11 (1)✓	
	13✓	12, 13 (1)✓	

(a)

Prime Implicants

$P_1 = \{2, 3, 10, 11\}$
$P_2 = \{4, 5, 12, 13\}$
$P_3 = \{8, 10\}$
$P_4 = \{8, 12\}$
$P_5 = \{3, 7\}$
$P_6 = \{5, 7\}$

(b)

Figure 19 Tabular determination of the prime implicants of $f = \Sigma(2, 3, 4, 5, 7, 8, 10, 11, 12, 13)$.

Is the converse also true? That is, if the minterm numbers differ by a power of 2, are the minterms adjacent? The answer is no! As an example, minterms 10 (1010) and 6 (0110) differ by $4 = 2^2$, a power of 2; but they are not adjacent. Their binary values differ in two positions, not just one. This fact has implications for the procedure we are in the process of describing; it means that testing for adjacency should never be carried out in decimal form, but always in binary representation.

Ranking by Index

If the binary values of two minterms differ in one position, that means the number of 1-bits differs by 1. Let the *index* of a minterm be defined as the number of 1's in its binary representation. That means that if the minterms are classified by index, then only those whose indices differ by 1 can possibly be adjacent. But what about the converse? If the indices of two minterms differ by 1, is it necessarily true that the binary values differ in one *position*? The answer is again no. The indices of minterms 8 and 3, for example, are 1 and 2, respectively. Yet their binary values differ in three positions. Thus, the condition that the decimal numbers of two minterms differ by a power of 2 and the condition that the number of 1's in their binary representations differ by 1 are *necessary* conditions, but *not sufficient*.

Let us now turn to the tabular procedure for a given function. The first step is to list the minterm numbers, grouped by index, in the first column of a table. The example introduced in the preceding subsection is used in the table in Figure 19.

The next step is to test those minterms that *might* be adjacent on the basis of their indices to see if they actually are. Each minterm number in the group with index 1 is checked against each one in the group with index 2. The 1-cubes formed by two adjacent 0-cubes are identified by listing the pairs of minterm numbers followed (in parentheses) by the value of the power of 2 in which their binary values differ, as shown in the third column. The 1-cube formed by minterms 4 and 12, with a difference of 8, for example, is labeled 4, 12 (8). Since those 0-cubes are covered by a higher-order k-cube, and therefore cannot be prime implicants, they are checked off.

Note that even though minterm 4 (index 1) differs from minterm 3 (index 2) by a power of 2, m_4 and m_3 are not adjacent; they differ in three binary positions. Furthermore, 4 in a lower-index group is a larger number than 3 in the next-higher-index group. The generalization of these observations is this: There is no need to check for adjacency a minterm having a lower index with one having a higher index if the latter is a smaller number; they cannot be adjacent.

When all 1-cubes involving minterms with indices 1 and 2 are formed, then index-2 minterms are checked with index-3 minterms for adjacency. However, the 1-cubes formed by such adjacencies are listed separately from the previous set. The process is continued until all indices are exhausted. In the present example there are no minterms with higher indices, so this step terminates.

What to do next is evident: we form 2-cubes from adjacent pairs of 1-cubes. The numbers in parentheses following the minterm numbers in the 1-cubes are the powers of 2 corresponding to the binary position in which the two minterms that form the 1-cube differ. The 1-cube 4, 12 (8), for example, has the binary representation -100. The binary value of 12 has a 1 in the 8 $(=2^3)$ position while the binary value of 4 has a 0 there. If this 1-cube is to be adjacent to another 1-cube, that other one must also have an 8 in parentheses, which means a - in the 2^3 position of its binary value.

At the same time, in the remainder of their bits, the two 1-cubes must differ in exactly one position. These two conditions are confirmed in two steps by comparing each 1-cube in the first category with those in the next category. First, the numbers in parentheses must agree; then the first of the two minterms in each 1-cube must differ by a power of 2 if the two 1-cubes are to be adjacent. (If all minterms are listed in ascending order within each k-cube, then checking the first minterm is enough; the others will automatically differ by the same power of 2.)

Starting with 1-cube 2, 3 (1), for example, the only 1-cubes in the next group to check are ones having (1) in parentheses: 10, 11 (1) and 12, 13 (1). To check whether the first minterms of the pairs of 1-cubes differ by a power of 2, we note that 2 and 10 do $(10 - 2 = 8 = 2^3)$, but 2 and 12 do not. So only (2, 3, 10, 11) form a 2-cube. To identify the positions in which their bits differ, we list both 1 and 8 in parentheses, as shown in the last column of the table.

Then the next 1-cube, 2, 10 (8) in the first group, is checked with those in the next group. The only candidates with which it can form a 2-cube are those with 8 in parentheses: 3, 11 (8) and 5, 13 (8). For 3, 11 (8), the first minterms (2 and 3) are adjacent, but not so for 5, 13 (8). The 2-cube thus formed is 2, 3, 10, 11 (1, 8); it is the same as the one already found and so need not be listed again. At each step, 1-cubes that form a 2-cube are checked off as having been covered and, hence, as not being prime implicants. The process of forming 2-cubes is continued until all 1-cubes have been exhausted. In the present example, only two 2-cubes exist. Eight of the 1-cubes are wholly covered by them, leaving four 1-cubes not covered. All of these 1-cubes are also prime implicants, making a total of six, as listed in the table.

The description just given has been lengthy, but the actual process is relatively simple. A few steps are repeated over and over; as already stated, the process can be implemented by a computer program.

Table 1 Summary of Prime Implicant Determination

1. Given a switching function, obtain a minterm list; group the decimal minterm numbers in a column in ascending order of index (number of 1's).
2. Test for adjacency each minterm having a lower index value with each one having the next-higher index value, starting with the lowest index; they are adjacent if their decimal values differ by a power of 2 and their binary values differ in only one bit position. List the 1-cubes formed by adjacent minterms, giving both minterm numbers in ascending order, and the power-of-2 difference between them in parentheses. Repeat until all index values are exhausted. Eliminate each minterm covered by a 1-cube since it is not a prime implicant. Separate the sets of 1-cubes formed by each index pair.
3. Starting with the lowest index pair, repeat for the groups of 1-cubes. Check for adjacency all 1-cubes in a group formed by an index pair with those in the group formed by the next index pair; they will be adjacent only if they have the same number in parentheses and if the first minterm numbers differ by a power of 2. List the 2-cube formed by adjacent 1-cubes, giving all minterm numbers in ascending order and, in addition to the number already in parentheses, the power-of-2 difference in the first listed minterm of the two 1-cubes. Again, 1-cubes covered by a 2-cube are eliminated to indicate that they cannot be prime implicants. Separate the sets of 2-cubes formed by each group of 1-cubes.
4. Repeat for each set of k-cubes until no higher-order cubes can be formed. All k-cubes not eliminated represent prime implicants. Number them, starting with the highest-order ones.

The process of finding all prime implicants by the tabular method is summarized in Table 1. The process generates *all* prime implicants of a specified switching function. From this set, we still need to select a minimal set, with the fewest literals, that covers all minterms. That is the subject of a subsequent section.

Incompletely Specified Functions

The preceding procedure, which applies to completely specified functions, can be easily extended to incompletely specified functions. In carrying out the algorithm for that case, all the don't-cares are taken as 1's. This will lead to a relatively large number of prime implicants, including some that cover only don't-cares. Although this requires some redundant effort, the subsequent step of selecting a minimal set among the prime implicants will guarantee that a minimal expression does not include such all-don't-care prime implicants, as the next section will show.

EXAMPLE 7

Given the following function, the objective is to generate all the prime implicants:

$$f(A, B, C, D, E) = \Sigma(0, 4, 5, 7, 9, 12, 13, 14, 15, 23, 31) + \Sigma d(3, 6, 10, 16, 20)$$

The table constructed by listing, in accordance with their indices, all don't-cares as if they were 1's is shown in Figure 20. Since 0 is a minterm in this example, index 0 also appears. The process outlined in Table 1 is carried out. All those unchecked k-cubes are prime implicants. Confirm all details of this table. ∎

Index	0-Cubes	1-Cubes	2-Cubes	3-Cubes
0	0✓	0, 4 (4)✓ 0, 16 (16)✓	0, 4, 16, 20 (4, 16)P_2	4, 5, 6, 7, 12, 13, 14, 15 (1, 2, 8) P_1
1	4✓ 16✓	4, 5 (1)✓ 4, 6 (2)✓	4, 5, 6, 7 (1, 2)✓ 4, 6, 12, 14 (2, 8)✓	
2	3✓ 5✓ 6✓ 9✓ 10✓ 12✓ 20✓	4, 12 (8)✓ 4, 20 (16)✓ 16, 20 (4)✓ 3, 7 (4)P_4 5, 7 (2)✓ 5, 13 (8)✓ 6, 7 (1)✓	4, 5, 12, 13 (1, 8)✓ 5, 7, 13, 15 (2, 8)✓ 6, 7, 14, 15 (1, 8)✓ 12, 13, 14, 15 (1, 2)✓	
3	7✓ 13✓ 14✓	6, 14 (8)✓ 9, 13 (4) P_5 10, 14 (4) P_6 12, 14 (2)✓	7, 15, 23, 31 (8, 16) P_3	
4	15✓ 23✓	7, 15 (8)✓ 7, 23 (16)✓		
5	31✓	13, 15 (2)✓ 14, 15 (1)✓ 15, 31 (16)✓ 23, 31 (8)✓		

Figure 20 Tabular procedure for prime-implicant determination of the function
$f = \Sigma(0, 4, 5, 7, 9, 12, 13, 14, 15, 23, 31) + \Sigma d(3, 6, 10, 16, 20)$.

Selection of a Minimal Expression

Once the prime implicants have been determined, the next step is to select the smallest number of them that together cover all minterms and have the least number of literals. This process is simplified by constructing a *prime implicant table*, or *chart*, in which the prime implicants constitute the rows and the minterms constitute the columns. If prime implicant P_j covers minterm m_i, a check mark is placed at the intersection of the row corresponding to P_j and the column corresponding to m_i.

Completely Specified Functions

Because there are differences, though small, we will first treat completely specified functions, starting with an example.

EXAMPLE 8

Prime implicants for $f = \Sigma(2, 3, 4, 5, 7, 8, 10, 11, 12, 13)$ were established in Figure 19. The resulting prime implicant chart is shown in Figure 21. The higher-order prime

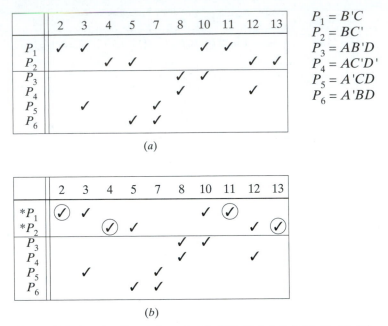

$$P_1 = B'C$$
$$P_2 = BC'$$
$$P_3 = AB'D$$
$$P_4 = AC'D'$$
$$P_5 = A'CD$$
$$P_6 = A'BD$$

Figure 21 Prime implicant chart for $f = \Sigma(2, 3, 4, 5, 7, 8, 10, 11, 12, 13)$.

implicants are listed first and the set of 2-cubes is separated from the 1-cubes. Note that the number of minterms covered by a k-cube is 2^k. That's how many check marks there should be in a row for a completely specified function. Confirm all the details of this table.

Since each essential prime implicant covers a minterm that is not covered by any other prime implicant, a minimal set *must* include the essential prime implicants. To identify these, each column is scanned vertically. If there is a column with a single check mark, that column corresponds to a distinguished minterm, and the row in which it appears is an essential prime implicant. As shown in Figure 21*b*, when a minterm is covered by a prime implicant that is selected, a circle is placed around the check mark.

In Figure 21, there are four distinguished minterms but just two essential prime implicants, indicated by asterisks. The minterms covered by the essential prime implicants don't need to be covered by any other prime implicant. To identify the already covered minterms, a check mark is placed above those minterm numbers at the head of the columns. The only minterms not covered by P_1 and P_2 are 7 and 8. Scanning those columns shows that minterm 7 is covered by either P_5 or P_6, and minterm 8 is covered by either P_3 or P_4.

Since the number of literals in each of these is the same, the selection of any combination of these prime implicants will lead to a minimal expression:

$$f = P_1 + P_2 + P_3 + P_5 = P_1 + P_2 + P_3 + P_6$$
$$= P_1 + P_2 + P_4 + P_5 = P_1 + P_2 + P_4 + P_6$$

The product-of-literals form for a prime implicant corresponding to a k-cube is obtained as follows. First write the binary value of any one of the minterm num-

	✓ 0	✓ 4	✓ 5	✓ 7	✓ 9	✓ 12	✓ 13	✓ 14	✓ 15	✓ 23	✓ 31
*P_1		✓	⊘	✓		⊘	✓	✓	✓		
*P_2	⊘	✓									
*P_3						✓			✓	⊘	⊘
P_4						✓					
*P_5					⊘		✓				
P_6								✓			

Figure 22 Prime implicant chart for incompletely specified function in Figure 20.

bers covered by the k-cube; then replace with - the bit in each position where one of the other minterms in the k-cube has the complementary value. Thus, in 1-cube $P_4 = (8, 12)$ the binary value for 8 is 1000. The binary value for 12 differs from this only in the 2^2 position, leading to 1-00. Assuming the variables are A, B, C, D, the second variable, B, is missing in the prime implicant, and the product is $AC'D'$. Verify the expressions for the other prime implicants given in the figure. ∎

Handling Don't-Cares

When determining the set of all prime implicants for an incompletely specified function, we saw earlier that the don't-cares are treated as if they are 1's. However, it is not *required* that don't-cares be covered by a minimal expression. Hence, in setting up a prime implicant table for incompletely specified functions, the don't-cares are not listed among the columns, even though some prime implicants do cover don't-cares. If a prime implicant that covers a don't-care ends up in a minimal expression, this don't-care takes on the value 1. The don't-cares that are not covered by any of the final prime implicants must have been specified as 0.

The procedure is illustrated using the example whose prime implicants were determined in Figure 20. That function has five don't-cares. They are not listed as minterms in the prime implicant table constructed in Figure 22. Observe from rows P_1, P_2, P_3, and P_4 that a row corresponding to a k-cube no longer requires 2^k check marks. There are four essential prime implicants, and together they cover all the minterms. Hence, the unique minimal expression is

$$f = P_1 + P_2 + P_3 + P_5 = A'C + B'D'E' + CDE + A'BD'E$$

Confirm the expressions for the prime implicants.

9   MULTIPLE-OUTPUT CIRCUITS

The circuits that have been treated so far have had just a single output. However, in many practical circuits there is more than one output dependent on the same set of inputs. A simple example is the following minterm expressions that describe a circuit with three inputs and two outputs:

$$f_1(a, b, c) = \Sigma(0, 1, 3), \qquad f_2(a, b, c) = \Sigma(0, 4, 5)$$

These two functions can be treated independently using the methods described earlier in the chapter, and a realization utilizing three 2-input NAND gates can be obtained for each function. Verify this by constructing the logic maps and deriving the implementations.

The functions are not independent, however, because they share a minterm. If f_1 is implemented as $a'b'c' + a'c$ and f_2 is implemented as $a'b'c' + ab'$, then the output of the 3-input gate $(a'b'c')$ can be used twice, once for each output. This implementation contains four 2-input NAND gates and one 3-input NAND gate. Identify this realization in the logic map and draw the logic gate schematic diagram. The savings in this particular example may not be great, but in many other cases, substantial economy can result.

The task of minimizing multiple-output circuits is normally left to computer-aided design tools. That subject is beyond the scope of this book, and we will not pursue the matter further. Nevertheless, most of the circuits presented in subsequent chapters are multiple-output by nature. Among these are the programmable-logic devices described in Chapter 4.

CHAPTER SUMMARY AND REVIEW

This chapter laid the foundations for methods of representing and implementing logic functions. Implementations considered here are those using primitive gates, the precursor for MSI circuits and implementations with PLDs, to be discussed in subsequent chapters. The following topics were included.

- Minterms, minterm lists, and sum-of-products implementations
- Maxterms, maxterm lists, and product-of sums form
- Logical (Karnaugh) maps and adjacency
- Logical adjacency and map adjacency
- Grouping minterms on a map
- Cubes of order k, or k-cubes
- Irreducible and minimal logical expressions
- One function covering another
- Implicants of a switching function
- Prime implicants
- Essential prime implicants
- Equivalent expressions
- Minimal sum-of-products expressions
- Minimal product-of-sums expressions
- Implementations of switching functions:

    - Two-level implementations
    - AND-OR implementation
    - NAND implementation
    - OR-AND implementation
    - NOR implementation
    - Multilevel implementations

- Analysis of logic circuits
- Incompletely specified functions and don't-cares
- Magnitude comparators
- Prime implicant determination: Quine-McCluskey algorithm

- Ranking by index: completely specified functions
- Incompletely specified functions
- Obtaining a minimal expression

- Timing diagrams
- Hazards
- Hazard-free circuits

PROBLEMS

Remember to save the solutions to early parts of the problems so that you can refer to them when you are ready to do the later parts.

1 For each truth table given in Figure P1:

   a. Construct the minterm list.
   b. Construct the logic map (K-map).
   c. Write the canonic sum-of-products expression.
   d. Specify all the prime implicants and note any that are essential.
   e. Find at least one minimal s-of-p expression from the map.
   f. Write the canonic product-of-sums expression.
   g. Find at least one minimal p-of-s expression using the map.

$x\ y\ z$	f_1	f_2	f_3	f_4
0 0 0	1	1	0	1
0 0 1	1	0	1	1
0 1 0	0	0	1	1
0 1 1	0	1	1	0
1 0 0	0	0	0	0
1 0 1	1	1	1	1
1 1 0	0	1	1	0
1 1 1	1	0	0	0

Figure P1

2 Construct a logic map for each of the following functions.

   a. $f_1 = y + xz' + y'z$
   b. $f_2 = xy + x'z' + yz'$
   c. $f_3 = ABD' + A'BD + A'C'D + AC + BCD + B'C'D$
   d. $f_4 = A'B + A'CD' + AC'D + B'CD' + BC'D$

For each case, find a minimal sum-of-products expression.

3 Two variations of a three-variable logic map were given in Figure 3. Now it's your turn.

   a. Determine an ordering of the variables $wxyz$ in a four-variable map, and the ordering of their logical values, such that each cell is adjacent to four other cells.
   b. Given $f = (wx' + y'z)(x + w'y')$, construct a logic map using the pattern of part *a*.
   c. From the map, determine minimal s-of-p and p-of-s expressions representing this function.

4 The following function of four variables is given:

$$f = \Sigma(0, 3, 4, 6, 7, 9, 12, 13, 14, 15)$$

a. With the use of a map, find the set of all prime implicants and specify those that are essential.

b. Specify the 1-cells in each map and find the minimal s-of-p expressions.

c. Repeat parts *a* and *b* for the complement, $f\,'$.

d. Convert the expression for $f\,'$ to an expression for f:
   i. Using De Morgan's theorem once
   ii. Using De Morgan's theorem twice

e. Repeat parts *a* to *d* for each of the following functions:

$$f_1 = \Sigma(1, 5, 6, 9, 10, 11, 12, 13)$$
$$f_2 = \Sigma(1, 2, 4, 6, 7, 14, 15)$$
$$f_3 = \Sigma(0, 3, 5, 7, 11, 13, 14, 15)$$
$$f_4 = \Sigma(1, 2, 4, 6, 8, 9, 10, 11)$$

f. Obtain a sum-of-products implementation of each preceding function.

g. Obtain a product-of-sums implementation of each function.

5 The following switching function is given:

$$f(A, B, C, D) = \Sigma(0, 1, 2, 4, 6, 10, 11, 12, 13, 14)$$

a. With the use of a logic map, find all the prime implicants and specify any that are essential.

b. Find as many distinct minimal expressions for f as you can.

c. Find minimal expressions for the complement, f', directly from the map.

d. Repeat the preceding parts for each of the functions in Problem 4.

6 The following function of four variables is given:

$$f = \Sigma(1, 3, 5, 6, 7, 9, 11, 12, 13)$$

a. With the use of a logic map, find all possible minimal s-of-p expressions.

b. Repeat for p-of-s expressions. Which implementation has fewer terms and literals?

c. From the map, find a minimal s-of-p expression for f'.

7 With the use of a logic map, find minimal s-of-p expressions for the following functions. In each case, find the distinguished 1-cells in the map.

a. $f_1 = \Sigma(0, 2, 3, 4, 6, 8, 10, 11, 12, 14)$
b. $f_2 = \Sigma(0, 4, 6, 8, 12, 14)$
c. $f_3 = \Sigma(1, 3, 7, 9, 12, 13, 14, 15)$
d. $f_4 = \Sigma(0, 2, 6, 7, 8, 9, 13, 15)$
e. $f_5 = \Sigma(4, 5, 6, 9, 11, 13, 14, 15, 20, 21, 22, 25, 27, 29, 31)$

Also find a minimal s-of-p expression for the complement of each function.

8 **a.** Using a logic map, find all minimal p-of-s expressions for the functions in Problem 7.

b. Repeat for the complement of each function.

9 **a.** A switching function of four variables, $f(w, x, y, z)$, is to equal the product of two other functions, f_1 and f_2, of the same variables: $f = f_1 f_2$. The functions f and f_1 are given:

$$f = \Sigma(4, 7, 15)$$
$$f_1 = \Sigma(0, 1, 2, 3, 4, 7, 8, 9, 10, 11, 15)$$

Find the number of fully specified functions f_2 (without don't-cares) that will satisfy the given condition. Select the simplest of these functions.

b. A logic circuit is to be designed to realize the following function f.

$$f(A, B, C, D) = A'B + B'C + BD'$$

Either by manipulating this expression or from logic maps of f or f', arrive at a form whose implementation includes any number of AND and OR gates but only a single inverter.

c. A logic function f has n prime implicants, $P_1, \ldots, P_n$. It is found that the pairwise product $P_i P_j = 0$ for all pairs. Show that the function has a unique minimal sum-of-products expression.

d. A function of four variables is to have eight minterms and no essential prime implicants. Show a possible logic map for such a function.

10 Using the tabular method, find minimal s-of-p expressions for f:

    a. In Problem 4
    b. In Problem 5

11   **a.** Repeat Problem 6 using the tabular method.
    b. Repeat Problem 7 using the tabular method.

12 The logic map in Figure P12 has no essential prime implicants.

    a. Find the maps of two other such four-variable functions that have the same number of minterms. Find a minimal s-of-p expression in each case.
    b. Find the map of a four-variable function with nine minterms that has the same property.

Figure P12

13 Given the incompletely specified function of four variables

$$f = \Sigma(0, 1, 2, 4, 7, 12) + \Sigma d(8, 10, 15)$$

    a. Find a minimal s-of-p expression.
    b. Find a minimal p-of-s expression.
    c. Use the distributive law to convert your part b answer to a s-of-p expression. Explain why it is not the same as the answer to part a.

14 The following incompletely specified functions are given. With the use of a logic map, find minimal s-of-p and p-of-s expressions for each.

    a. $f_1 = \Sigma(2, 4, 6, 10, 12) + \Sigma d(0, 8, 9, 13)$
    b. $f_2 = \Sigma(4, 5, 7, 12, 14, 15) + \Sigma d(3, 8, 10)$

   c. $f_3 = \Sigma(0, 6, 7, 10, 12) + \Sigma d(2, 8, 9, 15)$
   d. $f_4 = \Sigma(1, 2, 3, 4, 5, 11, 18, 19, 20, 21, 23, 28, 31) + \Sigma d(0, 12, 15, 27, 30)$
   e. $f_5 = \Sigma(5, 7, 12, 13, 15, 23, 24, 27, 28, 31) + \Sigma d(8, 16, 18, 21, 26, 29)$

15 Repeat Problem 14 using the tabular method.

16 The four inputs to the logic circuit shown in Figure P16 represent a binary-coded decimal (BCD) digit in the order $x_3 x_2 x_1 x_0$. Those input combinations that do not correspond to a BCD code word for a decimal digit are known never to occur. The output z is to be 1 if and only if the input word represents an even digit.

   a. Complete a logic map for the given condition.
   b. From the map, find a minimal s-of-p expression.
   c. Draw a two-level AND-OR implementation from this map; then convert it to an all-NAND circuit.
   d. Suppose that only two-input NANDs are available. Construct a circuit using only these gates.
   e. Find a minimal p-of-s expression.
   f. Construct a two-level OR-AND implementation from the map; then convert it to an all-NOR circuit.
   g. Draw a circuit implementation if only two-input NORs are available.

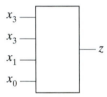

Figure P16

17 Suppose that the input word in Problem 16 represents a digit that is either 0 or a power of 2. Repeat steps *a* through *g* for this case.

18 Suppose that the input word in Problem 16 represents a digit that is prime (0 and 1 are not considered prime). Repeat steps *a* through *g* for this case.

19 Suppose that the input word in Problem 16 corresponds to the decimal digit 0, 2, 3, 5, or 8. Repeat steps *a* through *g* for this case.

20 In the setup of Problem 16, suppose that the forbidden combinations representing decimal numbers from 10 to 15 are no longer forbidden.

   a. Repeat Problem 17 for the power-of-2 detector.
   b. Repeat Problem 18 for the prime number detector.

For each case, make note of the essential prime implicants and the minterms (or maxterms) that distinguish them.

21 Suppose the number of inputs in Problem 16 is increased to five and the set of inputs represents a decimal number in binary code, not BCD. Repeat Problem 18 for the prime number detector.

22 A set of logic circuits is to have an even number of input lines: four, six, or eight. Half of the inputs are labeled x_i, the other half y_i. Each set of x_i inputs represents a decimal number (in binary code $X = x_1 x_0$, $x_2 x_1 x_0$, or $x_3 x_2 x_1 x_0$), as does each set of y_i inputs to form Y in a similar order. The purpose of each circuit is to compare the two incoming numbers X and Y and to produce output $z = 1$ if and only if $X > Y$. Find minimal s-of-p and p-of-s expressions for each case.

23 A certain university with excess professors uses five of them to grade each exam, with each professor giving the exam a *pass* (1) or a *fail* (0). To pass an exam, a student must receive either four or five passes. The exam is failed if a student receives either four or five fails. The exam must be retaken if the number of passes is either two or three. A logic circuit is to be designed, having five x_i inputs and two outputs y and z. Both outputs should equal 1 if the exam is passed; both outputs should be 0 if the exam is failed; the outputs should be $yz = 10$ if the exam is to be repeated.

    a. Find a minimal s-of-p expression for z.
    b. From the maps of y and z, deduce a relationship between the two outputs. From that relationship, find a p-of-s expression for y.
    c. Confirm that this expression is minimal.
    d. Use the distributive law (and anything else) in part b to obtain a s-of-p expression for y.
    e. Confirm by the tabular method that this is a minimal expression.
    f. Modify the expressions for the two outputs so as to make as many common terms as possible, thus reducing the number of gates as much as possible.

24 The prime implicant table of a function is shown in Figure P24. Every prime implicant of the function is included in the rows of the table. Every minterm (for which $f = 1$) is listed in the column headings, although two of the minterms are not known. (The order of those minterm numbers is not necessarily as shown.)

    a. Find the minterm numbers m_a and m_b. If there is more than one possibility, give them all. Show a logic map of the function.
    b. Find an expression for each of the unspecified prime implicants.

Prime	Minterms						
Implicants	1	3	4	6	9	m_a	m_b
$P_1 = B'D$	✓	✓			✓		
$P_2 = A'BD'$			✓	✓			
$P_3 =$				✓		✓	
$P_4 =$			✓				✓
$P_5 =$	✓						✓
$P_6 =$					✓		

Figure P24

25 Given a function represented by an expression $f(\{x_i\})$, with $i = 1, 2, \ldots, n$, we define the *dual function* f_d as represented by the expression obtained by interchanging the logical operations of addition and multiplication. (If the constants 0 and 1 appear explicitly in the function, they are also interchanged in the dual. However, if we assume that the function has no redundancies, the constants will not appear explicitly.) Prove the following:

$$f_d = f'(\{x_i'\})$$

That is, the dual of a function results from complementing the expression representing it and also complementing each literal in the expression. (In this process, due attention must be given to ensuring that implied parentheses are honored.)

26 A three-variable switching function has minterms m_3 and m_5. If the literals in these minterms are complemented, what are the corresponding minterm numbers? Repeat for a four-variable function with minterms m_6 and m_{13}. Generalize these observations for a function with n variables. That is, what is the number of the minterm that results from complementing the variables in m_j?

27 In light of Problem 26, given the map of a function, interpret how the map of its dual can be constructed.

28 A function that is its own dual is called a *self-dual* function. In view of Problems 25 and 26, give two necessary conditions for an n-variable function to be self-dual.

29 **a.** Using the result of Problem 28, decide which of the functions whose maps are shown in Figure P29 might be self-dual.

b. Repeat for the function: $f(A,B,C,D) = \Sigma(0, 2, 3, 7, 9, 10, 14, 15)$.

(a)

yz \ x	0	1
00	1	
01	1	
11		1
10		1

(b)

yz \ wx	00	01	11	10
00				
01	1	1		1
11		1	1	1
10		1		1

(c)

yz \ wx	v = 0: 00	01	11	10	v = 1: 10	11	01	00
00	1			1	1		1	1
01			1	1			1	
11	1	1			1			1
10	1	1	1		1	1		

Figure P29

30 Given a switching function g, prove that the following function is self-dual: $f = x_i g + x_i' g_d$, where x_i is a switching variable that may or may not be a variable in g.

31 Using the result of Problem 30, construct some self-dual functions using each of the g functions below.

a. $g = x + y$
b. $g = x'y$
c. $g = A + B'C$ (First use B as variable x_i, then use C.)

32 Using a logic map, show that a minimal p-of-s expression representing a self-dual function can be obtained from a minimal s-of-p expression by interchanging the sum and product operations.

33 Let $f(A, B, C, D) = f_1(A, B, C, D) + f_2(A, B, C, D)$, where $f_1 = B'D' + A'C'D' + AB'C'$. Find f_2 such that f will be the simplest self-dual function:

34 **a.** Assuming that both the variables and their complements are available as inputs, find an implementation of the following expression, without manipulating it:

$$f = xy + (w + z)(w' + z')$$

Note the number of levels, the overall propagation delay, and the maximum fan-in of each gate. Analyze your circuit and confirm that the output is the same as in the given expression.

b. Find five other expressions that represent the same function as the one in part *a* and implement each one. For each one, note the fan-in of each gate and specify the maximum fan-in. Compare the expressions in terms of number of levels and propagation delay, assuming equal gate delays.

c. Convert each circuit in parts *a* and *b* to NAND-only circuits. In each case, note if the number of gates has increased.

d. Construct a table whose rows are the implementations in *a* and *b*. Make columns corresponding to number of gates, number of internal connections, maximum fan-in of any gate, and longest delay from an input to the output. Compare these for the various implementations.

e. Analyze each circuit to obtain expressions for the output. Confirm that each expression represents the same function as in the originally given expression.

35 Repeat Problem 34 for each of the following expressions.

    a. $f = C(A' + DE) + D'(B'E' + A'BE)$
    b. $f = A(B' + CD') + B(AC' + D) + C'(A'B + AB'D)$

36 The number of minterms in a function having n variables is 2^n. Show that the largest number of terms in a minimal sum-of-products expression representing any such function is 2^{n-1}, half the number of minterms.

37 Using at most one additional inverter, convert each circuit in Figure 13 in the text to:

    a. A NAND-only circuit
    b. A NOR-only circuit

38 A 2-bit comparator is a logic device with two pairs of input lines—x_1, x_0, and y_1, y_0—and one output, z. The combinations $X = x_1x_0$ and $Y = y_1y_0$ represent binary numbers. The output is to be 1 whenever $X \geq Y$.

    a. Using a logic map, write a sum-of-products expression for the output.
    b. Obtain a circuit realization; then convert it to a NAND-only circuit.
    c. Also write a product-of-sums expression. Note which expression has fewer terms.
    d. Obtain a circuit realization for this case also; convert it to a NOR-only circuit.
    e. Draw a timing diagram depicting the output for the cases $X < Y$, $X > Y$, and $X = Y$ (use appropriate values of X and Y).

39 As discussed in section 7, a *magnitude comparator* is a combinational circuit whose $2n$ inputs are the bits of two n-bit binary words, A and B. It has three outputs:

    G, which is 1 when $A > B$
    E, which is 1 when $A = B$
    L, which is 1 when $A < B$

    a. When $n = 2$ bits, the input words are $A = A_1A_0$ and $B = B_1B_0$. Set up truth tables for G, E, and L, taking each bit of each input word as a variable. (This comparator is a four-input, three-output circuit.)
    b. From these tables construct logic maps for G, E, and L.
    c. Obtain a switching expression for E as an output in terms of G and L as inputs. Repeat for G in terms of E and L. Repeat for L in terms of E and G.
    d. Obtain a minimal implementation for the three outputs G, E, and L from the logic maps, utilizing the results of part c.
    e. Modify the result in part b, treating the circuit as a multiple-output circuit. That is, common terms are created among the outputs to the extent possible so as to reduce the overall gate count.

40 A 4-bit magnitude comparator is to be constructed from two 2-bit ones. One of these will compare the larger bits of each word and one will compare the smaller bits. The G, E, and L outputs from the comparator of the smaller bits become additional inputs to the comparator of the larger bits. To distinguish them from the overall G, E, and L outputs, they will be labeled G_{in}, E_{in}, and L_{in}.

    a. Draw a block diagram, showing one block each for the 2-bit comparators, with appropriate input and output terminals explicitly shown and labeled.
    b. Under what conditions will the result of comparing the smaller bits have an influence on the overall result? Explain.
    c. If all 2-bit units are to be identical modules, specify for the comparator of the smaller bits what values the inputs G_{in}, E_{in}, and L_{in} should have in order for the comparator to operate properly.

41 Two 2-bit binary numbers are given: $X = x_1 x_2$ and $Y = y_1 y_2$. The sum is $Z = CBA$, where C is the carry.

  a. Draw a logical map for each of the sum digits: C, B, A.
  b. From the maps, write expressions for the sum digits in s-of-p form.
  c. Rewrite these expressions to utilize XOR gates as far as possible.
  d. Assume each gate has a small delay. Is there any particular input sequence that causes the circuit to have more total delay than any other? Draw a timing diagram of the sequence.

42 Treat the circuit in Problem 41 as a multiple-output circuit and obtain a realization that reduces the number of gates to the extent possible.

43 A circuit is to have four inputs $wxyz$ and four outputs $ABCD$. The output set is to represent a BCD number that is to be the input number incremented by 1. Thus, if the input is 0111, the output will be 1000. (Suppose the input is 1011; what will the output number be?)

  a. Draw the logical maps for each output.
  b. Write the minterm list for each.
  c. From each list, write a minimal s-of-p expression.
  d. Draw a two-level logic diagram realizing each expression.
  e. Convert the minterm lists in b to maxterm lists.
  f. Draw a two-level logic diagram for each expression in e.

44 **a.** Repeat Problem 43, treating it as a multiple-output circuit. Use the relationships between the multiple outputs to minimize the total number of gates.
  b. Convert the realization to NAND-only.
  c. Convert the realization to NOR-only.

45 **a.** A four-input, one-output combinational circuit is to be a BCD prime number detector. Following the procedures in Exercise 18 in the text, obtain a minimal s-of-p circuit.
  b. Also obtain a minimal p-of-s circuit.

46 Two 2-bit binary numbers $A = a_1 a_0$ and $B = b_1 b_0$ constitute the inputs to a four-input logic circuit. The output is a 4-bit binary number $C = c_3 c_2 c_1 c_0$, which is to be the product of A and B.

  a. Construct logic maps for each output in terms of the four input variables. (It might help to construct a truth table first, but that is not necessary.)
  b. From the maps, find minimum s-of-p expressions for each c_i.
  c. Construct a circuit implementation of each output; make a note of the circuit complexity.

47 The prime implicant chart for an incompletely specified function is shown in Figure P47.

  a. Determine the essential prime implicants, if any; then construct a reduced chart if necessary.
  b. If any rows are dominated by others, specify them and remove them. Then determine the secondary essential prime implicants, if any.
  c. A column m_i is said to dominate another column m_j if m_i is dominated by all prime implicants that cover m_j, and possibly by other prime implicants also. For example, column 24 dominates column 17. Show that, if a prime implicant in the minimal s-of-p expression covers the dominated column, it will necessarily cover the dominating column. Hence the dominating column can be removed.

d. Use the result of part *c* to find a further reduced chart.
e. Complete the determination of a minimal s-of-p expression.

	6	9	10	12	17	20	24
P_1		✓	✓	✓			
P_2					✓	✓	✓
P_3	✓			✓			
P_4				✓			✓
P_5	✓		✓	✓			
P_6		✓			✓		✓
P_7						✓	

Figure P47

48 A k-bit comparator is represented by C_k in Figure P48. It compares two k-bit numbers $X_k = x_1 x_2 \dots x_k$ and $Y_k = y_1 y_2 \dots y_k$, with outputs G_k and S_k as follows:

$$G_k S_k = 10 \quad \text{if } X_k > Y_k$$
$$= 01 \quad \text{if } X_k < Y_k$$
$$= 00 \quad \text{if } X_k = Y_k$$

A $(k + 1)$-bit comparator is to be designed using a k-bit comparator package and another logic unit L as shown in the figure. (That is, G_{k+1} and S_{k+1} satisfy the preceding conditions on G_k and S, with k replaced by $k + 1$.)

a. Find logic expressions for outputs G_{k+1} and S_{k+1} in terms of all the inputs to L. (*Hint:* Consider the possible values of G_k and S_k and the resulting values of G_{k+1} and S_{k+1}.)
b. Assume that packages implementing L are available, as well as the constants 0 and 1. Show what the inputs to an L package would have to be so that it would serve as a comparator of 1-bit numbers.
c. Show a block diagram implementing C_3 (a 3-bit comparator) using only L packages.

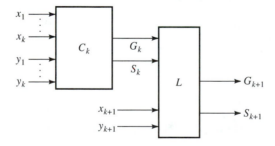

Figure P48

49 A switching circuit is to have four input lines. What it receives on these lines are known to be binary-coded decimal words. The output is to be 1 whenever the BCD word received corresponds to 0, 2, 3, 5, or 8; otherwise the output is 0. Design a two-level minimal circuit.

50 The diagram in Figure P50 represents a partially implemented logic circuit. The outputs are

$$f_1 = \Sigma(4, 9, 11, 12, 13)$$
$$f_2 = \Sigma(4, 5, 6, 7, 12, 13)$$

a. Construct a logic map of the function $x(A, B, C, D) = x_{max}$, which is to have the maximum number of minterms while still allowing the given outputs f_1 and f_2.

b. Construct maps of y and z corresponding to x_{max}, leaving the values unspecified wherever possible.

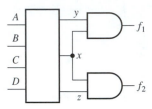

Figure P50

51 Suppose that the outputs in the combinational logic circuit in Figure P50 are to be

$$f_1 = \Sigma(0, 1, 3, 4, 5, 7, 11, 15)$$
$$f_2 = \Sigma(2, 3, 6, 7, 11, 15)$$

a. Of all possible intermediate functions x, find one, x_{min}, that has the fewest minterms.

b. Show the corresponding logic maps for y and z, assuming each has the maximum number of minterms.

52 The Lemon Logic Corporation has designed an IC package with multiple copies of a circuit, called a PU gate, with four input variables—A, B, C, D—and one output labeled PU. This gate implements the function

$$PU(A, B, C, D) = BC(A + D)$$

The design engineers at Lemon are investigating the possible implementation of switching functions using PU-OR logic. To help them out, design a circuit implementing the following function, using only three PU gates and an OR gate. (Assume that both a variable and its complement are available as inputs to PU.)

$$f(x_1, x_2\, x_3, x_4) = \Sigma(0, 1, 6, 9, 10, 11, 14, 15)$$

53 Carry out a research project to discover some properties of the Exclusive-OR function of three variables: $f = x \oplus y \oplus z$.

a. Find a Boolean expression for f in terms of x, y, and z.

b. From this, write f as a minterm list.

c. Write the terms in the minterm list as binary numbers. From an examination of these numbers, draw two general conclusions about the minterms of $x \oplus y \oplus z$.

54 Repeat Problem 53 for the Exclusive-OR of four variables, $f = w \oplus x \oplus y \oplus z$.

55 The objective of this problem is to design a circuit that uses a single full adder plus something else to add two n-bit binary numbers, one bit at a time. Describe what the "something else" would have to be, and illustrate with the numbers 1101 and 0110.

56 A multiplier has two pairs of input lines: a_1a_0 and b_1b_0. The two-digit inputs on these lines represent two-digit binary words. It has four output lines; the word $p_3p_2p_1p_0$ appearing on these lines represents the product of those numbers.

a. Write logical expressions for each of the product digits p_i.

b. Obtain an implementation for each digit.

c. To the extent possible, utilize common gates shared by more than one output.

57 A circuit is constructed with two XOR gates as follows. The first XOR gate has two external inputs, x_1 and x_2. The inputs to the second XOR are external input x_3 and the output of the first XOR. Remember the relationship of the inputs necessary for an XOR gate to yield a 1 output.

    a. Draw the resulting circuit and specify the relationships of the three inputs necessary to yield a 1 output.

    b. Extend the circuit by adding a fourth XOR gate, one of whose inputs is external and the other one being the output from the previous circuit. Again specify the relationship among the inputs necessary to yield a 1 output.

Additional XORs can be added in the same way, leading to the same relationship among inputs necessary to yield a 1 output. This structure is called a *daisy chain*, and it serves as an odd-parity detector. Verify this for the cases in *a* and *b*.

58 Sometimes a circuit is to have several outputs, all dependent on the same set of input variables. The procedures used in this chapter can be applied to implement each of the output functions independently of the others. However, sometimes it is possible to utilize *prime* implicants that are common among two or more of the output functions. In such cases the same gates can be used in the paths from the inputs to two or more of the outputs. Indeed, it might be profitable to select implicants that are common among several of the outputs even if they are not *prime* implicants. The trade-off here is to accept extra inputs to some gates at the benefit of fewer gates overall.

The following sets of functions are outputs depending on the same sets of inputs. In each case, consider two different implementations:

    Minimal sums of products independently implementing each function
    Sum-of-products implementations that utilize common gates among all three functions, or among pairs of them

Compare the numbers of gates, inputs, and SSI packages with those needed for implementing each of the functions independently.

    a. $f_1 = \Sigma(0, 1, 8, 9, 14, 15)$
       $f_2 = \Sigma(6, 7, 12, 13, 14, 15)$
       $f_3 = \Sigma(8, 9, 12, 13, 14, 15)$
    b. $f_1 = \Sigma(0, 1, 4, 8, 10, 15)$
       $f_2 = \Sigma(0, 1, 5, 6, 7)$
       $f_3 = \Sigma(1, 6, 9, 13, 14)$
    c. $f_1 = \Sigma(1, 3, 4, 5, 7, 9, 13, 18, 19, 20, 21, 26, 27)$
       $f_2 = \Sigma(4, 5, 6, 9, 12, 13, 14, 20, 21, 22, 23, 28, 29)$
       $f_3 = \Sigma(6, 7, 9, 11, 12, 13, 14, 15, 18, 19, 20, 21, 22, 23)$
    d. $f_1 = \Sigma(0, 4, 5, 11, 12) + \Sigma d(6, 10)$
       $f_2 = \Sigma(4, 8, 10, 12) + \Sigma d(0, 14)$
       $f_3 = \Sigma(4, 10, 12, 14) + \Sigma d(6, 7, 8)$
    e. $f_1 = \Sigma(1, 5, 7, 9, 20, 22, 29, 30) + \Sigma d(3, 4, 13, 21, 25, 28)$
       $f_2 = \Sigma(2, 6, 7, 14, 22, 25, 28, 30, 31) + \Sigma d(1, 12, 23, 27, 29)$
       $f_3 = \Sigma(6, 8, 12, 14, 22, 24, 26, 30) + \Sigma d(4, 10, 17, 18, 28)$

Chapter 4

Combinational Logic Design

The foundations for the design of digital logic circuits were established in the preceding chapters. The elements of Boolean algebra (two-element "switching algebra") and how the operations in Boolean algebra can be represented schematically by means of gates (primitive devices) were presented in Chapter 2. How switching expressions can be manipulated and represented in different ways was the subject of Chapter 3, which also presented various ways of implementing such representations in a variety of circuits using primitive gates.

With all of the tools for the purpose now in hand, we will be concerned in this chapter with the design of more complex logic circuits. Circuits in which all outputs at any given time depend only on the inputs at that time are called *combinational* logic circuits. The design procedures will be illustrated with important classes of circuits that are now universal in digital systems.

The approach taken is to examine the tasks that a combinational logic circuit is intended to perform and then identify one or more circuits that can perform the task. One circuit may have some specific advantages over others, but it may also have certain deficiencies. Often one factor can be improved, but only at the expense of others. Some important factors are speed of operation, complexity or cost of hardware, power dissipation, and availability in prefabricated units. We will take up a number of different operations that are useful in different contexts and show how appropriate circuits can be designed to carry out these operations.

1 BINARY ADDERS

One of the most important tasks performed by a digital computer is the operation of adding two binary numbers.[1] A useful measure of performance is speed. Of course, speed can be improved by using gate designs that favor speed at the

---

[1]As discussed in Chapter 1, subtraction of two numbers is included in the meaning of addition, since subtraction is performed first by carrying out some operation on the subtrahend and then adding the result. (What operation is first performed depends on the type of computer—either inverting the subtrahend or taking its two's complement, as discussed in Chapter 1.)

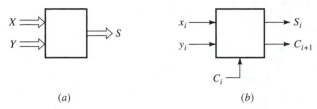

Figure 1 Binary addition. (*a*) General adder. (*b*) Full adder of two 1-bit words.

expense of other measures, such as power consumption (using advanced Schottky, for example, versus low-power Schottky designs). But for the *logic* designer, the important question is how to design an adder to increase the speed, regardless of the type of gate used. It may be that increased speed can be achieved at the expense of increased circuit complexity. That is, there might be several designs, each characterized by a certain speed and a certain circuit complexity. A judgment must be made as to the acceptable trade-offs between them.

A symbolic diagram representing a binary adder is shown in Figure 1*a*. Each open arrowhead represents multiple variables; in this case the inputs are two binary numbers. If each number has n digits, then each line shown really represents n lines. The sum of two n-bit numbers is an $(n + 1)$-bit number. Thus, S (sum) represents $n + 1$ output lines. If this circuit were designed by the methods of Chapter 3, we would require a circuit with $n + 1$ output functions, each one dependent on $2n$ variables. The truth table for each of the output functions would have 2^{2n} rows. Since n could easily be in the range 20–40, a different approach is obviously needed.

Full Adder

An alternative approach for the addition of two n-bit numbers is to use a separate circuit for each corresponding pair of bits. Such a circuit would accept the 2 bits to be added, together with the carry resulting from adding the less significant bits. It would yield as outputs the 1-bit sum and the 1-bit carry out to the more significant bit. Such a circuit is called a *full adder*. A schematic diagram is shown in Figure 1*b*. The 2 bits to be added are x_i and y_i, and the *carry in* is C_i. The outputs are the *sum* S_i and the *carry out* C_{i+1}. The truth table for the full adder and the logic maps for the two outputs are shown in Figure 2.

The minimal sum-of-products expressions for the two outputs obtained from the maps are

$$S_i = x_i'y_iC_i' + x_iy_i'C_i' + x_i'y_i'C_i + x_iy_iC_i \tag{1a}$$

$$\begin{aligned} C_{i+1} &= x_iy_i + x_iC_i + y_iC_i \\ &= x_iy_i + C_i(x_i + y_i) \end{aligned} \tag{1b}$$

(Make sure you verify these.) Each minterm in the map of S_i constitutes a prime implicant. Hence, a sum-of-products expression will require four 3-input AND gates and a 4-input OR gate. The carry out will require three AND gates and an

C_i	X_i	Y_i	S_i	C_{i+1}
0	0	0	0	0
0	0	1	1	0
0	1	0	0	1
0	1	1	1	0
1	0	0	1	0
1	0	1	0	1
1	1	0	1	1
1	1	1	0	1

Figure 2 Truth table and logical maps of the full adder. (*a*) Truth table.
(*b*) S_i map. (*c*) C_{i+1} map.

OR gate. If we assume that each gate has the same propagation delay t_p, then a two-level implementation will have a propagation delay of $2t_p$.

In the map of the carry out, minterm m_7 is covered by each of the three prime implicants. This is overkill; since m_7 is covered by prime implicant $x_i y_i$, there is no need to cover it again by using it to form prime implicants with m_5 and m_6. If there is some benefit to it, we might use the latter two minterms as implicants without forming prime implicants with m_7. The resulting expression for C_{i+1} becomes

$$C_{i+1} = x_i y_i + C_i(x_i'y_i + x_i y_i') = x_i y_i + C_i(x_i \oplus y_i) \tag{2}$$

(Confirm this result.) We already have an expression for S_i in (1*a*), but it is in canonic sum-of-products form. It would be useful to seek an alternative form for a more useful implementation.

Exercise 1   With the use of switching algebra, confirm that the expression for the sum in (1*a*) can be converted to

$$S_i = x_i \oplus y_i \oplus C_i \tag{3} \blacklozenge$$

Using the expressions for S_i and C_{i+1} containing XORs, confirm that we can obtain the implementation of the full adder shown in Figure 3*a*. Notice that the circuit consists of two identical XOR-AND combinations and an additional OR gate. The circuit inside each dashed box is shown in Figure 3*b*; it is named a *half adder*. Its only inputs are the 2 bits to be added, without a carry in. The two outputs are (1) the sum of the 2 bits and (2) the carry out.

Assuming that an XOR gate (implemented in a two-level circuit) has a propagation delay of $2t_p$, the full adder in Figure 3*a* has a propagation delay of $4t_p$, both for the sum and for the carry. (Verify these claims.)

We will observe in the following section that the overall speed in the addition of two n-bit binary numbers depends mainly on the speed with which the carry propagates from the least significant bit to the most significant bit. Hence, reducing the delay experienced by the carry of a full adder is a significant improvement. This is an incentive in seeking other implementations of the full adder. In some of the cases in Problem 1 at the end of the chapter, additional implementations of the full adder are

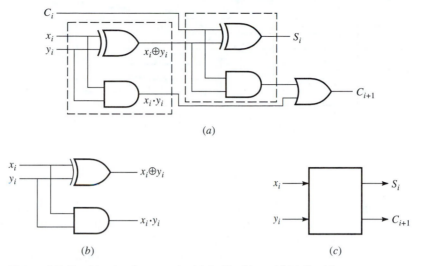

(a)

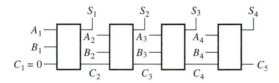

(b)   (c)

Figure 3 Full adder implemented with half adders. (a) Full adder.
(b) Half adder. (c) Half adder schematic diagram.

Figure 4 Four-bit ripple-carry adder.

proposed in which the propagation delay for the carry is $2t_p$ instead of $4t_p$. Henceforth, for a full adder, we will assume that the propagation delay of the carry is $2t_p$.

Ripple-Carry Adder

The problem of adding two multidigit binary numbers has the following form. Two n-bit binary numbers are available, with all digits being presented in parallel. The addition is performed by using a full adder to add each corresponding pair of digits, one from each number. The full adders are connected in tandem so that the carry out from one stage becomes the carry into the next stage, as illustrated for the case of four-digit numbers in Figure 4. Thus, the carry *ripples* through each stage. For binary addition, the carry into the first (least significant) stage is 0. The last carry out (the *overflow* carry) becomes the most significant bit of the $(n + 1)$-bit sum.

Since the carry of each full adder has a propagation delay of $2t_p$, the total delay in carrying out the sum of two n-bit numbers is $2nt_p$. Not every pair of two n-bit numbers will experience this much delay. Take the following two numbers as an example:

$$101010$$
$$010101$$

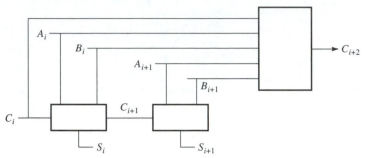

Figure 5 Carry-lookahead circuit schematic.

Assuming that the carry into the first stage is zero, no carries are generated at any stage in taking the sum. Hence, there will be no carry ripple, and so no propagation delay along the carry chain.

However, to handle the general case, provision must be made for the worst case; no new numbers should be presented for addition before the total delay represented by the worst case. The maximum addition speed, thus, is limited by the worst case of carry propagation delay.

Carry-Lookahead Adder

In contemplating the addition of two n-digit binary numbers, we were appalled by the thought of a single combinational circuit with all those inputs. So we considered the repeated use of a simpler circuit, a full adder, with the least possible number of inputs. But what is gained in circuit simplicity with this approach is lost in speed. Since the speed is limited by the delay in the carry function, some of the lost speed might be regained if we could design a circuit—just for the carry—with more inputs than 2 but not as many as $2n$. Suppose that several full-adder stages are treated as a unit. The inputs to the unit are the carry into the unit as well as the input digits to all the full adders in that unit. Then perhaps the carry out could be obtained faster than the ripple carry through the same number of full adders.

These concepts are illustrated in Figure 5 with a unit consisting of just two full adders and a carry-lookahead circuit. The four digits to be added, as well as the input carry C_i, are present simultaneously. It is possible to get an expression for the carry out, C_{i+2}, from the unit by using the expression for the carry of the full adder in (2).

For reasons which will become clear shortly, let's attach names to the two terms in the carry expression in (2), changing the names of the variables to A and B from x and y in accordance with Figure 5. Define the *generated carry* G_i and the *propagated carry* P_i for the ith full adder as follows:

$$G_i = A_i B_i \tag{4a}$$

$$P_i = A_i \oplus B_i \tag{4b}$$

Inserting these into the expression for the carry out in (2) gives

$$C_{i+1} = A_i B_i + C_i (A_i \oplus B_i) = G_i + P_i C_i \tag{5}$$

A carry will be *generated* in the ith full adder (that is, $G_i = 1$) if A_i and B_i both equal 1. But if only one of them is 1, a carry out will not be generated. In that case, however, P_i will be 1. (Confirm this.) Hence, the carry out will be $C_{i+1} = C_i$. We say that the carry will be *propagated* forward.

The expression for the carry out in (5) can be updated by changing the index i to $i + 1$:

$$C_{i+2} = G_{i+1} + P_{i+1}C_{i+1} = G_{i+1} + P_{i+1}(G_i + P_iC_i)$$
$$= G_{i+1} + P_{i+1}G_i + P_{i+1}P_iC_i \tag{6}$$

The last expression can be interpreted in the following way. A carry will appear at the output of the unit under three circumstances:

- It is generated in the last stage: $G_{i+1} = 1$.
- It is generated in the first stage, $G_i = 1$, and propagated forward: $P_{i+1} = 1$.
- The input carry C_i is propagated through both stages: $P_i = P_{i+1} = 1$.

Obviously, this result can be extended through any number of stages, but the circuit will become progressively more complicated.

Exercise 2   Extend the previous result by one more stage and write the expression for C_{i+3}. Then describe the ways in which this carry out can be 1. Confirm your result using the general result given next.   ◆

Extending the design to j stages, the expression in (6) becomes

$$C_{i+j+1} = G_{i+j} + P_{i+j}G_{i+j-1} + P_{i+j}P_{i+j-1}G_{i+j-2} + \cdots + (P_{i+j}P_{i+j-1} \cdots P_i)C_i \tag{7}$$

This expression looks complicated, but it is easy to interpret. Since the carry out $C_{i+j+1} = 1$ if any one of the additive terms on the right is 1, the carry out from the unit will be 1 for several possibilities. Either it is generated in the last (jth) stage of the unit, or it is generated in an earlier stage and is propagated through all succeeding stages, or the carry into the unit is propagated through all the stages to the output.

The greater the number of full-adder stages included in a unit, the greater the improvement in speed—but also the greater the complexity of the carry-lookahead circuit. There is an obvious trade-off between the two. Consider a unit of four stages. This unit is to add two 4-bit words A and B. Each stage can be considered as having a sum circuit (S) and a separate carry circuit (C). The sum circuit of each stage has as inputs the carry from the preceding stage and the corresponding bits of the A and B words. The inputs to the carry network of each stage consist of *all* the bits of the A and B words up to that stage and the carry—not just from the preceding stage, but from the input to the whole unit. Thus, if the first stage is stage i, the inputs to the carry circuit of stage $i + 2$ are: A_i, A_{i+1}, A_{i+2}, B_i, B_{i+1}, B_{i+2}, and C_i.

Exercise 3   Draw a schematic diagram for a three-stage unit using rectangles to represent the sum and carry circuits of each stage. (Let the first stage be 1 instead of the general i.)   ◆

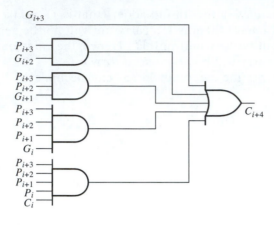

Figure 6 Four-stage carry-lookahead circuit.

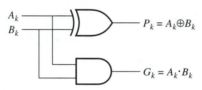

Figure 7 Half adder for generated and propagated carries.

A circuit implementation of the carry network of the last stage in a four-stage unit is shown in Figure 6. Except for C_i, the carry into the unit, all other inputs to the AND gates are generated carries and propagated carries from the various stages of the unit. These generated and propagated carries are produced by the half-adder circuits in Figure 7.

A semi-block diagram of the four-stage carry-lookahead adder is shown in Figure 8. (Note that pins that carry the same label in different subcircuits are assumed to be connected.) Since each propagated carry P_{i+j} is the output of an XOR gate, the overall propagation delay of the carry circuit having the design of Figure 7 is $4t_p$. However, all generated and propagated carries, G_{i+j} and P_{i+j}, of all units become available within $2t_p$ after the two words are first presented for addition, as evident from Figure 6. Hence, in all carry-lookahead units besides the first, the propagation delay of the carry network is only $2t_p$.

Exercise 4 Suppose that a carry-lookahead adder is to have k 4-bit units to carry out the addition of two $4k$-bit words. From the preceding discussion, from the diagram of Figure 8 implementing each unit, and from a consideration of the first and last units, determine the propagation delay of this adder in terms of t_p, the propagation delay through one gate. (Don't peek at the answer until you do the work.)
Answer [2]

---

[2]The sum of the delays through (a) the carry circuit of each unit ($2t_p$ each), (b) the sum circuit of the last unit ($2t_p$) since it depends on having the carry from the last unit, and (c) the extra delay in getting the carry from the first unit.   Total delay $= (k + 1 + 1)2t_p = (2k + 4)t_p$ ◆

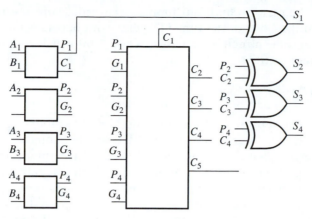

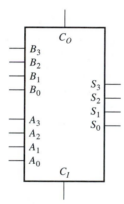

Figure 8 Schematic diagram of 4-bit carry-lookahead adder.

Figure 9 High-speed adder: 4-bit words.

If an adder has eight 4-bit units, the propagation delay through a carry-lookahead adder will be $20t_p$. The corresponding ripple-carry adder will have a propagation delay of $4 \times 8 \times 2t_p = 64t_p$. Thus, the carry-lookahead adder will have an advantage of 320 percent in speed over the ripple-carry adder. All is not gravy, however: the speed advantage has been paid for in the cost of the added hardware.

Exercise 5 From a count on the number of gates in each implementation, estimate the hardware disadvantage (in percent) of the carry-lookahead adder compared with the ripple-carry adder. Compare the disadvantage with the 320 percent speed advantage. ◆

The circuits described here are available in IC packages. A single full adder, for example, is available as a unit. A ripple-carry adder, as illustrated in Figure 4, and a carry-lookahead adder for 4-bit words, as shown in Figure 8, are available as MSI packages.

Externally, a package consisting of a ripple-carry adder of 4-bit words would look the same as a package consisting of a carry-lookahead adder of 4-bit words. The block diagram in Figure 9 illustrates such a package. There are nine inputs: the carry in and four inputs per word. There are five outputs: the carry out and the 4 bits of the sum. (The carry out becomes the most significant bit of the sum if the circuit is used just to add 4-bit words, and not as part of an adder of longer words.)

Binary Subtractor

In Chapter 1 two representations of signed binary numbers were studied: one's complement and two's complement. Recall that when numbers are represented in one of the complement forms, the only special treatment needed in the addition of a negative number with another positive or negative number is in the final carry out. Thus, the adders studied in the previous section are suitable for the addition of complement numbers if some additional circuitry is used to process the final carry out. Also, binary subtraction can be performed using the same adder circuits by negating the subtrahend.

Two's-Complement Adder and Subtractor

Recall from Chapter 1 that when the addition of 2 two's complement binary numbers produces a final carry, it can be ignored. However, it is necessary to detect the overflow that can occur when the result of the addition is out of range.[3] In Chapter 1 it was concluded that an arithmetic overflow could be detected if the carry in and carry out of the most significant bit position are different. Thus, the overflow can be detected with one additional Exclusive-OR gate. The two's complement adder is not much different from the binary adder for unsigned numbers.

What about subtraction? We already suggested that subtraction should be carried out by complementing the subtrahend and adding. So the task is to design a circuit whose output is the two's complement of the input, and use its output as one input to an adder. Such a circuit can be designed easily, but why should a system contain some hardware dedicated to addition and other hardware dedicated to subtraction? If the only difference between these two circuits is a circuit that computes the two's complement, then why not design a circuit where either addition or subtraction can be selected with one additional input? When this additional input is, say, 0 the circuit performs addition, and when the input is 1 the circuit performs subtraction. It sounds easy; a representation of the circuit can be derived using the techniques of Chapter 3, but an elegant solution exists that we describe next.

Examine the truth table of the Exclusive-OR operation and notice that it can be viewed as a conditional inverter. If one input is 0, then the output is identical to the second input. If one input is 1, then the output is the complement of the second input. This is convenient for producing the complement of an input to our adder/subtractor circuit when we want to perform subtraction. However, to compute the two's complement of a binary number we have to add

---

[3]The range of binary numbers having n binary digits represented in two's complement form is $-2^{n-1} \le m \le 2^{n-1} - 1$.

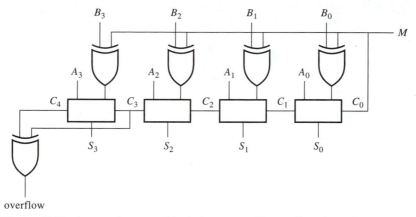

Figure 10 Two's complement adder/subtractor with overflow detection.

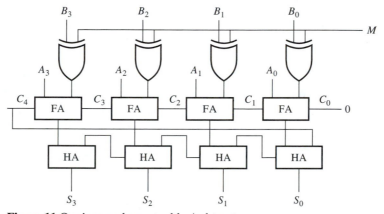

Figure 11 One's complement adder/subtractor.

1. Any ideas on how to do this without additional gates? (Think about it before you continue.)

The full adder for the least significant bit has a carry input signal that can be utilized to add the required 1. The design of our two's complement adder/subtractor circuit is complete; a version for adding 4-bit numbers is shown in Figure 10. If the control signal M is 0, then the circuit performs $A+B$; however, if M is 1, the circuit performs $A - B$.

One's-Complement Adder and Subtractor

To perform subtraction in one's complement we can use the Exclusive-OR circuit used in the two's complement adder/subtractor. The only difference is that we do not want to inject a carry into the least significant bit. One's complement addition requires the addition of 1 to the sum when a carry out from the most significant bit position occurs. This can be accomplished using multiple half adders as shown in Figure 11. Overflow detection for one's complement addition is left as a problem for you.

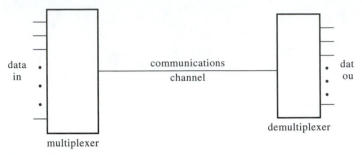

Figure 12 A data communication problem.

Two's complement addition is the most common method implemented in modern computers due to its reduced circuit complexity compared with one's complement.

This is as far as we will go with the addition of multibit words; other adder circuits are left for the problem set.

2   MULTIPLEXERS

Many tasks in communications, control, and computer systems can be performed by combinational logic circuits. When a circuit has been designed to perform some task in one application, it often finds use in a different application as well. In this way, it acquires different names from its various uses. In this and the following sections, we will describe a number of such circuits and their uses. We will discuss their principles of operation, specifying their MSI or LSI implementations.

One common task is illustrated in Figure 12. Data generated in one location is to be used in another location; A method is needed to transmit it from one location to another through some communications channel.

The data is available, in parallel, on many different lines but must be transmitted over a single communications link. A mechanism is needed to select which of the many data lines to activate sequentially at any one time so that the data this line carries can be transmitted at that time. This process is called *multiplexing*. An example is the multiplexing of conversations on the telephone system. A number of telephone conversations are alternately switched onto the telephone line many times per second. Because of the nature of the human auditory system, listeners cannot detect that what they are hearing is chopped up and that other people's conversations are interspersed with their own in the transmission process.

Needed at the other end of the communications link is a device that will undo the multiplexing: a *demultiplexer*. Such a device must accept the incoming serial data and direct it in parallel to one of many output lines. The interspersed snatches of telephone conversations, for example, must be sent to the correct listeners.

A digital multiplexer is a circuit with 2^n data input lines and one output line. It must also have a way of determining the specific data input line to be selected at any one time. This is done with n other input lines, called the *select* or *selector* inputs, whose function is to select one of the 2^n data inputs for connec-

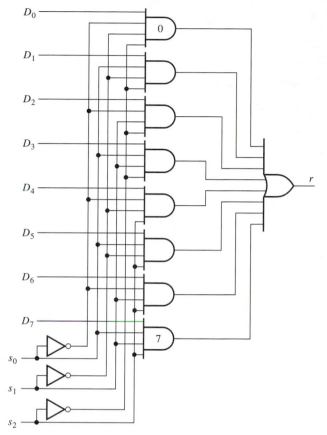

Figure 13 Multiplexer with eight data inputs.

tion to the output. A circuit for $n = 3$ is shown in Figure 13. The n selector lines have $2^n = 8$ combinations of values that constitute binary *select numbers*.

Exercise 6 Write expressions for each of the AND gate outputs in terms of the s_i and D_i inputs, confirming that the multiplier of D_k is the binary equivalent of k. ◆

When the selector inputs have the combination $s_2 s_1 s_0 = 011$, for example, the outputs of all AND gates except the one to which data line D_3 is connected will be 0. All other inputs to that AND gate besides D_3 will be 1. Hence, D_3 appears at the output of the circuit. In this way, the select inputs whose binary combination corresponds to decimal 3 have selected data input D_3 for transmittal to the output.

Standard MSI packages are available as multiplexers. Figure 14a shows the circuit for a package containing two separate multiplexers for $n = 2$. Practical considerations not included in Figure 13 account for some of the features of this circuit. The *enable* input E, for example, is used to control the period of time that the multiplexer is operative. Thus, when the value of E is 1, the output will be 0 no matter what the values of the select inputs. The circuit will be operative only when the corresponding enable input is 0. (In other circuits, the

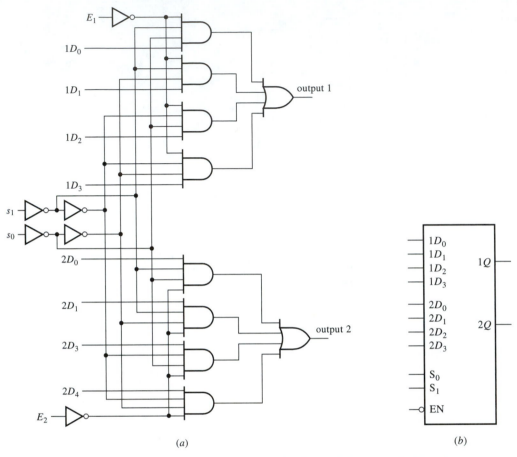

(a)

(b)

Figure 14 (*a*) Dual four-input multiplexer with enable. (*b*) Dual four-input multiplexer with single enable.

enable signal is not inverted; in such cases, the circuit is operative when $E = 1$, just the opposite of the case shown in Figure 14*a*.)

In addition, note from the figure that both the selector signals and their complements are inputs to AND gates. The signal inputs themselves are obtained after two inversions. This is especially useful if n is large. In this way, the circuit that produces the select inputs has as load only a single gate (the inverter) rather than several AND gates. In Figure 14*a* the select inputs are common to both multiplexers, but each has its own enable. In other designs, the enable can also be common. A schematic diagram of a dual four-input multiplexer (MUX) with a single enable is shown in Figure 14*b*.

The preferred gate form for many IC logic packages (for example, the 74LS00 and the 74LS10) is the NAND gate. Since the multiplexer design in either Figure 13 or 14 is a two-level AND-OR circuit, a direct replacement of all AND and OR gates by NAND gates will maintain the logic function, as discussed in the preceding chapter. In this way, the actual implementation of the multiplexer is carried out with NAND gates.

Multiplexers as General-Purpose Logic Circuits

It is clear from Figures 13 and 14 that the structure of a multiplexer is that of a two-level AND-OR logic circuit, with each AND gate having $n + 1$ inputs, where n is the number of select inputs. It appears that the multiplexer would constitute a canonic sum-of-products implementation of a switching function if all the data lines together represent just one switching variable (or its complement) and each of the select inputs represents a switching variable.

Let's work backward from a specified function of m switching variables for which we have written a canonic sum-of-products expression. The size of multiplexer needed (number of select inputs) is not evident. Suppose we choose a multiplexer that has $m - 1$ select inputs, leaving only one other variable to accommodate all the data inputs. We write an output function of these select inputs and the 2^{m-1} data inputs D_i. Now we plan to assign $m - 1$ of these variables to the select inputs; but how to make the assignment?[4] There are really no restrictions, so it can be done arbitrarily.

The next step is to write the multiplexer output after replacing the select inputs with $m - 1$ of the variables of the given function. By comparing the two expressions term by term, the D_i inputs can be determined in terms of the remaining variable.

EXAMPLE 1

A switching function to be implemented with a multiplexer is

$$f(x, y, z) = \Sigma(1, 2, 4, 7) = x'y'z + x'yz' + xy'z' + xyz$$

Since the function has three variables, the desired multiplexer will have $3 - 1 = 2$ select inputs; half of the dual four-input MUX of Figure 14 will do. The expression for the multiplexer output is

$$f = s_1's_0'D_0 + s_1's_0D_1 + s_1s_0'D_2 + s_1s_0D_3$$

There are no restrictions on how to assign the selector inputs to the variables of the given function; let $s_1 = x$ and $s_0 = y$ arbitrarily. Then

$$f = x'y'D_0 + x'yD_1 + xy'D_2 + xyD_3$$

Comparing this with the original expression for the given function leads to

$$D_0 = D_3 = z$$
$$D_1 = D_2 = z'$$

The original function is thus implemented with a four-input multiplexer. ∎

There are five other ways that the two select inputs could have been assigned to two of the three switching variables. No conditions need to be satisfied by the choice, so it is arbitrary. However, the specific outcome obtained for the D_i inputs depends on that initial choice.

[4]For a set of $m - 1$ variables, there are $m!$ ways of assigning $m - 1$ quantities to specific variables.

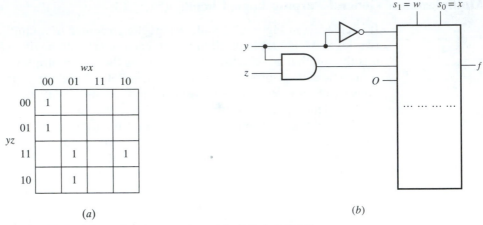

	wx			
yz	00	01	11	10
00	1			
01	1			
11		1		1
10		1		

(a)

(b)

Figure 15 Multiplexer implementation of $f = \Sigma(0, 1, 6, 7, 11)$.

Exercise 7 In the problem of Example 1, choose $s_1 = z$ and $s_0 = x$. Determine the D_i.
Answer[5]

Exercise 8 For practice, choose each of the remaining possible ways of assigning select inputs to the switching variables, and then determine the required D_i; specify the external gates needed. ◆

To implement a switching function of m variables, we have seen that a multiplexer of $m - 1$ select inputs will work. It might be possible in some cases that even a smaller multiplexer can be used. It should be expected that, when possible, this savings in MUX complexity must come at some other cost.

EXAMPLE 2

The function of four variables whose map is shown in Figure 15 is to be implemented by a multiplexer. One with $4 - 1 = 3$ select variables is always possible. However, let's explore the possibility of using a multiplexer with only two select variables to implement this function.

Arbitrarily assign the two select inputs s_1 and s_0 to w and x. The expression for the output of the multiplexer is the same one given in Example 1, since this one has the same dimensions. For $wx = s_1 s_0 = 00$, that expression reduces to D_0. But for the values $wx = 00$, the expression that covers the 1's in the map is $y'z'$ $+ y'z = y'$. Hence, $D_0 = y'$. Similarly, in the 01 column of the map, the expression reduces to D_1 and the map gives $yz + yz' = y$; hence, $D_1 = y$. In the same way, from the 11 column we find $D_3 = 0$ and from the 10 column $D_2 = yz$. (Confirm these.) The rather simple circuit is shown in Figure 15b. We find that to imple-

---

[5]$D_0 = D_3 = y, D_1 = D_2 = y'$ ◆

ment a certain specific function of four variables, a multiplexer of order lower than 3 can be used, at the cost of an additional AND gate (The inverter would be necessary even with a higher-order multiplexer, so it does not count as additional cost.) ∎

Exercise 9 In the preceding example, suppose that s_1 and s_0 are identified as y and z instead of w and x. Determine expressions for the data inputs in terms of w and x, and specify the external hardware that will be needed besides the multiplexer. Note the difference in complexity for the two choices of select inputs.
Answer[6]

In the implementation of an arbitrary switching function, different choices for the select inputs lead to different amounts of external hardware for a smaller-than-normal multiplexer. Unfortunately, short of trying them, there is no way to determine which choice will be most economical.

3 DECODERS AND ENCODERS

The previous section began by discussing an application: Given 2^n data signals, the problem is to select, under the control of n select inputs, sequences of these 2^n data signals to send out serially on a communications link. The reverse operation on the receiving end of the communications link is to receive data serially on a single line and to convey it to one of 2^n output lines. This again is controlled by a set of control inputs. It is this application that needs only one input line; other applications may require more than one. We will now investigate such a generalized circuit.

Conceivably, there might be a combinational circuit that accepts n inputs (not necessarily 1, but a small number) and causes data to be routed to one of many, say up to 2^n, outputs. Such circuits have the generic name *decoder*. Semantically, at least, if something is to be decoded, it must have previously been *encoded*, the reverse operation from decoding. Like a multiplexer, an encoding circuit must accept data from a large number of input lines and convert it to data on a smaller number of output lines (not necessarily just one). This section will discuss a number of implementations of decoders and encoders.

Demultiplexers

Refer back to the diagram in Figure 12. The demultiplexer shown there is a single-input, multiple-output circuit. However, in addition to the data input, there must be other inputs to control the transmission of the data to the appropriate data output line at any given time. Such a demultiplexer circuit

---

[6] $D_0 = D_1 = w'x', D_2 = w'x, D_3 = w \oplus x$; three AND gates and one XOR gate, in addition to a four-input MUX. ◆

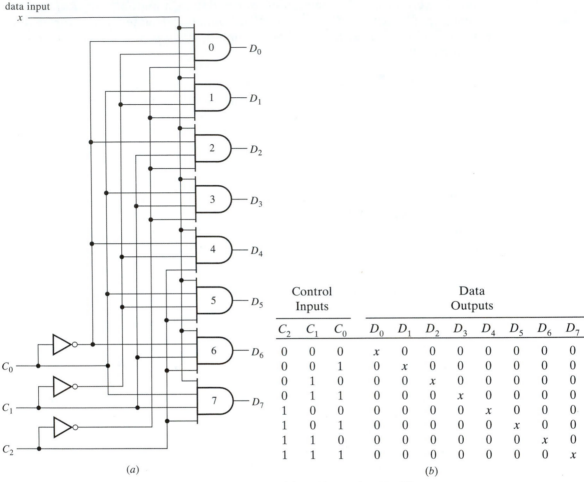

Control Inputs			Data Outputs							
C_2	C_1	C_0	D_0	D_1	D_2	D_3	D_4	D_5	D_6	D_7
0	0	0	x	0	0	0	0	0	0	0
0	0	1	0	x	0	0	0	0	0	0
0	1	0	0	0	x	0	0	0	0	0
0	1	1	0	0	0	x	0	0	0	0
1	0	0	0	0	0	0	x	0	0	0
1	0	1	0	0	0	0	0	x	0	0
1	1	0	0	0	0	0	0	0	x	0
1	1	1	0	0	0	0	0	0	0	x

(a)   (b)

Figure 16 A demultiplexer circuit (a) and its truth table (b).

having eight output lines is shown in Figure 16a. It is instructive to compare this demultiplexer circuit with the multiplexer circuit in Figure 13. For the same number of control (select) inputs, there are the same number of AND gates. But now each AND gate output is a circuit output. Rather than each gate having its own separate data input, the single data line now forms one of the inputs to each AND gate, the other AND inputs being control inputs.

When the word formed by the control inputs $C_2C_1C_0$ is the binary equivalent of decimal k, then the data input x is routed to output D_k. Viewed in another way, for a demultiplexer with n control inputs, each AND gate output corresponds to a minterm of n variables. For a given combination of control inputs, only one minterm can take on the value 1; the data input is routed to the AND gate corresponding to this minterm. For example, the logical expression for the output D_3 is $xC_2'C_1C_0$. Hence, when $C_2C_1C_0 = 011$, then $D_3 = x$ and all other D_i are 0. The complete truth table for the eight-output demultiplexer is shown in Figure 16b.

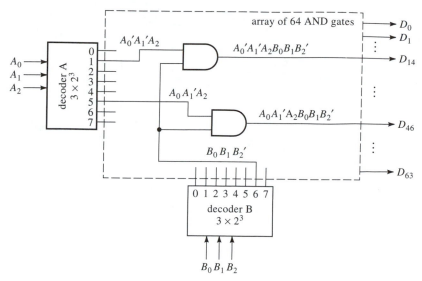

Figure 17 Design of a 6-to-2^6-line decoder from two 3-to-2^3-line decoders with an interconnection matrix of 64 AND gates.

n-to-2^n-Line Decoder

In the demultiplexer circuit in Figure 16, suppose the data input line is removed. (Draw the circuit for yourself.) Each AND gate now has only n (in this case three) inputs, and there are 2^n (in this case eight) outputs. Since there isn't a data input line to control, what used to be control inputs no longer serve that function. Instead, they are the data inputs to be decoded. This circuit is an example of what is called an *n-to-2^n-line decoder*. Each output represents a minterm. Output k is 1 whenever the combination of the input variable values is the binary equivalent of decimal k.

Now suppose that the data input line from the demultiplexer in Figure 16 is not removed but retained and viewed as an enable input. The decoder now operates only when the enable x is 1. Viewed conversely, an n-to-2^n-line decoder with an enable input can also be used as a demultiplexer, where the enable becomes the serial data input and the data inputs of the decoder become the control inputs of the demultiplexer.[7]

Decoders of the type just described are available as integrated circuits (MSI); $n = 3$ and $n = 4$ are quite common. There is no theoretical reason why n can't be increased to higher values. Since, however, there will always be practical limitations on the fan-in (the number of inputs that a physical gate can support), decoders of higher order are often designed using lower-order decoders interconnected with a network of other gates.

An illustration is given in Figure 17 for the design of a 6-to-2^6-line decoder constructed from two 3-to-2^3-line decoders. Each of the component decoders

[7]In practice, the physical implementation of the decoder with enable is carried out with NAND gates. In that case, it is the complements of the outputs in the circuit under discussion that are obtained, and the enable input is inverted before it is applied to the NAND gates. These are practical details that do not change the principles described here.

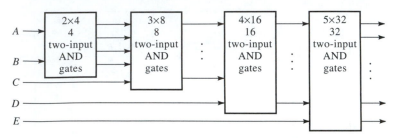

Figure 18 Design of tree decoder.

has eight outputs. Each of the outputs from the A decoder must be ANDed with each of the outputs from the B decoder to yield one of the 64 outputs from the complete decoder. Thus, in addition to the 8 three-input AND gates in each component decoder, there are 64 two-input AND gates in the interconnection network. Only two of these are shown explicitly in Figure 17.

Exercise 10   A 6-to-2^6-line decoder is to be designed using the structure of Figure 16. Specify the number of AND gates and the total number of input lines to all gates. Compare this with the design in Figure 17.   ◆

Tree Decoder

When higher-order decoders are designed in a hierarchy of several stages of lower-order ones, a practical difficulty with fan-out (number of gates driven by one terminal) results. (By a hierarchy of stages we mean, for example, two 3×8 stages to form a 6×64 decoder, as in Figure 17; then two 6×64 stages to form a 12×2^{12} decoder; and so on.) Even in Figure 17, each gate in the component decoders drives eight other gates. In the next level of the hierarchy, each of the outputs from the gates in the next-to-last level will have to drive 64 other gates.

This problem is overcome, but only partially, by the decoder design illustrated in Figure 18, called a *tree decoder*. The first stage is a 2-to-4-line decoder. A new variable is introduced in each successive stage; it or its inverse becomes one input to each of the two-input AND gates in this stage. The second input to each AND gate comes from the preceding stage. For example, one of the outputs of the second stage will be $AB'C$. This will result in two outputs from the next stage, $AB'CD$ and $AB'CD'$. This design does avoid the fan-out problem in the early stages but not in the later stages. Nevertheless, the problem exists only for the variables introduced in those stages. Any remedies required will have to be used for relatively few variables, as opposed to the large number needed by the design of Figure 17.

Decoders as General-Purpose Logic Circuits: Code Conversion

Since each output from an n-to-2^n-line decoder is a canonic product of literals, simply ORing all the outputs produces a canonic sum of products. And since every switching function can be expressed as a canonic sum of products, it fol-

Decimal Digit	Inputs: Excess-3				Outputs: Seven-Segment						
	w	x	y	z	S_1	S_2	S_3	S_4	S_5	S_6	S_7
0	0	0	1	1	1	1	1	1	1	1	0
1	0	1	0	0	0	0	0	1	1	0	0
2	0	1	0	1	1	0	1	1	0	1	1
3	0	1	1	0	0	0	1	1	1	1	1
4	0	1	1	1	0	1	0	1	1	0	1
5	1	0	0	0	0	1	1	0	1	1	1
6	1	0	0	1	1	1	0	0	1	1	1
7	1	0	1	0	0	0	1	1	1	0	0
8	1	0	1	1	1	1	1	1	1	1	1
9	1	1	0	0	0	1	1	1	1	0	1

Figure 19 Excess-3 to seven-segment code conversion.

lows that every switching function can be implemented by an n-to-2^n-line decoder followed by an OR gate. (If 2^n exceeds the fan-in limitation of the OR gate, additional levels of OR gates will be needed.) Indeed, if more than one function of the same variables is to be implemented, the same decoder can be used, with each function having its own set of OR gates.

One major class of logic circuits is known as a *code converter*. This is a circuit that accepts as inputs the digits of a word that expresses some information in a particular code and that yields as outputs the digits of a word in a different code. (See Chapter 1 for an introduction to codes.) We will illustrate the use of a decoder as a code converter by designing a circuit to convert from excess-3 code to seven-segment code. (These codes were given in Figure 4 and Exercise 12 in Chapter 1; they are repeated here in Figure 19.)

Assume that a 4-to-16-line decoder is available. Since there are only 10 valid excess-3 code words, only 10 of the 16 AND gate outputs ever become 1. So only those 10 outputs from a 4-to-16-line decoder will be used. They are indicated in Figure 19 by their decimal equivalents.

Figure 19 is the truth table for each of seven output functions (the S_i) in terms of the four input variables. The circuit external to the decoder will consist of seven OR gates, one for each segment. Only one decision needs to be made: Which outputs from the decoder should become inputs to each OR gate? This is answered for each segment by listing the minterm numbers corresponding to each code word for which that segment output has the value 1. The minterm lists for the outputs corresponding to some of the segments are as follows:

$$S_3 = \Sigma(3, 5, 6, 8, 10, 11, 12)$$
$$S_4 = \Sigma(3, 4, 5, 6, 7, 10, 11, 12)$$
$$S_5 = \Sigma(3, 4, 6, 7, 8, 9, 10, 11, 12) \tag{8}$$
$$S_6 = \Sigma(3, 5, 6, 8, 9, 11)$$

Only one of the OR gates (the one for S_6) is shown in Figure 20; there should be six others. Then, when an excess-3 code word corresponding to a decimal

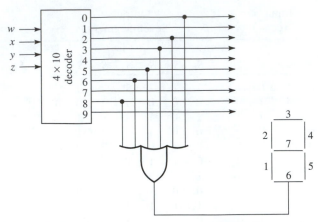

Figure 20 Excess-3 to seven-segment code converter.

digit appears at the input, the appropriate segments will light up, displaying the digit.

Exercise 11   Write the minterm lists for the three segments whose minterm lists were not given in (8). Confirm the inputs to the OR gate in Figure 20.   ◆

4   READ-ONLY MEMORY (ROM)

A circuit for implementing one or more switching functions of several variables was described in the preceding section and illustrated in Figure 20. The components of the circuit are

- An $n \times 2^n$ decoder, with n input lines and 2^n output lines
- One or more OR gates, whose outputs are the circuit outputs
- An interconnection network between decoder outputs and OR gate inputs

The decoder is an MSI circuit, consisting of 2^n n-input AND gates, that produces all the minterms of n variables. It achieves some economy of implementation, because the same decoder can be used for any application involving the same number of variables. What is special to any application is the number of OR gates and the specific outputs of the decoder that become inputs to those OR gates. Whatever else can be done to result in a general-purpose circuit would be most welcome.

The most general-purpose approach is to include the maximum number of OR gates, with provision to interconnect all 2^n outputs of the decoder with the inputs to every one of the OR gates. Then, for any given application, two things would have to be done:

- The number of OR gates used would be fewer than the maximum number, the others remaining unused.
- Not every decoder output would be connected to all OR gate inputs.

This scheme would be terribly wasteful and doesn't sound like a good idea.

Instead, suppose a smaller number, *m,* is selected for the number of OR gates to be included, and an interconnection network is set up to interconnect

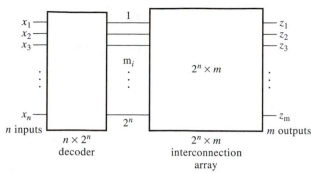

Figure 21 Basic structure of a ROM.

the 2^n decoder outputs to the m OR gate inputs. Such a structure is illustrated in Figure 21. It is an LSI combinational circuit with n inputs and m outputs that, for reasons that will become clear shortly, is called a *read-only memory* (ROM). A ROM consists of two parts:

- An $n \times 2^n$ decoder
- A $2^n \times m$ array of switching devices that form interconnections between the 2^n lines from the decoder and the m output lines

The 2^n output lines from the decoder are called the *word* lines. Each of the 2^n combinations that constitute the inputs to the interconnection array corresponds to a minterm and specifies an *address*. The *memory* consists of those connections that are actually made in the connection matrix between the word lines and the output lines.

Once made, the connections in the memory array are permanent.[8] So this memory is not one whose contents can be changed readily from time to time; we "write" into this memory but once. However, it is possible to "read" the information already stored (the connections actually made) as often as desired, by applying input words and observing the output words. That's why the circuit is called *read-only* memory.[9]

Before you continue reading, think of two possible ways in which to fabricate a ROM so that one set of connections can be made and another set left unconnected. Continue reading after you have thought about it.

The one-time "writing" into memory can be done as follows:

- A ROM can be almost completely fabricated except that none of the connections are made. Such a ROM is said to be *blank*. Forming the connections for a particular application is called *programming* the ROM. In the process of programming the ROM, a *mask* is produced to cover those connections that are not to be made. For this reason, the blank form of the ROM is called *mask programmable*.[10]

---

[8]In certain designs, it is possible for the connections to be *erasable;* this will be described shortly.

[9]Although "memory" appears in its name, a ROM does not have memory in the usual sense. As will be described in Chapters 5 and 6, memory is a characteristic of sequential, but not combinational, circuits.

[10]The mask, requiring minute attention, is expensive to produce. Hence, mask-programmable ROMs are used only when the cost is justified by very large production runs.

x_1	x_2	z_1	z_2	z_3
0	0	1	0	1
0	1	0	1	0
1	0	1	1	1
1	1	0	0	1

(a)   (b)

Figure 22 A ROM truth table and its program.

- A ROM can be completely fabricated such that *all* potential connections have been made. Such a ROM is also said to be blank. Programming the ROM for a specific application in this case consists of *opening* those connections that are unwanted. In this case, the blank ROM is said to be *field programmable* (designated PROM). The connections are made by placing a *fuse* or *link* at every connection point. In any specific application, the unwanted connections are opened or "blown out" by passing pulses of current through them. A measure of PROM cost is the number of fusible links, $2^n \times m$.[11]

Once a ROM has been programmed, an input word $x_1 x_2 \ldots x_n$ activates a specific word line corresponding to the minterm formed by the specific values of the x_i. The connections in the output matrix result in the desired output word.

EXAMPLE 3

Figure 22a gives the truth table for the interconnection matrix of a $2^2 \times 3$ ROM. The truth table leads to the ROM program represented by the solid dots at the intersections of the input and output word lines in Figure 22b. Each input word defines an output word, as required by the truth table. If the input word is 01 (corresponding to minterm m_1), for example, only output line z_2 will be activated because that is the only connection with m_1 in the connection matrix. Hence, the output word will be 010, as confirmed also from the truth table. (Confirm from the truth table that the rest of the program is correct.)   ∎

Exercise 12   A ROM is to be programmed to implement the conversion from excess-3 to seven-segment code whose table was given in Figure 19. ROMs come in standard sizes, and $m = 7$ is not one of them. The next larger standard size is $m = 8$. Hence, the truth table will have six more rows and one more column than shown in Figure 19. (Specify what the entries in the truth table will be for these extra rows and column.) Draw the appropriate number of crossing lines for the input and output words. Using the truth table, program the ROM by putting dots at the appropriate intersections of the two words.   ◆

---

[11]Some PROMs are fabricated so that it is possible to restore them to their blank condition after they have been programmed for a specific application; these are *erasable* PROMs, or EPROMs. They have some clear advantages over the nonerasable kind, but their cost is correspondingly higher.

In Exercise 12 the number of entries in the truth table (which corresponds to the number of links between the input and output words) is $2^n \times m = 16 \times 8 = 128$. Of these, fully half represent don't-cares. There are cases far worse than this; sometimes as few as 1 percent of the links are used, resulting in considerable "waste" in such ROM implementations. Another implementation that avoids this waste would be most welcome. That's the subject of the next section.

5 OTHER LSI PROGRAMMABLE LOGIC DEVICES

One way of looking at the ROM discussed in the previous section is as a device with a specific structure (a set of AND gates and a set of OR gates) that a designer can use to achieve desired outputs by making a few modifications. We might say that the ROM has been "programmed" to produce its specific outputs. There are other structures that have this property, namely, programmability. A generic name for them is *programmable* (or programmed) *logic device* (PLD).

The ROM implements logic functions as sums of minterms. For n input variables there are 2^n minterms and, hence, 2^n AND gates, each one with n inputs. As just discussed, in a number of important logic functions, many of the AND gates and the links connecting them to the output OR gates are unused. We will now discuss two implementations in which some of this "waste" is avoided.

Programmed Logic Array (PLA)

The canonic sum-of-products implementation of a logic function is wasteful in two ways: in the number of AND gates used (as many as there are minterms, 2^n) and in the number of inputs to each AND gate (n). Suppose we contemplate a reduced (possibly minimal) sum-of-products implementation. Given a logic function of n variables, the largest number of terms in a minimal sum-of-products expression representing this function is 2^{n-1}—just half the number of minterms. (See Problem 36 in Chapter 3.) That means a savings of 50 percent in AND gates for the worst single-output case. Since there will be a reduced set of inputs to the AND gates, this saving in gates is paid for by the need to program not only the outputs of the AND gates but their inputs as well. The structure of the circuit that results is called a *programmable* (or *programmed*) *logic array* (PLA). It is illustrated in Figure 23 for the case $n = 3$ input variables, $m = 4$ output functions, and four AND gates.

The diagram in Figure 23 is not a circuit diagram but a schematic diagram. A single line is shown to represent all inputs to each AND and OR gate. The number of input lines to each AND gate should be $2n$, twice the number of inputs, to accommodate the possibility of connecting each variable or its complement to each AND gate. The number of input lines to each OR gate should equal the number of AND gates, say p. (For simplicity and without fear of confusion, even the gate symbols can be omitted.) The programmed connections between the inputs and the AND gates, and between the AND-gate outputs and the OR gates for a specific set of output functions are shown by the heavy dots at the intersections.

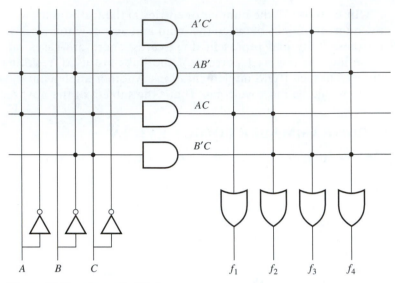

Figure 23 Structure of a PLA.

Maps of the four output functions and minimal sum-of-products expressions are shown in Figure 24. In this example, a total of only four product terms covers all functions, so only four AND gates are needed in the implementation. Two sets of lines must be programmed: the input lines and the output lines. To do this, we construct a *programming table* as follows:

- The implicants (product terms) are listed as row headings.
- In one set of columns, the headings are the input variables; this part of the table must provide the information that tells which variables (or their complements) are factors in each implicant.
- In a second set of columns, the headings are the output functions; this part of the table must provide the information that indicates the output gate to which each implicant (AND-gate output) is directed.

In the first set of columns, if a variable (uncomplemented) is present in a particular row, the corresponding entry is 1; if its complement is present, the entry is 0. If neither is present, the entry can be left blank, but it is preferable to show some symbol instead; a dash is often used.

In the second set of columns, corresponding to the output functions, if a particular function covers a particular implicant, then the corresponding entry is 1; otherwise it could be left blank, but it is customary to enter a dot. To illustrate, consider row 4. Since the implicant is $y'z$, the entry in column z is 1, that in column y is 0, and that in x is a dash. In the output columns, only f_1 does not cover implicant $y'z$; hence, the entry will be 1 in every column in row 4 except the f_1 column, where the entry is •. Confirm the remaining rows.

Once the programming is done, fabricating the links (connection points) in a PLA is carried out in a similar manner as for the ROM. The PLA is either mask programmable or field programmable (FPLA). In the case of the FPLA, with p = the number of AND gates, there will be $2np$ links at the inputs and mp

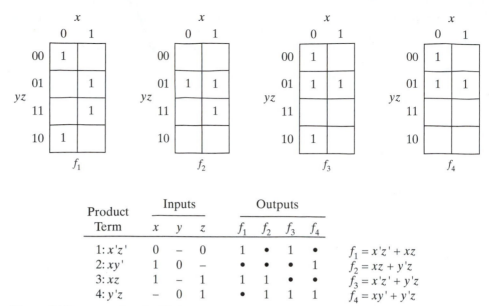

Product Term	Inputs			Outputs				
	x	y	z	f_1	f_2	f_3	f_4	
1: $x'z'$	0	–	0	1	•	1	•	$f_1 = x'z' + xz$
2: xy'	1	0	–	•	•	•	1	$f_2 = xz + y'z$
3: xz	1	–	1	1	1	•	•	$f_3 = x'z' + y'z$
4: $y'z$	–	0	1	•	1	1	1	$f_4 = xy' + y'z$

Figure 24 Programming the PLA.

links at the outputs. For the example in Figure 23, the number of links is $4(6 + 4) = 40$. Only 16 of these are to be kept, meaning that, during field programming, 24 links are to be blown out. Typical PLAs have many more inputs, outputs, and AND gates than are shown in the example in Figure 23. (IC type 82S100, for example, has $n = 16$, $m = 8$, and $p = 48$.)

When a set of switching functions is presented for implementation with a PLA, a design goal would be reduction in p (the number of AND gates). The economy achieved is not derived from a reduction in the production cost of gates. (The production cost of an IC is practically the same for one with 40 gates as it is for one with 50 gates.) Rather, the removal of one AND gate eliminates $2n + m$ links; the main source of savings is the elimination of a substantial number of links due to the elimination of each AND gate. On the other hand, reduction of the number of AND gates to a minimum does not mean that each function should be minimized or that all implicants should be *prime* implicants. The implicants should be chosen so that as many as possible of them are common to many of the output functions.

Programmed Array Logic (PAL)

A ROM has a large number of fusible links ($m \times 2^n$) because of the large number (2^n) of AND gates. Programming of links is performed only on the outputs from the AND gates. In a PLA, the number of links is drastically reduced by reducing the number of AND gates. The latter is done by changing the expression representing the switching function from a canonic sum-of-products form to a sum of products with fewer terms. The price paid is the need to program not only the outputs from the AND gates, but also the inputs to the AND gates. What other possibility for programming is there beyond the two cases of (a) programming the outputs of the AND gates and (b) programming both the inputs

Product Term		Inputs												Outputs					
Number	Function	1	2	3	4	5	6	7	8	9	10	11	12	1	2	3	4	5	6
1														•	•	•	•	•	1
2														•	•	•	•	•	1
3														•	•	•	•	•	1
4														•	•	•	•	•	1
5														•	•	•	•	1	•
6														•	•	•	•	1	•
7	$x_1 x_2' x_5 x_7 x_{11}' x_{12}$	1	0	–	–	1	–	1	–	–	–	0	1	•	•	•	1	•	•
8														•	•	•	1	•	•
9														•	•	1	•	•	•
10														•	•	1	•	•	•
11														•	1	•	•	•	•
12														•	1	•	•	•	•
13														1	•	•	•	•	•
14														1	•	•	•	•	•
15														1	•	•	•	•	•
16														1	•	•	•	•	•

Figure 25 Programming table for a PAL example.

and the outputs? We're sure you answered, "programming only the inputs." This is a possibility, but is it worthwhile?

In the case of the ROM, there is no need to program the inputs because, for *any* function of *n* variables, there will be the same (large) number of AND gates. In the same way, if the number of OR gates at the output could be fixed, then programming the outputs of the AND gates could be avoided.

In many circuits with multiple outputs, even though the outputs are functions of a large number of input variables, the number of product terms in each output is small. Hence the number of AND gates that drive each OR gate is small. In such cases, permanently fixing the number of OR gates and leaving only the programming of the AND gate inputs for individual design might make economic sense. The resulting circuit is called *programmed array logic* (PAL).[12] The number of fusible links in a PAL is only $2np$. Standard PALs for a number of low values of p exist. For example, the PAL16L8 has a maximum of 16 inputs and 8 outputs.

A programming table for a PAL is similar to the one for a PLA. A case with six outputs is illustrated in Figure 25. A ROM with 12 input variables would require $2^{12} = 4096$ AND gates. However, let's assume that for some possible cases, the canonic sum-of-products expression can be reduced to 16 implicants, only one of which is shown in Figure 25. The entries in the table would have the same meanings as those for the PLA. However, for the PAL, the output columns would be fixed by the manufacturer on the basis of the number of AND gates already connected to each OR gate.

In the present case, two of the output OR gates are each driven by four AND gates; the remaining four OR gates are each driven by two AND gates. For any

---

[12]PAL is a registered trademark of Advanced Micro Devices.

given design problem, the first step is to obtain an appropriate sum-of-products expression, just as in the case of a PLA implementation. The input connections are indicated in the table as in the case of the PAL: an entry is a 1 if a variable appears uncomplemented in an implicant, a 0 if it appears complemented, and a dash if it does not appear at all. This is illustrated for one row in Figure 25. The number of fusible links in this example is $2 \times 12 \times 16 = 384$. This is 20 percent fewer than the number of links of a PLA having the same dimensions. Typically, however, PLAs have many more AND gates and so, for a PAL, the number of links would typically be many times more than the number for a comparable PLA.

Exercise 13   Suppose two of the rows of inputs in Figure 25 are as follows:

$$0\ 1\ 0 - 0 - - 1 - - - -$$
$$1\ 0\ 1 - - 0 - - 1\ 1 - -$$

What are the corresponding product terms?   ◆

Further attention will be devoted to PLDs in Chapter 8. Attention will also be given there to the use of hardware description languages in designs using PLDs.

CHAPTER SUMMARY AND REVIEW

In Chapter 3, designs were carried out with primitive gates in SSI circuits. This chapter advanced the design process to more complex circuits implemented in MSI units. The topics included were

- Binary adder
- Full adder
- Ripple-carry adder
- Carry-lookahead adder
- Binary subtractor
- Two's complement adder and subtractor
- One's complement adder and subtractor
- Multiplexer
- Data input
- Select input
- Implementation of general-purpose logic circuits with multiplexers
- Demultiplexer
- Data input lines
- Control input lines
- Decoder
- $n \times 2^n$-line decoder
- Tree decoder
- Implementation of general-purpose logic circuits with decoders
- Code conversion
- Read-only memory (ROM)
- $n \times 2^n$ decoder
- $2^n \times m$ interconnection array
- Programming a ROM

- Mask-programmable ROM
- Field-programmable ROM
- Programmable logic device (PLD)
- Programmed logic array (PLA)
- Programmed array logic (PAL)

PROBLEMS

1   a. Analyze each of the full adder circuits shown in Figure P1 and write expressions for the output of each intermediate gate.
   b. Obtain logic expressions for the sum and carry circuit outputs.
   c. Verify that these expressions are equivalent to the sum and carry functions in equations (1) in the text.

2   a. A 4-bit carry-lookahead adder is to be designed. In equation (7) in the text for the carry function, let $i = 0$ and let j range from 0 to 4. Write the resulting expressions for C_1, C_2, C_3, and C_4.
   b. Construct the logic diagram for the 4-bit carry-lookahead whose schematic diagram is given in Figure 8.

3 A 4-bit binary number $Y = y_3y_2y_1y_0$ is to be multiplied by a 3-bit binary number $X = x_2x_1x_0$. Use two 4-bit adders and other gates that you might need to implement this operation, and draw the corresponding diagram.

4 Prove formally that if the propagate variable P_i for a carry-lookahead adder is defined as $A_i + B_i$ instead of $A_i \oplus B_i$, the sum and carry outputs of the adder will still be computed correctly. (Give an informal proof also.) Which definition is better for implementation purposes?

5 Design a circuit for overflow detection in the one's complement adder/subtractor shown in Figure 11.

6   a. Show the connections on a schematic diagram of a dual four-input multiplexer for implementing the sum and carry functions of a full adder.
   b. Repeat using a 3-to-2^3-line decoder.

7 Realize each of the following functions using an 8×1 multiplexer.

   a. $f = \Sigma(0, 1, 10, 11, 12, 13, 14, 15)$
   b. $f = \Sigma(0, 3, 4, 7, 10)$
   c. $f = \Sigma(0, 3, 4, 6, 7, 8, 12)$
   d. $f = \Sigma(1, 2, 5, 8, 11, 12, 14)$

8 Realize each of the functions in Problem 7 using half of a dual 4×1 multiplexer and the minimum number of external gates.

9 Repeat Problem 7 using a 3-to-2^3-line decoder.

10 Use a dual four-input multiplexer to implement each of the following pairs of functions with the fewest external gates.

   a. $f_1 = \Sigma(0, 4, 5, 7, 9, 11)$, $f_2 = \Sigma(2, 3, 5, 6, 10, 13)$
   b. $f_1 = \Sigma(0, 4, 7, 10, 12, 14, 15)$, $f_2 = \Sigma(2, 7, 8, 9, 12, 13, 14, 15)$

11   a. Show how to connect a 4-bit MSI adder to serve as a BCD-to-excess-3 code converter.
   b. Repeat using a 4-to-10-line (BCD-to-decimal) decoder and four AND gates.

12 Design a BCD-to-decimal decoder using two 2-to-4-line decoders and a minimum of interconnecting AND gates.

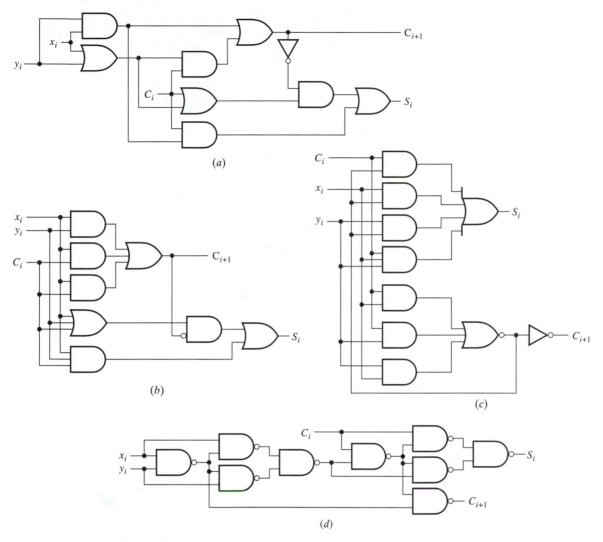

Figure P1

13 A circuit is to accept two 2-bit binary numbers x_1x_0 and y_1y_0 and emit the product as a 4-bit binary number $z_3z_2z_1z_0$. (Review binary multiplication in Chapter 1 if you need to.)

   a. The result is to be achieved by a (possibly) multilevel circuit with two-input gates. Determine appropriate expressions for each output. How many levels of gates does each output have?
   b. Design a circuit using a 4-to-2^4-line decoder with external OR gates.

14 Examine late editions of manufacturers' data books.

   a. What is n for the largest n-to-2^n-line decoders?
   b. Note what the standard sizes of ROMs are.
   c. What are some representative dimensions of a PLA chip?
   d. What are some representative dimensions of a PAL?
   e. Is there a BCD adder in a single MSI package?

15 A switching function of n variables is to be implemented by an n-to-2^n-line decoder followed by an external OR gate. The physical gate available for this purpose has both an OR and a NOR output. (It is an ECL gate.) For practical reasons (to avoid fan-in problems), it would be best to try to reduce the number of inputs to an external gate.

    a. Describe how to implement the function using the available physical gate if the number of minterms contained in the function is more than $2^{n-1} = 2^n/2$.

    b. Illustrate with the following function:

$$f = \Sigma(0, 1, 2, 3, 4, 5, 7, 8, 9, 10, 13, 14, 15)$$

16 **a.** Design a BCD-to-decimal decoder using the minimal number of two-input AND gates.

    b. Repeat, using two 2-to-4-line decoders and a few interconnecting AND gates.

17 **a.** Use two identical n-to-2^n-line decoders with enable inputs to construct an $(n+1)$-to-2^n-line decoder without enable. Show how the outputs are obtained.

    b. Illustrate with two 2-to-4-line decoders.

18 Design an octal to binary encoder. This is a circuit with 8 inputs, x_i, and 3 outputs, z_i. Only one of the outputs is 1 at any one time. Octal digit k is represented by $x_k = 1$.

19 A decimal-digit code converter from 2-out-of-5 to seven-segment code is to be designed. A number of different possibilities are to be explored, assuming that only valid code words will occur as inputs.

    a. Draw a circuit diagram using a complete 5×2^5 decoder design.

    b. Assuming a design using discrete gates:

        i. Draw a circuit for a sum-of-minterms design. (This would constitute a partial decoder.)

        ii. The AND gates in the preceding design are five-input gates. Is it possible to use the same structure but with two-input gates? Justify your answer.

        iii. Carry out a minimal sum-of-products design that uses 11 AND gates and 7 OR gates, each with no more than three inputs.

        iv. Consider a minimal product-of-sums design. Is this more economical than the minimal sum-of-products design?

        v. Now suppose that, in addition to valid code words, invalid ones can also occur. Modify the best of the preceding designs so that, whenever there is an invalid code word, the symbol E (for error) is displayed.

20 The code converter in Problem 19 is to be designed with a ROM. The closest-size ROM available is a $2^5 \times 8$. Construct the required programming table. Specify the number of links.

21 The code converter in Problem 19 is to be implemented with a PLA. A 5×8 PLA with 12 AND gates is available. Draw a programming diagram for implementing the desired code converter. Specify the number of links.

22 **a.** Suppose the circuit in Problem 13 is to be implemented with a $2^4 \times 4$ PROM. Show the programming table and draw an appropriate diagram.

    b. Suppose instead that the circuit is to be implemented by a 4×4 PLA with 10 AND gates. Show the programming diagram (in the form of Figure 23 in the text). Compare the number of links with those of the PROM implementation. Construct the programming table in the form of Figure 25 in the text.

    c. Now suppose that the circuit is to be implemented by a PAL. Construct the programming table in the form of Figure 25 in the text.

23 A combinational circuit having three inputs and six outputs is to be designed. The output word is to be the square of the input word.

    a. Design the circuit using a ROM that has the smallest possible dimensions. Construct the truth table and specify the number of links.

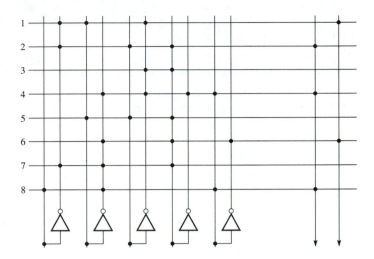

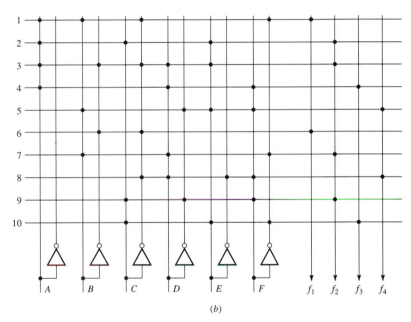

(b)

Figure P24

b. Design the circuit using a PLA with the fewest number of product terms. Construct the programming diagram and specify the number of links.

24 The programming diagrams for two PLAs are shown in Figure P24.

a. Write the equations of the outputs realized by each PLA. Specify the number of links.
b. The same functions are to be implemented with a ROM. Specify the dimensions of the ROM and the number of links. Set up its programming table.
c. The same functions are to be implemented with a PAL. Is it possible to do so? If so, set up the programming table and specify the number of links. If it is not possible, explain why not.

25 (Review Chapter 1 on Hamming codes if you need to.) Using an n-to-2^n-line decoder (for an appropriate n) and any additional logic:

    a. Design the error-correcting logic for a single-error-correcting Hamming code assuming 3 *message* bits in each code word. The outputs of the circuit should be

- E, indicating that an error has been detected
- IV, indicating that the MSG output is invalid (obviously, IV is 0 when no error, or only a single error, has occurred)
- MSG, a 3-bit output that contains the corrected transmitted message in the cases of zero and one error

    b. Design the single-error-correcting and *double-error-detecting* (SEC-DED) logic for an error-correcting Hamming code extended by the addition of a parity bit over all (that is, message and parity) positions. Assume 3 *message* bits in each code word. The output signals and their meanings are to be the same as in part *a*.

26 Explain in words the behavior of the diagram in Figure P26. (The open-headed arrows represent multiple-bit inputs and outputs.)

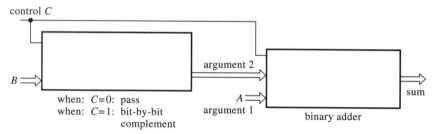

Figure P26

27 A microprocessor (μp) outputs three control signals that have the meanings given in the following table. (No knowledge of μp is necessary to solve this problem.)

R'	W'	M/I'O'	
0	1	1	μp wants to read memory
1	0	1	μp wants to write to memory
0	1	0	μp wants to read an input/output device
1	0	0	μp wants to write to an input/output device
1	1	×	μp wants none of the preceding operations

    a. Design a logic circuit using a suitable multiplexer and minimal additional logic to transform these three signals into the following four signals, each representing an operation:

    $(M R)'$, $(M W)'$, $(IO R)'$, $(IO W)'$

    When any of the operations is desired (not desired), the value of the corresponding signal is to be 0 (1).

    b. Design a multiplexer implementation to perform the inverse transformation.

28 The 4-bit lookahead unit shown in Figure P28a receives generate and propagate variables from units 0 through 3 comprising a similar group. It also receives C, the carry input to unit

0 of the group. It computes $C_0, C_1,$ and $C_2,$ which are the carry outputs from units 0, 1, and 2, respectively. It also computes the generate and propagate variables, G and $P,$ for the whole group. The carry outputs are generated in parallel, not in ripple fashion.

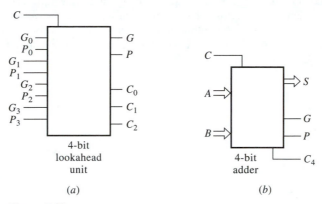

(a)   (b)

Figure P28

    a. Derive equations for all the outputs, and show the implementation.

    b. Using 4-bit lookahead units of the above type and 4-bit adders of the type shown in Figure P28b, draw the logic diagram for a 48-bit adder using a single-level lookahead. (The open arrows represent multiple-bit inputs and outputs—in this case, 4 bits. $A,$ for example, stands for a vector of 4 bits: $A_0, A_1, A_2, A_3.$)

    c. Repeat part b using two levels of lookahead, in which the G and P outputs of the first-level lookahead units feed the G_i and P_i inputs of the second-level lookahead units. Compare with respect to speed with the design of part $b.$

29 This problem concerns the design of a 4-bit lookahead subtractor (Figure P29). The 4-bit vector B $(B_3B_2B_1B_0)$ is to be subtracted from 4-bit vector $A.$ The borrow input C_0 is 1 if and only if the next lower unit is borrowing a 1 from this unit. The 4-bit vector D is the difference output, and C_4 is the borrow output. G and P are generate and propagate variables from the whole unit.

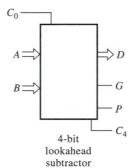

Figure P29

    a. Give an expression for each output and show the implementation.

    b. As in Problem 28, there will be more than one way to define the propagate variable. Give these definitions and compare the differences in their implementation.

    c. Suppose that multiple-bit subtraction is to be carried out. For this purpose, can 4-bit lookahead units, of the type described in Problem 28 in the context of addition, be used with 4-bit lookahead subtractors of the type defined here? Justify your answer.

    d. Using 4-bit subtractors of the type described in this problem, and also suitable 4-bit lookahead units, design a 24-bit lookahead subtractor.

30 An 8-input priority encoder (Figure P30) has eight request inputs: $I(7\ldots0)$. A logic 1 on any of these lines denotes the presence of a request from the corresponding source for some service. The priority varies from the highest for 7 to the lowest for 0. Output LR (Local Request) is 1 if and only if there is at least one request among the eight I inputs. If EI (Enable Input) is 1, the encoder identifies the request having the highest priority and outputs its 3-bit address on $A(0\ldots2)$. If no request is active, it outputs a zero address. If the encoder is not enabled (EI = 0), it outputs zeros on A. EO (Enable Output) is 1 if and only if the encoder is enabled (EI = 1) *and* there is no request among the eight I inputs.

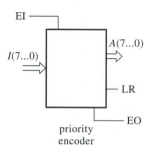

Figure P30

a. Derive expressions for each output and simplify.
b. Design a 48-input priority encoder using 8-bit priority encoders of the type described in this problem and minimal additional logic. Use a ripple configuration.
c. Considering the enable signals, EI and EO, as the equivalent of carry signals, derive expressions for the generate and propagate variables for the eight-input priority encoder. As in Problem 29, give two expressions for the propagate variable and pick the "better" one. Does it require extra logic to compute the generate and propagate variables, or are they available from the outputs of the eight-input priority encoder described here?
d. Using suitable 4-bit lookahead units, design a lookahead implementation for a 48-input priority encoder and compare its speed with the design in part *b*.
e. Suppose that the eight-input priority encoder has *disable* signals, DI and DO, instead of enable signals EI and EO. Repeat parts *c* and *d* considering the disable signals as the equivalent of carry signals.

31 A BCD-to-seven-segment decoder has "blank" signals, BI and BO, to help suppress leading 0's in integer displays and trailing 0's in fraction displays. When BI is 1, if the input digit is 0, all outputs should be 0; that is, the digit will be blanked. When BI is 0, there is no blanking, but then BO is a blank signal to the next digit. A diagram is shown in Figure P31*a*.

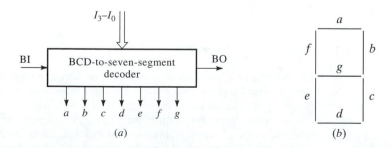

Figure P31

    a. Give expressions for the outputs BO, *a*, and *f*.

    b. Design an 8-bit display with four digits each for the integer and fractional parts. The least significant integer digit should never be blanked, even if the integer part of the number is 0.

    c. Considering sluggish human response times, the ripple implementation in part *b* should be adequate. However for pedagogical purposes, suppose you wanted to design a lookahead implementation of the display, so that each digit would settle into the blanked or unblanked state faster. Treating BI and BO as carry signals, give expressions for the generate and propagate variables for this decoder.

    d. Suppose that, instead of the BI and BO pins, the decoder has DBI ("don't blank input") and DBO ("don't blank output") pins. Treat these as the carry signals this time, and repeat part *c*.

32 Prove formally that if the propagate variable P_i for a lookahead adder is defined as the Boolean sum of A_i and B_i instead of their Exclusive-OR, the sum and carry outputs of the adder will still be computed correctly. Give an informal proof also. Which definition is better for implementation purposes?

33 A 4-bit data selector has four data inputs, $D_3...D_0$, and two select inputs, $s_1 s_0$. The output z is one of the data inputs as selected by the select inputs. Thus, $z = D_2$ when $s_1 s_0 = 10$.

    a. Draw an AND-OR diagram of the data selector.

    b. Another circuit consists of two XOR gates. The inputs to XOR1 are two signals *A* and *B*. The inputs to XOR2 are the output of XOR1 and a third signal *C*. Draw this circuit and write its output in terms of *A*, *B*, and *C*.

    c. Choose the select inputs and the data inputs in part *a* in terms of *A*, *B*, and *C* so that the circuits in parts *a* and *b* will have the same outputs. If there is more than one choice, show all of them.

34   **a.** Design a BCD adder using a ROM (and any other logic needed), assuming only legal BCD words are used as input. Specify the dimensions of the ROM and show a schematic diagram.

    b. Describe the programming table and illustrate it (at least partially).

    c. Specify the number of links.

Chapter 5

Sequential Circuit Components

In combinational circuits, the outputs at any given time depend only on the inputs at that time, not on the history of past inputs. This implies

- Either that the gates in the circuit respond to inputs without delay, or
- That the time intervals between successive inputs are so long compared with the response time of gates that the output responses to a set of past inputs have already occurred before the next input comes along.

There are situations, however, when an output depends not only on the present inputs but on the condition of the circuit at the time those inputs arrive. The condition of the circuit at any given time, in turn, depends on the history of the past inputs. This means that there must be a mechanism for *storing* the information conveyed by the sequence of past inputs. This chapter is concerned with devices that store past inputs—either a single past input or a sequence of them.

1 DEFINITIONS AND BASIC CONCEPTS

Storage of information about past inputs implies *memory* of those past inputs. In this chapter we will introduce a number of specific circuits that serve as cells (often called *primitive* cells) for the storage of a single past input. The interconnection of such memory cells, together with combinational circuits, can deal with situations where an output depends on the present input and a finite string of past inputs. This discussion triggers the following definition:

> *A digital circuit is a sequential circuit if its outputs at any given time are functions both of the external inputs at that time and of sequences of past inputs.*[1]

On this basis, models of a sequential circuit can be constructed, as shown in Figure 1.

---

[1]A variation of this general case will be discussed in Chapter 7.

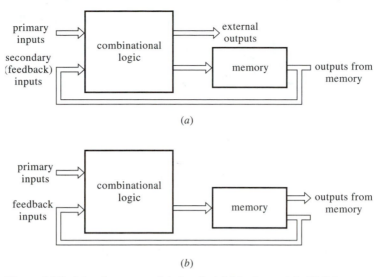

Figure 1 Models of a sequential circuit. (*a*) Mealy model. (*b*) Moore model.

There are two varieties of inputs to the part of the circuit consisting of combinational logic. There are inputs from the external world, called the *primary* inputs. There are also *secondary* inputs that describe the condition (or state) in which the circuit is found at the arrival of the present inputs. These secondary inputs are fed back from the memory. Both sets of inputs are processed by combinational logic. In the model referred to as a *Mealy machine,* there are two varieties of output from the combinational logic: outputs to the outside world and outputs that represent new information to be stored in memory. The memory incorporates into the previous information stored there the new information brought in by the latest inputs. The outputs from memory are fed back as *secondary inputs* to the combinational logic. Causing the *contents* of the memory unit to be thus modified is called *writing in memory.* The reverse process—acquiring as an output the information currently stored in a memory unit—is called *reading from memory.*

In the Mealy model shown in Figure 1*a*, the external outputs depend both on inputs from the outside world and on the feedback inputs from memory. There are, however, digital circuits in which the output does not depend directly on external inputs. Rather, the external inputs cause changes in memory, after which external outputs are emitted from the memory. As in the Mealy machine, another set of outputs from memory become feedback inputs to the combinational logic. The model of such a sequential circuit is shown in Figure 1*b* and is called a *Moore machine.* In subsequent chapters we will deal with both models.

Timing difficulties are inherent in switching circuits. With variability both in the time of arrival of inputs and in the time delay inherent in gates, it may be difficult to keep track of what is a past input and what is a present input. To illustrate, suppose an input variable changes its value at some time. After some

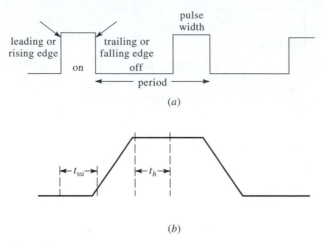

Figure 2 (*a*) A periodic pulse train. (*b*) A "realistic" clock pulse.

propagation delay, the response to this change will be felt both at the input to the memory and at the output terminals. Hence, the contents of the memory will change. Since, in both models, there is feedback to the input, this change may cause a further change in the output. From the time of the initial change in input and the final change in output there is a period of instability. If a further change in primary input takes place during this period of instability, the circuit might fail to give a reliable response because it isn't clear what the contents of memory are when the second input arrives. (There will be a *glitch*.)

A common approach to overcoming this problem is to introduce an extra signal in the form of a periodic pulse wave train called a *clock,* as shown in Figure 2a. (Such a clock signal is generated by what is known as an *oscillator,* almost universally a quartz crystal oscillator.) The memory part of the circuit in Figure 1 is constructed so that it is disabled during the time that the clock pulse is absent; activity in the circuit takes place only in the presence of the clock pulse. If the pulse is narrow enough, it is unlikely that more than one input change will take place during the time that the clock pulse is *on*.[2] In fact, in many circuits the mere initiating of the clock pulse (the rising edge) enables the circuit—the so-called edge-triggered case, to be described in a later section. No subsequent activity will take place in the circuit, no matter how wide the clock pulse, until the onset of the next clock pulse. (The edge triggering can also be done by the trailing edge of the clock pulse.)

The type of circuit just described is called a *clocked* sequential circuit. Several clocked sequential circuits can be interconnected, but there would be no end of confusion if the clocks enabling the respective circuits were different. If all enabling clocks in such interconnected circuits are the same, then activity in all circuits will occur synchronously. The result is called a *synchronous* se-

---

[2]Figure 2 is appropriate to the "1-high" system used here; for the "1-low" system, the pulse train would be negative going.

quential circuit, because the parts of the circuit are all synchronized by the same clock, the system clock.

One period of the clock waveform includes an interval of time when the clock pulse is 1 ("high" in positive logic; some would say "active high") and another interval of time when it is 0 ("low" in positive logic). If these two intervals are equal, the signal is a square wave. (It is difficult to conceive of a square wave as a sequence of pulses.) If the high interval is much shorter than the low interval, the signal is a pulse train of positive-going pulses. On the other hand, if the high interval is much *longer* than the low interval, the signal is a pulse train of negative-going pulses. (To get a feel for this before continuing, draw a waveform where the duration of the low interval is 10 times the duration of the high interval and vice versa.)

The important features of the clock signal are the following:

- Duty cycle
- Clock frequency
- Sharpness of the edges
- Stability of the frequency and the waveform

The *duty cycle* is the fraction of the clock period in which the clock signal is high. For a square wave, the duty cycle is 0.5; a duty cycle of less than 0.01 (1 percent) is not uncommon. The frequency is particularly important. Since the timing of all activity in a clocked circuit is based on the occurrence of a clock pulse, the rapidity with which the circuit performs its operations depends on the frequency. Frequencies on the order of 1 MHz (1 million operations per second) are considered slow in computer operations; 5 to 50 MHz is common. Contemporary microprocessors have clock frequencies as high as 600 MHz.

The last two characteristics of the clock signal depend on the quality and design of the oscillator. Although they are important, logic designers have no control over such characteristics; they must accept what oscillator designers provide.

Ideally, the sides of clock pulses rise and fall in zero time. But real life deviates from the ideal in two ways: (a) The sides rise and fall in nonzero time, and (b) the slopes of the sides do not change from zero to nonzero (or vice versa) in zero time. Neglecting (b) results in pulses that are trapezoidal rather than rectangular, as in Figure 2b. More realistically, the slopes at both the rising and falling corners of the clock pulse change gradually, so that the "corners" are curved. (See Figure 22 in Chapter 2.) We will neglect this curvature in the clock and other signals in the discussion that follows.

Exercise 1

    a. Estimate the approximate value of the duty cycle in Figure 2a.

    b. Suppose that the duty cycle of a pulse train is a reasonable 1 percent and that its frequency is a reasonable 10 MHz. How long in nanoseconds is one period? Suppose that one period is to occupy 4 cm on a horizontal axis; draw two cycles of such a pulse train to scale. How long in nanoseconds is the duration of the clock pulse? How many centimeters is the width of the clock pulse?   ◆

In this chapter, a considerable number of circuits will be introduced that serve as elements or cells of memory, starting with the simplest one. Each new circuit introduced will be justified by outlining the shortcomings of the preceding one and showing how the new circuit improves on it or has a useful feature that preceding ones do not have. By the nature of this material, the chapter is rather descriptive; there is little in the way of mathematics or analysis. The purpose here is not to enable you to engage in the design of such devices, but to help you learn enough of their characteristics to use them with facility in the design of larger systems.

2   LATCHES AND FLIP-FLOPS

As indicated in Figure 1, one of the needs in a sequential circuit is the storage in memory of information about the present condition of the circuit as a result of past inputs. A *bistable* device, which can exist in one of two stable conditions, can store one piece of information. A light switch, for example, can be in either of two positions. Information about the light being *on* or *off* is contained in the position of the switch. In digital circuits, the most common elements of memory are electronic devices called *flip-flops*. Several varieties of flip-flops will be considered here, each useful under certain circumstances.

The subject will be introduced using the circuit in Figure 3a, which shows two *cross-coupled* inverters. Note that the complex-looking interconnection in Figure 3a can be redrawn as in Figure 3b. ("Unwrap" the preceding circuit and confirm this.) The input of each inverter is the output of the other, without any external input. This is a digital example of a bistable device. To confirm this claim, suppose that the top output Q in Figure 3a has the logic value 1. Since this is the input to the bottom inverter, the latter will have output $x = 0$. But x is also the input to the top inverter, confirming the assumed output $Q = 1$. (Now you do some work: Assume that $x = 1$ and confirm that $Q = x' = 0$, verifying that $x = (x')' = 1$, as assumed.) Thus, the output of this circuit can remain stable in either of the values $Q = 0$ or $Q = 1$. Hence the terminology *bistable*. The value that the output actually takes on depends on a possible external input, as shown in Figure 3c. Each of the two inverter outputs is the complement of the other one.

SR Latch

The bistable circuit in Figure 3c has two sources of data for the input to the first inverter: the external input and the feedback from the output of the second inverter. It is difficult to write a new value into the circuit because the feedback is designed to hold the existing value. The feedback and the external input fight to determine the value stored in the memory element. Figure 4a shows the design of a memory element called an *SR latch*. The design uses two NAND gates preceded by inverters.[3] (Another design, using two NOR gates, is left for you to analyze in an exercise.) What happens in this circuit without the inverters will be left for you to figure out in a problem. The ad-

---

[3]There would be no inverters if negative logic were used or if the *S* and *R* inputs were asserted low in mixed logic; see Chapter 2.

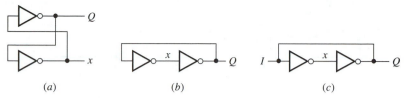

Figure 3 A bistable circuit of inverters.

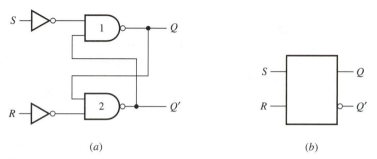

Figure 4 *SR* latch design. (*a*) With NANDs. (*b*) Schematic symbol.

vantage of the circuit in Figure 4*a* compared with the one in Figure 3*c* is that the external inputs can influence the state of the memory without fighting the feedback in the circuit.

The novel feature, different from anything you have seen in combinational circuits but found in the circuit of Figure 4*a*, is cross-coupling, or feedback, from the output of each gate to an input of the other gate. The two external (primary) inputs are called *set (S)* and *reset (R)*.

Based on experience with the cross-coupled devices in Figure 3, we will assume that each NAND gate output is the complement of the other. In the literature, such outputs are universally designated Q. However, when external connections are made from a memory cell output to other devices, these external variables can be given other designations. We will often use y for this purpose, sometimes even when discussing a single memory cell. A schematic diagram is shown in Figure 4*b*. The bubble on the lower output implies that this signal is the complement of the one labeled Q. If the bubble is shown, it isn't really necessary to provide any other label, such as Q' shown in Figure 4*b*. Some might even misinterpret this as a double negation.[4]

The gates shown in Figure 4 are assumed to be real, physical gates, not simply pictorial representations of the Boolean NAND operation. At first it might seem surprising that a mere interconnection of a couple of gates can lead to something other than a strictly combinational circuit. The answer lies (a) in the feedback provided by the cross-coupling and (b) in the delay of signals going

---

[4]Consult a number of logic design books and notice that some books place the labels Q and Q' *inside* the rectangle representing the device. What do you think of the propriety of using a label Q' and following it with a bubble?

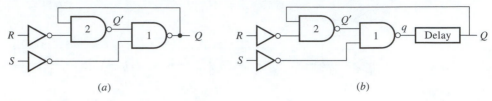

Figure 5 *SR* latch with feedback and delay emphasized.

through real gates. The feedback feature can be emphasized by redrawing the latch in the manner shown in Figure 5a.

Although there is propagation delay through each real, physical gate, let us temporarily assume — *as a digression in this paragraph only* — that all the delays in the circuit of Figure 5a are lumped together in one place, as in Figure 5b; the gates in the latter circuit are assumed to be ideal and, hence, respond instantly. Such a model for treating the propagation delay of gates would give sufficiently accurate results. The effect of a change in the *S* or *R* input is felt immediately in the output labeled *q* (to distinguish it from the eventual output, *Q*) in Figure 5b. It isn't until after a delay, however, that this effect is passed on to *Q* and fed back to the input.

In subsequent discussions, the delay block in Figure 5b will not be shown explicitly. All gates will be assumed to be real, physical gates; thus, the presence of propagation delay in any gate will be assumed implicitly. Where it makes a difference, the analysis will take this delay into account.

When the *SR* latch is unexcited (that is, both *S* and *R* equal to 0), it is a bistable device, able to maintain an output of $Q = 1$ or $Q = 0$ indefinitely, just like the cross-coupled inverters in Figure 3. To change the output of the latch requires that the *S* or *R* input signal become 1. The condition of the latch, indicating whether the output is 0 or 1, is referred to as its *state*. (This terminology will be extended later to circuits containing any number of storage elements; the state of the circuit will refer to the collection of output values of all the memory-storing devices.)

Exercise 2   One at a time, assume the latch is in each of the two states ($Q = 1$, $Q = 0$). For each state, assume inputs $SR = 00$ (both inputs 0) and confirm that the *SR* latch will remain in whatever state it happens to be in.[5]   ◆

Suppose the latch is in one of its two states, $Q = 0$ or $Q = 1$, called the *present state*. Now the *S* and/or *R* inputs change to a new combination of values; after a delay corresponding to the propagation delay through the physical gates, there should be a transition to a new (but not necessarily different) state, called the *next state*. If present time is labeled t_n, the time at the occurrence of a transition is labeled t_{n+1}. The corresponding sequences of states will be indicated by one of the following notations:

---

[5]Note that the notation *SR* is not intended to be the AND of inputs *S* and *R*, but rather just an easy way of identifying the sequence of values that follows, that is, 00. This practice can be confusing; we will minimize the confusion whenever it is likely by explicitly showing the AND operation in the usual way: *S•R*.

S	R	Q	Q+
0	0	0	0
0	0	1	1
0	1	0	0
0	1	1	0
1	0	0	1
1	0	1	1
1	1	0	×
1	1	1	×

(a)

S	R	Q+
0	0	Q
0	1	0
1	0	1
1	1	×

(b)

Figure 6 Transition table for the *SR* latch.

$$Q(t_n) \rightarrow Q(t_{n+1})$$
$$Q^n \rightarrow Q^{n+1}$$
$$Q \rightarrow Q^+ \tag{1}$$

The last of these is the simplest, and we will use it the most often.

What we need to do now is work out what the next state of the *SR* latch will be for any combination of inputs and any present state. Let's start with present state $Q = 0$ and $SR = 00$ in Figure 4a. (Exercise 2 should have convinced you that this combination of inputs and present state is possible.) Now suppose SR becomes 10; that is, input S changes to 1. One of the inputs of gate 1 goes to 0; hence, Q becomes 1. Now both inputs of gate 2 are 1; so Q' goes to 0. Since this is fed back to gate 1, you should determine if it influences the output of gate 1. (It doesn't.) Hence, the next state is $Q^+ = 1$. The S input has been passed on to the output. We say that the latch has been *set*.

Exercise 3 Follow the same process just described to determine the next state Q^+ when SR becomes 01 from the preceding value of 00. The answer is given in what is called the *transition table* shown in Figure 6a. Note that when $R = 1$ (while $S = 0$), the output has gone to 0; we say the latch has been *reset*. ◆

The case $SR = 11$ is anomalous. In this case, each of the two NAND gates has an input that is 0. It would follow that both Q and Q' would become 1, an inconsistent outcome! Furthermore, what would be the consequence if SR subsequently went from 11 to 00? The inputs to both gates would be 1 and, hence, both outputs should go to 0—another inconsistency. But, more important, even if the inputs *could* change at exactly the same time, since it is unlikely that the propagation delays of the two gates are exactly the same, one of the outputs will reach 0 first.[6] Because of the feedback, the other gate output should then become 1. Thus, out of many copies of this latch, some will have $Q = 1$ in these circumstances and others $Q = 0$, an uncertainty that is unacceptable. Hence, the input combination $SR = 11$ cannot be tolerated.

---

[6]The two gates seem to be racing each other to see which one will first have a 1 output. For this reason, this situation is referred to as a *race* condition. This matter will be discussed at greater length in Chapter 7.

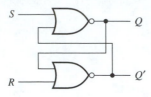

Figure 7 NOR gate design of *SR* latch.

The *SR* = 11 condition can be eliminated by imposing the logical requirement that $S \cdot R = 0$. Since *SR* = 11 is not to occur, we don't care what the output is for such a combination of inputs, as shown in Figure 6. Hence, the complete transition equation for the *SR* latch obtained from the transition table in Figure 6*a* is

$$Q(t + 1) = S(t) + R'Q(t), \quad S(t) \cdot R(t) = 0$$
$$Q^+ = S + R'Q, \qquad\qquad S \cdot R = 0 \tag{2}$$

(Confirm this result by writing a sum of minterms, including don't-cares, and reducing it using Boolean algebra. Does the notation $SR = 0$ in (2) unambiguously imply Boolean AND, and not a juxtaposition of *S* and *R*?)

Now examine the transition table in Figure 6*a* from a different point of view. When either $S = 1$ or $R = 1$, but not both, the next state takes on the value 0 or 1 *independent of the present state.* That is, the next state depends only on the inputs, not on the present state. This means that, for these particular input combinations, the circuit behaves as a combinational circuit. But notice also that for the combination $SR = 00$, the next state *does* depend on the present state. All this information on the transitions is unambiguously presented in the reduced transition table of Figure 6*b*.

Exercise 4   A student studying digital logic, seeing that the design of the *SR* latch using NAND gates requires inverters at the inputs, came up with an idea: What would happen if the NAND gates were replaced by NOR gates, but without the inverters? So she constructed the diagram in Figure 7. Construct a transition table for this circuit similar to the one in Figure 6. By comparing the two tables, confirm that this design also yields an *SR* latch.   ◆

Timing Problems and Clocked *SR* Latches

The design of the *SR* latch in Figure 4*a* does not include a clock. The inclusion of a clock as a separate input could cause all state changes in a circuit to take place simultaneously. It might also result in all state transitions being made reliably, without the output uncertainty described in section 1.

A design of a *clocked SR latch* with NAND gates is shown in Figure 8*a*. The two gates to the left of the latch are called the *steering gates.* The clock signal is one of two inputs to each steering gate. When the clock signal *C* is 0, the *S* and *R* inputs have no influence on the state *Q*. In this case the circuit is equivalent to a latch with $S = R = 0$. (Confirm it.) When the clock signal is 1, on the other hand, the behavior of the circuit reduces to that of the *SR* latch in Figure 4*a*. Confirm the transition table shown in Figure 8*c*.

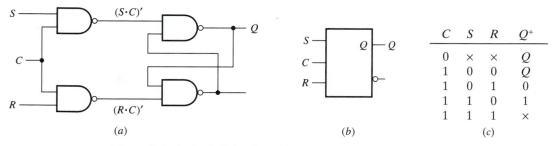

Figure 8 A clocked *SR* latch and its transition table.

Exercise 5

a. Using the transition table in Figure 8*c*, construct a four-variable map with *C*, *Q* as two variables and *S*, *R* as the other two. The entries in the map are to be the next states. (Complete your own map before you confirm it using the completed map below.)

b. From the map, obtain a minimal expression for the next state.

Answer

$$CQ$$

		00	01	11	10
SR	00		1	1	
	01		1		
	11		1	×	×
	10		1	1	1

◆

Note that for $C = 0$, the expression for Q^+ in the answer to Exercise 5 reduces to $Q^+ = Q$; this tells us that the next state is the same as the present state. So long as the clock remains at 0, no state transition will take place. For $C = 1$, on the other hand, Q^+ reduces to the next-state expression for the *SR* latch in (2), as it should. The latch is said to be *transparent* when $C = 1$, because the outputs respond to changes in the inputs. For future reference, the expression for Q^+ in Exercise 5 is given here.

$$Q^+ = C'Q + CS + R'Q \tag{3}$$

A symbol, or block diagram, for the clocked *SR* latch is shown in Figure 8*b*. Note that the presence of the clock does not change the indeterminacy of the output for the condition $S = R = 1$.

JK Latch

Proper operation of the *SR* latch requires that both inputs not be 1 simultaneously. This is a headache and causes practical problems. A design change that overcomes this difficulty would be most welcome. A modified design using two

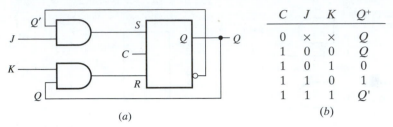

C	J	K	Q^+
0	×	×	Q
1	0	0	Q
1	0	1	0
1	1	0	1
1	1	1	Q'

(a)   (b)

Figure 9 *JK* latch.

AND gates whose outputs play the role of S and R is shown in Figure 9a. One of the two inputs to each of these AND gates is fed back from the *SR* latch outputs. The other two (external) inputs are labeled J and K, respectively. The expressions for S and R are $S = JQ'$ and $R = KQ$. (Confirm these.) Inserting these into (3) gives the transition equation as

$$Q^+ = C'Q + C(JQ') + Q(KQ)'$$
$$= C'Q + CJQ' + K'Q \qquad\qquad (4)$$
$$= JQ' + K'Q \qquad\qquad \text{for } C = 1$$

Using the expressions $S = JQ'$ and $R = KQ$ leads to $S{\cdot}R = JKQQ' = 0$; thus, the condition that S and R never simultaneously be 1 is automatically satisfied.

It is evident that the clocked *SR* latch has no advantage over the *JK* latch in synchronous circuits, so they are seldom used as memory devices in such circuits.[7]

Master-Slave Latch

The clocked *JK* latch overcomes some of the timing problems and the prohibition against simultaneously high inputs in the *SR* latch, but another timing problem remains. It is true that the occurrence of a clock pulse initiates a state transition on the basis of the J and K signals present at the time. Suppose that J and K are both 1 when a clock pulse arrives. According to (3) or the transition table in Figure 8c, the transition is $Q^+ = Q'$, and so the state changes. After the transition is completed, the new state is fed back to the inputs of the steering gates.

If the delay in this process is relatively small and the clock pulse is still present, a further transition will take place in accordance with the transition table. This process will continue until the clock pulse goes to 0. The latch output, therefore, will be uncertain, depending on the width of the clock pulse relative to the propagation time through the latch. This is another example of a race condition, this time a race to beat the clock. Such an uncertainty in the final state cannot be tolerated.

---

[7]However, they do find use in control systems and other sequential circuits that are not synchronous, to be discussed in Chapter 7.

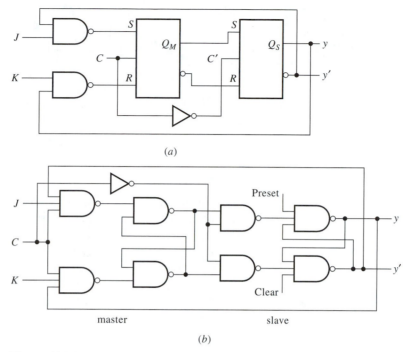

(a)

(b)

Figure 10 Master-slave latch.

A Possible Design

How can this problem be fixed? What we need, once a clock pulse has initiated a transition, is some way to prevent the transition from being completed until the pulse has ended. Several designs that perform this function exist; one is shown in the schematic diagram in Figure 10a.

This circuit is called a *master-slave latch*. The unit on the right (called the *slave*) has an inverted clock signal compared with the one on the left (called the *master*). That is, when the master clock is 1 the slave clock is 0, and vice versa. In the actual design shown in Figure 10b, the master is like the *JK* design of Figure 9 except that the feedback into the steering gates of the master is taken from the output of the slave instead of from its own output. Study this diagram carefully and note its features compared with those of the *JK* latch.

When the input clock signal turns low, the master is disabled but the slave is enabled (its clock goes high). During this interval, the master output does not change, but the slave output makes a transition to whatever the master output was at the start of this interval. At the end of the low interval of the clock, the master and slave outputs are in the same state, the present state. Now, when the next clock pulse arrives, the slave is disabled and its output does not change. Equation (3) is still valid for the state transition of the master. Since the slave output is not changing, however, the present state, fed back to the master, stays fixed, no matter how long the clock stays high. At the end of the clock pulse, the slave is enabled and the state of the master (the next state) is transferred to the slave. It appears that the timing problem is solved!

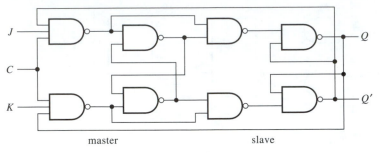

<figure>
master　　　　　　　　slave
</figure>

Figure 11 Alternative master-slave latch.

It is sometimes necessary at the beginning of an operation (such as counting) to set the initial states of the latches and flip-flops in the circuit by external means. A mechanism for doing this is shown in Figure 10*b* in the form of two direct inputs to the slave from the outside (Preset and Clear) that take precedence over the usual inputs. These inputs are never allowed to be 0 simultaneously.[8]

Exercise 6　Determine the state Q_s of the slave for combinations of Preset (PS) and Clear (CL) inputs other than 00.
Answer[9]

An Alternative Master-Slave Design

The design in Figure 10 is not the only possibility; another design for a master-slave *JK* latch is shown in Figure 11.[10] In this circuit, the inverted clock inputs to the steering gates of the slave in the preceding circuit are removed and replaced by the outputs from the steering gates of the master. That's the only change. Confirm that no state transition occurs in the slave when $C = 1$ and that $Q_{\text{slave}}^+ = Q_{\text{master}}$. Hence, the circuit behaves like a master-slave latch.

Note the sequence of events in this circuit. Changes of logic values (levels) occur on the *J* and *K* lines from time to time. At the leading edge of a clock pulse a state transition is initiated. The output of the master takes on a value appropriate to the present state of the slave and the *J* and *K* values at this time. Nothing happens to the slave output until the trailing edge of the clock pulse. When that occurs, after an appropriate propagation delay, the slave output acquires its new value. Since the clock pulse triggers the entire state transition (initiation and completion), we say this latch is *pulse-triggered*.

Unfortunately, a problem still remains with the master-slave pulse-triggered *JK* latch. Even though the slave output cannot change in the presence of the clock pulse, if a change in the *J* or *K* input should occur during this time, the master output *will* undergo another transition. Then, after the trailing edge of the clock

---

[8]In some cases, the enabling and disabling are done by negative-going Reset and Clear signals. In such cases, these incoming external signals must be inverted first. This is indicated by a bubble at the input.

[9]$Q_s = 1$ for PS = 0, CL = 1; $Q_s = 0$ for PS = 1, CL = 0; normal operation for PS = 1, CL = 1. ◆

[10]The designs presented here are included for pedagogical reasons; practical considerations often result in different actual, commercially available designs.

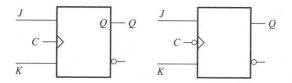

Figure 12 Schematic symbol of an edge-triggered flip-flop.

pulse, this value will be dumped into the slave. Hence, the transition made by the latch may not be the one made by the master upon arrival of the clock pulse—as it should be—but a subsequent transition. This, too, cannot be tolerated.

One possible solution is to reduce the duty cycle of the clock signal (the width of the clock pulse for a given frequency) as much as possible. The chance of a change in J or K during this shorter time will be correspondingly reduced. In any case, to avoid uncertainty in the output, input J and K values must be held stable during the entire interval of the clock pulse.

Edge-Triggering Parameters

A solution to the preceding timing and race problems is to abandon pulse triggering altogether and to design memory elements that are triggered just by an *edge* of the clock pulse. Such elements are said to be *edge-triggered* and are called *flip-flops*. In an edge-triggered flip-flop, once the information needed for a state transition (the present values of J, K, and the state) is *loaded,* during an interval of time surrounding an edge (either a leading or trailing edge), any further changes will have no effect; we say these effects have been *locked out.* In other words, a flip-flop does not exhibit transparency in either clock state; the outputs of the flip-flop respond to the state of the inputs at a clock edge.

A realistic shape for a clock pulse is shown in Figure 2*b*. (Look back at it.) Two intervals of time are shown at the leading edge: the *setup time* and the *hold time*. (Similar intervals obviously occur at the trailing edge.) The setup time extends from some instant before the initiation of the clock pulse to the start of the clock pulse. The hold time extends from the end of the clock pulse to some time after the clock pulse drop has been completed. For proper operation of the edge-triggered flip-flop, the values of the J and K inputs must remain stable from the beginning of the setup time to the end of the hold time. Since this entire interval is much shorter than the pulse width itself, it is much easier to ensure that no change in J or K will occur during this interval.

Edge triggering is indicated in a schematic diagram by a small triangle, called a *dynamic indicator,* placed at the clock terminal, as shown in Figure 12. If the triggering occurs at the leading edge, the symbol is the one shown in Figure 12*a*; the bubble in Figure 12*b* indicates that the triggering occurs at the trailing edge. Although no circuit diagram for an edge-triggered *JK* flip-flop will be given here, such circuits exist. (Their existence is assumed in what follows and in the problems.) Instead, we will discuss an edge-triggered design for the type of flip-flop to be described next.[11]

---

[11]A typical edge-triggered *JK* flip-flop is the SN74111. It has a setup time of 0 and a hold time of 30 ns. The total of setup and hold times for other flip-flops in TTL technology is also about 30 ns.

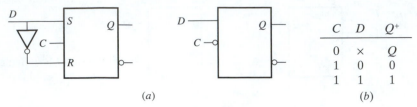

C	D	Q^+
0	×	Q
1	0	0
1	1	1

(a) (b)

Figure 13 D latch.

Delay (D) Flip-Flops

The condition required for proper operation of an SR latch in (2), that $S \cdot R = 0$ always, can be achieved automatically if the R input is obtained from the S input through an inverter so that $R = S'$. The result is a circuit with a single external input in which S is relabeled D (standing for *data*). A simple version is shown in Figure 13a. The clock pulse acts as a gate that, when high, permits the data on D to pass to the output. Viewed in this way, this circuit is sometimes called a *gated latch*. ("Latch" because the clock is viewed as a gate that lets the data through when it is enabled.)

The transition equation for the D latch can be obtained from that for the SR latch by setting $R = S'$ and then renaming S as D in (2). The result is

$$Q(t + 1) = D(t) \tag{5}$$

What this expression tells us is that the next state will be whatever the D input is during the interval of the clock pulse. (Hence, D can also stand for *delay*.) Or, looking backward in time from the present, whatever the D latch state is at the time of the present clock pulse, that's what its input was one clock pulse earlier. Note that, whereas the SR and JK latches have two inputs (excitations) each, not counting the clock, the D latch has only one.

Exercise 7 Using the transition equation for the SR latch, confirm the transition table given in Figure 13b. ◆

Although the D latch is created from the SR latch in Figure 13, the same behavior can be obtained if the SR latch is replaced by a JK latch. That is, K is forced to equal J' by means of an inverter. (Verify that $Q^+ = D$ if J is renamed D.) The D-type latch has the same timing problems as the JK latch. However, it is a simple and cheap device that finds application when it isn't necessary to synchronize all the latch transitions in a system.

Edge-Triggered D Flip-Flop

The solution mentioned for the timing problems of pulse-triggered latches is edge triggering. That is, for a D flip-flop, changes in the input D value will have no effect on the state except at the edge of the clock pulse, either the rising edge or the falling edge, depending on the design. The design of an edge-triggered D flip-flop is shown in Figure 14. Gates 5 and 6 constitute a basic SR latch.

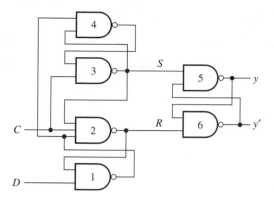

Figure 14 Edge-triggered D flip-flop design.

Expressions for the S and R inputs to this latch can be obtained by analyzing the remainder of the circuit, consisting of gates 1 through 4. Analyze the circuit to confirm the following:

$$S^+ = C(S + DR') \tag{6}$$
$$R^+ = CS'(D' + R)$$

When the clock signal is low ($C = 0$), these expressions show that both S^+ and R^+ are 0, independent of input D. According to (2), $Q^+ = Q$, no transitions take place, and y maintains whatever value it had: $y^+ = y$.

Now suppose C rises to 1; according to (6) and because S and R were both 0 just before that event, $S^+ = D$ and $R^+ = D'$. After a delay needed for this transition to occur, using these values for S and R in (2) gives $Q^+ = D + DQ = D$. To summarize, on the rising edge of the clock pulse (after an appropriate delay), the output of the circuit becomes whatever the input D was at the time of the rising edge.[12]

Suppose now that the clock pulse ends: $C = 0$. We already saw that the values of S and R will become 0 but that no transition in output will take place: $Q^+ = Q$. No further change in output will take place until the rising edge of the next clock pulse. The subsequent output (after a delay) will be whatever the input value is at that time.

Possible waveforms for the input signal (D) and the clock (C) are shown in Figure 15. The resulting output waveform is also shown, under the assumption that the flip-flop was initially set ($Q = 1$). The non-idealness of the clock pulse is neglected in this figure; attention will be given to this matter shortly. Both D and Q are 1 at the first rising edge of the clock (t_1), so, since $Q^+ = D = 1$, no transition takes place. (Confirm each point in this and the next paragraph.)

At the next rising edge (t_4), $Q^+ = D = 0$, and a transition takes place. At the next rising edge (t_7), D is again 0 so no transition in output will occur. As t_{11} approaches, $Q = 0$ and $D = 1$, so a transition to 1 again takes place at the rising edge of the clock pulse. At all preceding times when D changes value (t_5, t_6, t_8, t_9, t_{10}), the clock is stable, so no transitions in output take place.

---

[12]Functionally, "the value of D at the rising edge" means that D must hold its value during a time no less than the setup time plus the hold time surrounding the rising edge.

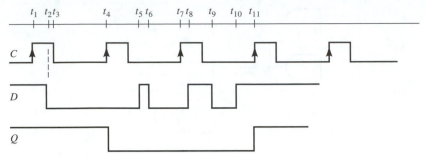

Figure 15 Edge-triggered D flip-flop waveforms.

As a final comment, what would happen if input D were to change during the interval over which the clock pulse rises? The answer is, chaos! That's because the "edge" is not really a sharp edge. It would not be fully clear what value the input had during this slow "edge," so the output would be uncertain. This is the reason for the requirement that the input remain stable (unchanging) over an interval of time spanning the setup and hold times of the flip-flop. In Figure 15, for example, the change in D from 0 to 1 near t_7 may not be permitted if the rise in the clock pulse has not been completed.

T Flip-Flop

A device that would be useful in digital systems is one whose output *toggles,* that is, whose output is replaced by its complement whenever there is an incoming signal. Such a device would be inherently single input. A device that exhibits such behavior is called a *toggle (T) flip-flop.* There is no need for an independent design because such a flip-flop is easily obtained from a JK flip-flop with certain connections at the terminals.

One possible design is shown in Figure 16a. The J and K inputs of a JK flip-flop are tied together and relabeled T. The transition table is easily obtained from that of the JK flip-flop in Figure 9. For $C = 1$, the transition equation in (4) reduces to

$$Q^+ = TQ' + T'Q = T \oplus Q \tag{8}$$

For $T = 1$ (when $C = 1$), $Q^+ = Q'$; that is, the new state is the complement of the old state—the state toggles. But if $T = 0$ when $C = 1$, then $Q^+ = Q$; *there is no toggling.* Since, in the presence of a clock pulse, toggling takes place only when $T = 1$ and not when $T = 0$, this circuit is somewhat deficient.

Exercise 8 Another design for a synchronous T flip-flop is shown in Figure 16c. The J signal is fed back from Q', and K is fed back from Q. (Remember that this is a block diagram; in the actual JK flip-flop circuit of Figure 9, this amounts to connecting the J and K terminals to logic 1.) Determine the transition equation for this circuit and compare it with that of Figure 16a for both $C = 0$ and $C = 1$.
Answer[13]

---

[13]Same.

◆

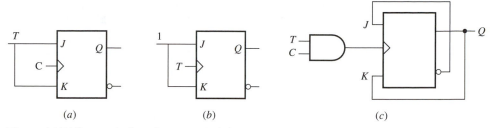

Figure 16 Different designs for a toggle (*T*) flip-flop.

A modification of the design in Figure 16*a* that overcomes the deficiency is shown in Figure 16*b*. The *J* and *K* terminals are again connected but are now permanently set at 1. The *T* input is now limited to being a pulse and is introduced into the clock terminal in place of the clock. The resulting flip-flop is *asynchronous, not clocked*. To determine the transition equation, set $J = K = 1$ and $C = T$ in (4). The result is the same as (8). Again, toggling occurs when $T = 1$, but now that's for every occurrence of the *T* pulse. That is, every time the input *T* pulse goes high, the output toggles.

We have briefly considered two *T* flip-flop designs. One is synchronous but does not toggle at each clock pulse; the other toggles at each input pulse but is not synchronous.

Flip-Flop Excitation Requirements

The preceding parts of this chapter can be summarized as follows. From a given flip-flop design, it is possible by analysis of the circuit to determine a state transition table—or the equivalent, a transition equation. From either of these, the next state can be determined for each present state and each input combination.

In sequential circuit design, to be discussed in Chapter 6, the flip-flop excitations are not known. Instead, for each design and each input combination, what are known are both the present state and the next state. From this information, we must *derive* the required excitation values that will result in the given transition.

EXAMPLE 1

Suppose a transition from $Q = 1$ to $Q^+ = 0$ is required for a *JK* flip-flop. If these values are inserted into the transition equation (4), the result will be $Q^+ = JQ' + K'Q$ or $0 = 1' \cdot J + 1 \cdot K' = K'$. So $K = 1$, independent of *J*. That is, if the transition $Q = 1 \rightarrow Q^+ = 0$ is desired, the required *JK* flip-flop inputs are $K = 1, J = $ don't-care. ∎

The excitation requirements for the other transitions can be determined in a similar way. The results for all flip-flops under consideration are shown in the excitation tables of Figure 17.

Transition Required		Inputs Needed	
Q	Q^+	S	R
0	0	0	×
0	1	1	0
1	0	0	1
1	1	×	0

$$Q^+ = S + R'Q, \quad S \cdot R = 0$$

(*a*) *SR* flip-flop

Transition Required		Inputs Needed	
Q	Q^+	J	K
0	0	0	×
0	1	1	×
1	0	×	1
1	1	×	0

$$Q^+ = J'Q + K'Q$$

(*b*) *JK* flip-flop

Transition Required		Input Needed
Q	Q^+	D
0	0	0
0	1	1
1	0	0
1	1	1

$$Q^+ = D$$

(*c*) *D* flip-flop

Transition Required		Input Needed
Q	Q^+	T
0	0	0
0	1	1
1	0	1
1	1	0

$$Q^+ = T \oplus y$$

(*d*) *T* flip-flop

Figure 17 Flip-flop excitation requirements.

Exercise 9 Using the appropriate transition equations for each of the flip-flops under consideration, confirm each of the tables of excitation requirements in Figure 17. ◆

3 REGISTERS

The following features distinguish a sequential circuit from a combinational circuit:

- The ability to store, in memory, information about the state of the circuit due to past inputs
- The utilization of this information to produce an output in response to new inputs

The basic units for the storage of 1 bit of information are flip-flops or latches. Just as logic gates can be interconnected to constitute larger units such as multiplexers, decoders, and full adders, so flip-flops can be organized into groups called *registers*. An *n*-bit register is a set of *n* (usually *D*-type) flip-flops, all with a common clock. Not only must the clock signal be common, but the flip-flops must all respond to the clock in the same way: all rising-edge triggered or falling-edge triggered. It can store *n* bits of information that may or may not be related.

Transferring information to a register is called *loading* the register. Conceptually, it is possible to transfer the information into all flip-flops in a

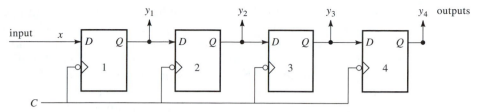

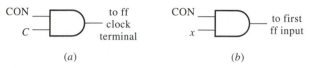

Figure 18 Shift-right shift register.

Figure 19 Shift-register input control.

register simultaneously (in parallel) or 1 bit at a time (serially). Similarly, the information can be transferred out of the register in parallel or serially. All four combinations of loading and reading out are possible: parallel-in/parallel-out, parallel-in/serial-out (you list the others).

Registers are available in MSI circuits. Information is processed in units of 2^n bits. (The higher this number, the faster the processing.) For illustrative purposes and ease of visualization in what follows, registers with fewer than eight (2^3) flip-flops will be used.

Serial-Load Shift Register

The schematic diagram for a 4-bit serial-in register is shown in Figure 18. The flip-flops shown are D-type, although they could also be JK. The dynamic indicator (triangle) indicates an edge-triggered flip-flop, and the bubble on the clock input terminal means that the triggering occurs at the trailing edge of the clock pulse. Let's make a number of observations by examining the diagram.

At each trailing edge of the clock pulse, the input on the x line is transferred to the output of the first flip-flop. Whatever the output of the first flip-flop is at that time is transferred to the output of the second flip-flop—and similarly, in a chain to the right, until the last flip-flop. (Would this scheme work if there were more flip-flops in this chain?) That is, the data is shifted with each clock pulse—hence the name *shift register*. If the only output available externally is the last one on the right (or on the left in the left-shift case), then the register is called a *serial-out* shift register. However, if each flip-flop output (each Q, not just Q_4) is available to be read externally, this is also a *parallel-out* register. Both possibilities are often available, and specific control signals are used to control which mode will be used in a particular application.

Because the transfer of data takes place with each clock pulse in the circuit of Figure 18, there is no control over the timing of data transfer. Such control can be achieved in one of two ways. One possibility is illustrated in Figure 19a, where CON is the shift-control input signal. Because CON is ANDed with the clock, the flip-flops in the shift register are enabled only when CON = 1. For a

k-bit register, CON should become high right after a trailing edge of the clock pulse and remain high for k clock periods. The register will operate as described above only for these k clock periods. Hence, a k-bit word will be transferred to the register. The shift-control signal changes to CON = 0 at the end of the k-bit word, thus disabling the flip-flops in the register until the control signal again goes high.

The major drawback of this scheme is that logic is performed with the clock signal. Since the gates with which the logic is performed (an AND gate in this case) have propagation delay, the clock signal will not reach all flip-flops in the system at the same time. Hence, the system may fail to perform all system functions synchronously—a highly undesirable result.

A simple way to overcome this input-control difficulty is shown in Figure 19*b*. Instead of ANDing the control signal with the clock, the signal is ANDed with the input x. Again, CON should go high right after a trailing edge of the clock pulse, remain high for k clock periods, and then go low. In this way, only a k-bit word will be transferred to the shift register. This approach to load control is illustrated next.

Parallel-Load Shift Register

The schematic diagram of a parallel-in, parallel-out register is shown in Figure 20. It is designed with edge-triggered *JK* flip-flops and uses the load-control method discussed above. Let's observe some of its features by studying the diagram.

1. The clock pulse is applied to each of the flip-flops through an inverter, so that the triggering is performed on the trailing edge of the clock pulse. The main purpose of the inverter, however, is to provide buffering, thus reducing the loading on the clock source: the clock won't have to drive all the flip-flops, only the inverter.
2. The input control scheme in Figure 20 is used with each control signal. This time, to avoid loading the control source, the signal is introduced through a buffer.
3. Finally, there is an independently operated CLEAR (CLR) input applied to each flip-flop in the register; it must be high for normal operation of the register. However, when it is necessary to override the function being carried out and to clear the contents of the register, this can be done, asynchronously, by setting CLR = 0. (What is the function of the buffer in this line?)

Although the flip-flops are *JK*, when the load-control input is $L = 1$, they are being used as D flip-flops because then $K = J'$. (Confirm this.) When $L = 0$, however, $J = K = 0$; hence, no flip-flop transitions will take place. The inputs I_j are loaded simultaneously into the corresponding flip-flops at the trailing edge of a clock pulse, provided the CLR input is 1 at that time and the load control input is also 1. The flip-flop outputs are all available externally, so this is a parallel-in, parallel-out register.

For simplicity, the registers shown here are 4-bit registers. Obviously, more bits can be accommodated by adding more flip-flops. The only difference between registers processing 16 (2^4)—or 32 (2^5) or 2^n for a higher value of n—bits and the simpler ones discussed here is the number of flip-flops.

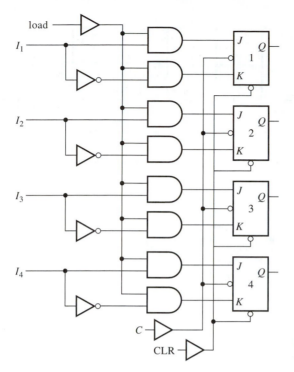

Figure 20 Parallel-load 4-bit register with load control.

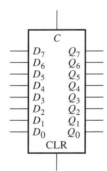

Figure 21 Symbol for an 8-bit register.

All such registers are available in MSI packages. Figure 21 shows a schematic symbol for an 8-bit parallel-in, parallel-out register using D flip-flops and a CLR input.

Parallel-To-Serial Conversion

Although all bits of a word may be available at the same time, it may be desirable to convert this information to serial form. This is the case, for example, if the data is to be transmitted to another location serially over a single-line communication channel. For this purpose, a shift register is needed that can be parallel loaded. A possibility is shown in Figure 22. The 4-bit shift register is made up of JK flip-flops, but they act as D flip-flops since $K_i = J_i'$. Each flip-flop has asynchronous CLEAR (CLR) and PRESET (PR) terminals also.

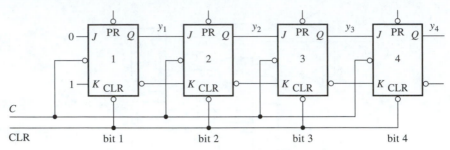

Figure 22 Four-bit shift register for parallel-to-serial conversion.

Clock Pulse	y_1	y_2	y_3	y_4	
		(input word)			(lsb in)
1	1	1	0	1	
2	0	1	1	0	(lsb out)
3	0	0	1	1	
4	0	0	0	1	

serial out

output word

Figure 23 Parallel-to-serial data conversion.

The data enters the register through the PRESET terminals. (Not shown are load-control units at each PRESET input.) The J and K inputs to the first flip-flop on the left are set permanently low (0) and high (1), respectively. This guarantees that after each clock pulse, the output of the first flip-flop will go low. Unless the flip-flop states are asynchronously set, this low value will propagate to the right and, after three more clock pulses, all flip-flops will be cleared. At a time determined by the load-control input, the register is loaded through the PRESET terminals.

Suppose that the word to be transmitted is 1101. At a given clock pulse, the load-control input enables the PRESET terminals, and all bits of the word are loaded simultaneously into the register. The result is shown in the first row of the table in Figure 23. PRESET is now disabled. At each successive clock pulse, the contents of the register shift to the right, leaving 0's on the left, as shown in the succeeding rows of the table. The output, appearing in the last column, is taken from the rightmost flip-flop, the least significant bit first.

Another application of such a parallel-to-serial conversion register is as follows. Suppose two binary numbers (that are to be added) are available in parallel form but are to be processed through a serial adder. The numbers can be converted to serial form and applied as serial inputs to the adder.

Universal Registers

Each type of register described in the preceding sections (right-shift, left-shift, parallel-load, parallel-read) has important applications. What would be of even greater value is a register that combines some or all of these features, a sort of universal register.

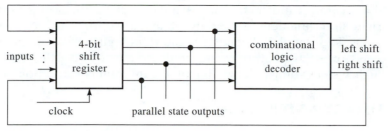

Figure 24 Universal register.

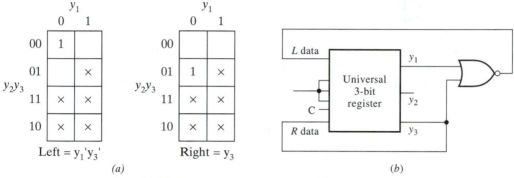

Figure 25 Modulo-3 counter using a universal register.

The schematic for a 4-bit register of this type is shown in Figure 24. There are four terminals for parallel loading and parallel reading, as well as serial left-shift and right-shift inputs. The purpose of the external decoder is to produce the bits that are to be shifted and to control the direction in which they are shifted.

One of the applications for this type of circuit is as a counter. Since 0 and 1 bits can be shifted in either direction at the clock pulse, a large number of different codes of varying length can be generated. Problems for carrying out such designs are provided in Chapter 6.

As a simple illustration, suppose a 3-bit universal register is to be used to design a modulo-3 counter according to the following code. (Is this a unit-distance code?)

$$\rightarrow 000 \rightarrow 001 \rightarrow 100 \rightarrow$$

Suppose the state variables are designated y_1, y_2, y_3, with y_1 representing the most significant bit. Starting with state $y_1 y_2 y_3 = 000$, a 1 bit is shifted to the left, then a 1 bit to the right, then a 0 bit to the left. Logic maps for the data to be shifted left and right can be constructed as shown in Figure 25a.

To illustrate, for present state 000 (cell 000 in the map), the bit to be shifted to the left is a 1 and to the right, a 0. Since shifting is specified for only three present values of the state variables, the entries in five of the map cells are don't-cares. (Confirm all entries in the maps.) Expressions for the left-shift and right-shift functions are easily obtained; confirm the ones given under the maps. The completed counter is shown in Figure 25b. The parallel-input terminals are inactive; all of them should be connected either to the low or to the high system voltage depending on the internal construction of the register.

CHAPTER SUMMARY AND REVIEW

This chapter introduced sequential circuits. It dealt with memory storage devices: flip-flops and registers. The following topics were covered.

- Timing problems and clocks
- Operation of the basic *SR* latch
- Feedback
- The concept of circuit state and its change to the next state
- State transition equations and transition tables
- Unclocked and clocked *SR* latches and the problem of simultaneous high inputs
- The *JK* latch and its transition equation
- The race problem and procedures to overcome it
- Master-slave design and edge triggering
- *D* flip-flops
- *T* flip-flops
- Excitation requirements for each type of flip-flop
- Shift registers
- Serial loading of registers
- Parallel loading of registers
- Conversion from parallel to serial loading and vice versa
- Universal registers

PROBLEMS

You will be able to carry out parts of many of these problems only after studying the later parts of the chapter. Save those parts of your solutions that you can carry out early; then complete the problems after you have studied the later material.

1 A latch is to be defined with inputs L and M (an LM latch). The table specifying the desired next state at a clock pulse is given in Figure P1a.

L	M	Q	Q^+		N	P	Q	Q^+		G	H	Q	Q^+
0	0	0	0		0	0	0	1		0	0	0	1
0	0	1	0		0	0	1	1		0	0	1	1
0	1	0	0		0	1	0	0		0	1	0	0
0	1	1	0		0	1	1	0		0	1	1	1
1	0	0	1		1	0	0	1		1	0	0	1
1	0	1	1		1	0	1	1		1	0	1	0
1	1	0	1		1	1	0	0		1	1	0	0
1	1	1	0		1	1	1	1		1	1	1	0
		(a)					(b)					(c)	

Figure P1 Transition tables for three latches. (*a*) *LM* latch. (*b*) *NP* latch. (*c*) *GH* latch.

a. Describe how the latch state is to change for each combination of L and M values. (For example, for $LM = 10$, the next state is 1, independent of the present state.)
b. Write expressions for the next state Q^+ in terms of L, M, and present state Q.
c. Construct the table of excitation requirements.
d. Construct a diagram of a circuit that realizes this latch using a clocked SR latch and any additional gates needed.
e. Repeat part d using a JK latch.

2 Repeat Problem 1 for the NP latch defined by the transition table in Figure P1b.
3 Repeat Problem 1 for the GH latch defined by the transition table in Figure P1c.
4 An AB latch is constructed from an SR latch as shown in Figure P4.

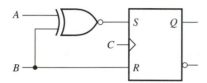

Figure P4

a. Obtain expressions for S and R in terms of A and B.
b. Write an expression for the next state Q^+ in terms of A, B, and present state y.
c. Construct the excitation requirements table for A and B.

5 The schematic diagram of a G flip-flop is shown in Figure P5. The transitions allowed in this flip-flop are

$$Q^+ = Q \text{ when } G = 0, \qquad Q^+ = 1 \text{ when } G = 1$$

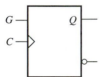

Figure P5

a. Write an expression for the next state Q^+ in terms of G and present state Q.
b. Construct the excitation requirements table for G.

6 The schematic diagram of an H flip-flop is shown in Figure P6. The transitions allowed are

$$Q^+ = 0 \text{ when } H = 0, \qquad Q^+ = Q' \text{ when } H = 1$$

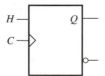

Figure P6

a. Write an expression for the next state Q^+ in terms of H and present state y.
b. Construct the excitation requirements table for H.

7 A D-type latch can be constructed from an SR latch and an inverter, forcing $R = S'$. Figure P7 shows another type of gated latch.

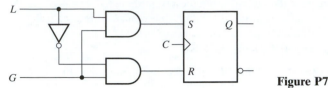

Figure P7

a. Find an expression for Q^+ in terms of G and L.
b. From this, determine expressions for Q^+ for $G = 0$ and $G = 1$.
c. Describe how G controls the transfer of information from L to the latch output.

8 In the clocked SR latch, the simultaneous condition $S = 1$, $R = 1$ results in an uncertain state. A variation on the SR latch is the *set-dominate* (SD) latch. For this latch, simultaneously high inputs result in setting the latch, that is, setting $Q = 1$.

a. Construct a transition table for the SD latch. From this, determine an expression for the next state.
b. Construct a logic diagram that implements this latch.

9 The circuit in Figure P9 has been proposed as a clock generator. It is assumed that a pulse is inserted by some external mechanism at the input of one of the three leftmost inverters, that is, at point A, B, or C.

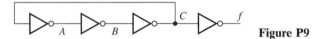

Figure P9

a. Assume that the inverters are ideal, with zero delay. Describe the resulting anomaly.
b. Now assume that each inverter has a 25 ns delay. Use timing diagrams to analyze the circuit and determine the output f. Specify the frequency of this clock.

10 Analyze the edge-triggered D flip-flop shown in Figure 14 in the text, proceeding by the steps outlined below. Assume the gate delay is 1 percent of the clock period and construct a timing diagram as your analysis proceeds.

a. Start with the clock low and find the logic values of all intermediate variables with $D = 0$. Verify that the flip-flop will be stable in either state.
b. Now assume that the D input changes from 0 to 1 with the clock still low. Again find the values of intermediate variables. Is there a change in the flip-flop state?
c. Next suppose that the clock goes from 0 to 1 while $D = 0$. Determine the changes in intermediate variables and flip-flop output. (It might be useful to use the gate delay as a unit of time for the purpose of tracing the changes in intermediate variables and flip-flop output.)
d. Repeat part c, this time with $D = 1$.

11 A ripple counter is made up of toggle (T) flip-flops in which the input is to the clock terminal while the J and K terminals are held at logic 1, as in Figure 16b in the text. The schematic diagram of a 3-bit ripple counter is shown in Figure P11. Since there is no system clock, this is not a synchronous counter. Assume that flip-flop outputs change, with a delay of d_f, on the rising edge of pulses on the clock terminal.

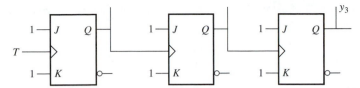

Figure P11

  a. Describe the operation of this ripple counter. In particular, assume that the count is to be used by some other circuit. Will the count sampled by this circuit always be correct?
  b. What is the maximum counting frequency of the 3-bit counter?
  c. In a 10-bit ripple counter, assume that $d_f = 40$ ns. Discuss the maximum rate at which the input signal (T) can change.

12 Assume that a JK flip-flop is available. Draw a diagram that includes a JK flip-flop and one or more logic gates to construct:

  a. A D flip-flop
  b. A T flip-flop

13 Assume that a D flip-flop is available. Draw a diagram that includes a D flip-flop and one or more logic gates to construct:

  a. A T flip-flop
  b. A JK flip-flop

14 Assume that a T flip-flop is available. Draw a diagram that includes a T flip-flop and one or more logic gates to construct:

  a. A D flip-flop
  b. A JK flip-flop

15 A sequential circuit is made of three toggle flip-flops with inputs labeled T_0, T_1, and T_2 and outputs Q_0, Q_1, and Q_2, respectively. Each of these is designed by using a JK flip-flop whose J and K inputs are both set at 1, leaving the clock terminal as the only external input. The input to flip-flop 1 on the left is a clock signal C. The input to the flip-flop on its right comes from the output of a two-input AND gate whose inputs are C and Q_0. The input to the last flip-flop comes from the output of a three-input AND gate whose inputs are C, Q_0, and Q_1. The flip-flop outputs constitute a word $Q_2Q_1Q_0$. Assume that all flip-flop outputs are 0 initially.

  a. Draw the appropriate circuit diagram using AND gates and T flip-flops.
  b. Write expressions for each flip-flop input.
  c. Using the property of toggle flip-flops, find the next-state values of flip-flop outputs upon the occurrence of the first clock pulse, and write the output word.
  d. Starting from the 000 state, repeat this for ten successive clock pulses and list the output words one under the other. Describe what happens after the eighth clock pulse.
  e. Assume the output words are code words. By examining the consecutive outputs, specify what the code is. What operation does this circuit seem to perform?
  f. Suppose a fourth toggle flip-flop is appended on the right, with the input to this flip-flop being the output of a four-input AND gate whose inputs are C, Q_0, Q_1, and Q_2. Draw the circuit and speculate on its nature. Repeat as in part d for 20 successive clock pulses and verify your speculation.

16 The circuit in Figure P16 performs a certain function. The objective of this problem is to determine what this function is.

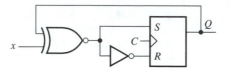

Figure P16

a. Determine an expression for the next state Q^+ in terms of input x and the present state Q.

b. Let the clock pulses be numbered $1, 2, 3, \ldots$ after the point at which the flip-flop is reset $(Q_0 = 0)$, and let x, Q, and Q^+ take on these subscripts to designate the clock pulse during which they occur. Derive expressions for Q_{i+1}^+, for $i = 1, 2, 3, 4$ in terms of the input values x_i. Starting from the reset state $(Q_0 = 0)$, determine the condition on the inputs that results in an output $Q^+ = 1$. Create an appropriate name for the circuit.

c. Draw a timing diagram that includes the clock and waveforms for x, S, and Q to verify your previous conclusions.

d. Suppose a JK flip flop is to be used to carry out the same function. Draw the circuit diagram.

17 A 3-bit shift register using D flip-flops has an input x and left-to-right states Q_0, Q_1, and Q_2. An output z is obtained from an XOR gate whose inputs are x and Q_2. The start of a timing diagram, showing a clock sequence and input x, is given in Figure P17.

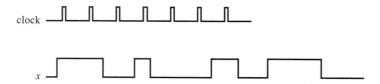

Figure P17

a. Draw the circuit.

b. Extend the clock pulse for six more periods, and complete the timing diagram by adding the waveforms for Q_0 to Q_2 and output z.

18 a. Lemon Logic (LL.com) has just patented a new Lemon flip-flop with one input, L, and outputs Q and Q', having the following properties: If $L = 0$, then $Q^{n+1} = 0$; if $L = 1$, then $Q^{n+1} = (Q^n)'$. Set up a table showing the excitation requirement for each state transition. Give your reasoning. Does anything appear to be anomalous?

b. Another Lemon design is a clocked LM latch. It is constructed by connecting the output of an XOR gate to the R input of a clocked SR latch. One input of the XOR is labeled L and is also connected to the S input of the latch; the other input is labeled M. Repeat part a.

19 Why are the setup and hold times of a flip-flop defined, respectively, relative to the beginning and end of a clock transition?

20 For the shift register shown in Figure 22, derive an expression for the delay that will determine its maximum clock frequency. The equation should be the sum of a few delay terms.

21 Explain why the *JK* master-slave latch is not considered an edge-triggered flip-flop. Design a master-slave latch of type *D, T,* or *SR* that is edge triggered.

22 **a.** Design a 4-bit parallel-load register that can accept its input from two different sources.

  b. Modify your design of part *a* so that the register can be loaded with new values or rotated right.

23 Design a 4-bit parallel-load register that can be loaded with new values, can be rotated left or right, or can preserve its existing value.

24 Design a 4-bit parallel-load register that can either be loaded with new values or increment its existing value.

25 Design a 4-bit serial register that can be loaded or cleared (all zeros) synchronously.

26 Refer to the *SR* latch design in Figure 4. A curious student wanted to know what would result if the two inverters at the inputs were removed. You are called in as a consultant.

Analyze the resulting circuit and construct a transition table for this design similar to Figure 6. Discuss your findings.

Chapter **6**

Synchronous Sequential Machines

A block diagram of a sequential circuit was shown in the introduction of Chapter 5. That chapter concentrated on one part of such a circuit: the memory devices, or flip-flops. Those circuits in which state transitions are controlled, or synchronized, by a clock are said to be *clocked,* or *synchronous,* sequential circuits. Other sequential circuits exist, called *asynchronous* circuits, in which state transitions are not synchronized by a clock. These are less common, although they do have important applications. We will postpone the discussion of such circuits to Chapter 7.[1]

A number of tools are used to describe the behavior of sequential logic circuits and to analyze and design them. We will introduce and develop such tools in this chapter. Included are formal procedures for the design of synchronous machines. Finally, we will concentrate on one class of such circuits and their design: circuits called *counters*.

1 BASIC CONCEPTS

The generic description of a problem requiring the design of a synchronous sequential logic circuit can be given as follows.

Design a digital circuit whose outputs are to take on specific values after a specific sequence of inputs has taken place.

Such a problem statement is very broad. What is clear is that

- There are to be certain sequences of inputs to the circuit.

---

[1]With or without adjectives to qualify it, the term *machine* is often used to designate a sequential circuit, as in the title of this chapter. Because such circuits can have only a finite number of states, they have also been called *finite-state machines*. Since finiteness is all that is possible in the physical world, this adjective is often dropped and the circuits in question are simply called *state machines*. "Machine" normally has the connotation of something physical. However, in the present usage, the term refers to an abstract entity described by mathematical, graphical, or tabular means, as we will describe in this chapter.

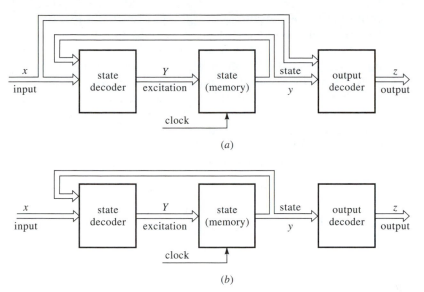

Figure 1 Models of Mealy and Moore machines.

- Different sequences of past inputs will constitute different states or conditions of the circuit.
- Specific outputs result only following a specified input sequence.

The general description does not specify whether the output will depend just on the past inputs or on the latest input as well.

These two possibilities lead to different structures, first explored in the mid-1950s by two people named Mealy and Moore. As mentioned in Chapter 5, in *Mealy machines* the outputs depend both on the present state (resulting from past inputs) and on the current input. The outputs in *Moore machines* depend only on the present state (resulting from past inputs). Attention will be devoted to both types.

Block diagrams of both Mealy and Moore models of sequential machines were given in Chapter 5, Figure 1. Somewhat refined versions are shown in Figure 1 here. The open arrowheads imply multiple variables. For example, input x stands for the set of variables $\{x_1, x_2, \ldots, x_n\}$. The combinational part of the circuit is broken down into two separate parts: the *state decoder* and the *output decoder*. The state decoders in both models accept as inputs both primary (external) inputs and the present state. In the Moore machine, however, the output decoder accepts only the present state to yield outputs. In the simplest case, there is no output decoder at all; the states themselves are the outputs. In a given machine, there may be some outputs that are Moore-type outputs and others that aren't. Such a machine must be classified as a *Mealy machine,* since at least *some* of its outputs depend not only on the state, but also on the inputs. Thus, the Mealy machine is the more general (and more common) type.

The behavior of synchronous sequential logic circuits can be described in a number of ways. At any given clock pulse, the state of the circuit is the *present state*. Signals present at the input terminals at that time are the *inputs*. This combination of present state and input results in two things: a transition to the *next state* and an *output*. At the next clock pulse, the process is repeated, except that the present state is now what the next state was at the preceding clock pulse. It is possible to conceive of this process as never ending; that is, the "next state" is never a state that had been previously encountered. In this case, the machine would be infinite—a peculiar machine indeed. Barring such an unlikely event, somewhere along the line the next state will be a previously encountered state. After this, the machine will retrace its steps over and over; no new states will be encountered.

Several means are available for illustrating the following sequence:

$$\text{present state/input} \rightarrow \text{clock pulse} \rightarrow \text{next state/output}$$

One of these means is graphical/diagrammatic; another is tabular. A third approach utilizes a chart not unlike a flow chart describing an algorithm. All are treated in this chapter.

State Diagram

The graphical/diagrammatic tool for describing sequential circuit behavior is known as a *linear graph*.[2] For each state of the circuit, there is a corresponding node in the graph. (The circle representing the node is made large enough that the symbol for the state can be written inside.) With the machine in any one state (node in the graph), at the occurrence of a clock pulse, there will be a state transition to the next state and there will be an output, both in accordance with the problem statement. For a single-input machine, two lines emanate from each node, one each for a 0 and for a 1 input. For two input variables, four lines emanate from each node, one for each input combination: 00, 01, and so on. (How many lines will emanate from each node if the number of input variables is *n*?) Along each line we write the input value and the corresponding output separated by a slash. The resulting graph is called a *state diagram*.

For some state machines it is known from the statement of the problem just how many distinct states the machine has. However, in general, the number of possible states is initially unknown. To establish the state diagram, we arbitrarily choose an initial state and label it, say, A. (State names can be anything convenient.) A state is identified by unambiguously specifying how it is reached. To say, for example, that "state S is reached when the input is 1" is inadequate,[3] because the statement does not *unambiguously* identify it: Does this 1 follow another 1? Is it the first 1 after a string of 0's? Does this 1 follow a string of two or more preceding 1's? One unambiguous specification would be: "State S is reached by the second of two input 1's after one or more 0's."

Because it is difficult to describe, in the abstract, both the state diagram and the tabular tool to be discussed next, we will continue this discussion in conjunction with an example.

---

[2]A linear graph is a set of *nodes*, or *vertices* (drawn as circles), interconnected by a set of directed lines (or arcs), that is, lines that have an orientation indicated by an arrowhead.

[3]Except for a trivial case that will be described shortly.

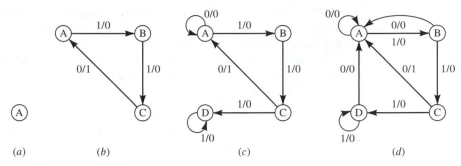

Figure 2 State diagram of sequence detector.

EXAMPLE 1

A synchronous sequential machine with a single input line x and a single output line z is to be designed. The specifications are as follows: the output is to be $z = 1$ if and only if the specific input sequence ...0110 occurs at consecutive clock pulses; otherwise $z = 0$. (Unless otherwise stated, the most recent bit of a sequence is the one on the right in all cases.) Suppose, for example, that an input sequence is ...0110110. The first 4 bits identified constitute an *acceptable* (that is, *output-producing*) input sequence. But the fourth bit starts another acceptable sequence that overlaps with the first. Hence, the output sequence will be ...0001001. (Confirm.) Such a machine is called a *sequence detector*.

Let's identify our initial state A as the state reached by an input $x = 0$ regardless of the preceding sequence of inputs. It doesn't matter if the preceding input bit is 1 or 0 for a 0 input bit to start an acceptable sequence. We start the state diagram by drawing a node labeled A. Following this, there are two possible strategies:

- We can explore the consequences (next state/output) resulting from each possible input starting in this state, and continue in this fashion with all the next states encountered along the way.
- We can assume an acceptable sequence and pursue the consequences (string of next states and outputs), adding states as needed. Then we return to each state encountered along the way to fill in the consequences of inputs that are not part of an acceptable sequence.

The second method is carried out in Figure 2. Starting from state A (reached by a 0 input), we assume an input sequence 110 to complete an acceptable sequence. The result is shown in Figure 2b. (Confirm that the two additional states shown must be introduced along the way. Also confirm each of the steps that follow.)

Starting at each state in Figure 2b, only one of the two possible inputs has been used so far. Now we fill in the other possibilities. From state A, an input of 0 leads back to state A/output 0. From state B, an input of 0 also leads back to A/output 0. But what is the next state if there is an input of 1 while the machine is in state C (which was reached by an input sequence 011)? It can't be to any of the three states reached so far (confirm this), so it must be to a new state D/output 0. The state diagram so far is shown in Figure 2c. Finally, from this new state, an input of 0 leads back to A/output 0 and an input of 1 leads back to D/output 0. The final diagram is shown in Figure 2d.

Study the last diagram. A is the state reached by the first bit in an acceptable sequence. The sequence is aborted if the next input is also 0. Now it is this last 0 that starts an acceptable sequence. Any number of additional inputs of 0 lead to the same result: the latest 0 becomes the first bit of an acceptable sequence. The last state, D, is an acceptable-sequence spoiler; it is reached by an input of 1 following a sequence ...011. ∎

Notice in Figure 2 that all but one of the arcs in the graph are labeled with an output of 0. A lot of clutter could be avoided if we adopt the convention that only outputs of 1 will be shown explicitly. When the output associated with a particular arc is 0 (or 00, 000, etc. for more output variables), henceforth it will not be shown explicitly on the state diagram.

In constructing a state diagram, there are generally two major decision points:

1. Choosing the initial state
2. When in a particular state, deciding whether the transition resulting from a particular input is to an existing state or to a new state not yet identified

In some (not all) sequential machines there is a specific *reset* state; the machine must be in this state at the starting time. In such cases, the initial state is predetermined. When there is no reset state, the initial state is chosen arbitrarily, as in the preceding example. Although the problem statement might guide the choice of initial state, different designers might choose different initial states.[4] No problem. Assuming there are no mistakes, two state diagrams constructed with different initial states will be *isomorphic;* that is, they will become identical by an appropriate interchange of state names.[5]

As for the second decision point, a transition to a new state rather than to an existing state will result in more states in the state diagram. Eventually, a circuit must be implemented. Generally speaking, more states mean more flip-flops, though not proportionately more. A circuit with n flip-flops, for example, will have 2^n states. Conversely, then, eight (2^3) states will require three flip-flops. But even as few as five states will still require three flip-flops. Thus, if a state diagram has five states already, increasing the number of states to as many as eight will not increase the number of flip-flops needed. Thus, introducing more states in a state diagram may simply mean introducing redundancies; these might be removable later.

Although in the preceding example we constructed a state diagram that describes all the state transitions of the desired machine, we still did not complete a design. Before tackling that task, we turn to the tabular tool for describing the behavior of a sequential machine.[6]

---

[4] They may also give the states different names, the placement of the nodes might be different, and the curvatures of the lines joining the nodes might be different. All of these are trivial matters.

[5] This assumes that no extra states are introduced, as will be described shortly.

[6] Suppose we reconsider the statement "State S is reached when the input is 1." The state diagram will then consist of just two states: S is the state reached by an input 1 and, say, T is reached by an input 0. Any other 0 inputs while in state T will return the machine to state T. A 1 input will send the machine back to state S. Any other 1 inputs will keep the machine in state S. Construct a state diagram for yourself. This is a trivial "machine."

PS	NS,z	
	$x = 0$	$x = 1$
A	A,0	B,0
B	A,0	C,0
C	A,1	D,0
D	A,0	D,0

Figure 3 State table of sequence detector in Example 1 obtained from its state diagram.

State Table

A table can be described by its row headings (or names) and its column headings. Its entries occur at the intersections of the rows and columns. To describe the operation of a synchronous machine, it is customary to choose the row headings as the present states and the column headings as the inputs.

Since two outcomes (next state, output) result from an input to the circuit when the circuit is in a particular state, it is conceivable to construct two separate tables. The entries in one of these would be the circuit outputs—hence it is called the *output table*. The entries in the other table would be the next states. Since the table is intended to show transitions from a present state to a next state, it might be tempting to call this table a state transition table. But in the preceding chapter, state transition table was the name given to the table that specifies the next state resulting from inputs to a flip-flop for each present state of the flip-flop. So we use a different name; it is called simply a *state table*. In the present usage, the circuit is not limited to a single flip-flop but encompasses an entire machine.

Remember from Figure 1*b* that the output of a Moore machine depends only on the present state. For such a machine, there will be only one output combination for each input combination; hence, a separate output table makes more sense. For a Mealy machine, on the other hand, we will combine the state and output tables into a single table in which the entries are both the next states and the resulting outputs, separated by a comma.[7] This will be illustrated for Example 1, whose state diagram was obtained in Figure 2. (Of course, once this table is available, it is always possible to separate the next-state part and the output part into two separate tables if there is a reason for doing so.)

Constructing a State Table from a State Diagram

Once a state diagram is available, the corresponding state/output table is easily constructed. The state diagram in Figure 2 has four states. Hence, the corresponding state/output table will have four rows. Starting in any state, the output and the next state can be obtained from the diagram and entered in the table. The result is given in Figure 3.

Exercise 1 Notice in Figure 2 that state A is reached from each state (including A) by an input 0 but with different outputs: 0 from states A, B, D and 1 from state C. Instead, let's suppose that state A in the state diagram of Figure 2 is

---

[7]It is also sometimes called a *flow table*.

identified as the state reached by an input 0 resulting in an output 1. Now start from Figure 2a and assume that a 0 input while in state A or state B leads to a new state E rather than back to A. Complete the resulting state diagram. (You don't need to take the following advice, but you won't get tangled up in crossing lines if you put A, B, and C in a row, with D under C and E under B.) Then, from the diagram, construct a state table.

Answer[8]

Examine states A and E in your table from Exercise 1. Both are reached by an input bit 0. Hence, starting from either state, the next states and outputs must be the same for each input bit. In this case, these two states can't be distinguished from each other. This is the basis for a definition:

> *Two states are said to be* indistinguishable *if, for each input combination, the resulting outputs and next states are the same.*

Actually, the next states need not be the same—only indistinguishable, as just defined.

On this basis, states A and E are indistinguishable. If all E next states in the table are replaced by A, and row E is eliminated, the table obtained in Exercise 1 will become the same table obtained in Figure 3.

There are formal procedures for extending the concept just defined. We will pursue this generalization in the following section and discuss ways of reducing the number of rows of a state table. Consequently, in constructing a state diagram in the design of a sequential circuit, there is no great need to worry about introducing redundant states; such states can be removed subsequently. On the other hand, there is no point in needlessly extending a state table, since effort will be needed later to reduce it. When in doubt while constructing a state table, by all means introduce a new state. However, restrain yourself if you are certain that the relevant conditions have already been identified by an existing state.

EXAMPLE 2

(*Note:* One of the topics treated in section 4 of Chapter 1 is the Hamming code. Review it if you need a refresher.) To an n-bit message, an additional k bits are added, making the parity of the resulting $(n + k)$-bit string either odd or even—our choice. In this example, let $n = 3$ and $k = 1$; let's choose odd parity. Suppose that a 4-bit string is to be received; the first 3 bits constitute the message, and the fourth bit is always 0 (equivalent to a blank). If the number of 1's in the 3-bit message is odd, the parity bit is to remain 0. If the number of 1's is even, a 1 bit is to be generated

---

[8]Now, only one arc of the graph with input 1 enters A (from C/output 1). Four arcs enter E (from A, B, D, and E, all with 0 in, 0 out. D is a spoiler state, entered by a string of three 1's starting at A; any further 1's while in state D will keep the state in D. An input 1 while in state E is the first 1 after one or more 0's; in that respect, an input 1 while in state E should lead to the same state as a 1 while in state A, namely, state B. ◆

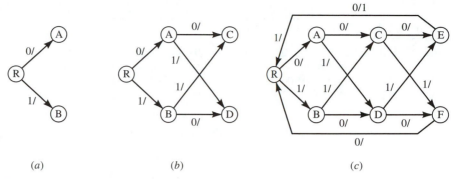

Figure 4 State diagram of parity-bit generator.

	NS,z	
PS	$x = 0$	$x = 1$
R	A,0	B,0
A	C,0	D,0
B	D,0	C,0
C	E,0	F,0
D	F,0	E,0
E	R,1	R,×
F	R,0	R,×

Figure 5 State table for parity-bit generator.

and inserted in the fourth position to make the parity of the 4-bit string odd. In either case, after the fourth bit, transition is to be to a reset state, in which the machine is ready to receive the next message sequence. Such a machine is a *parity-bit generator.* Our objective is to create a state table.

The first step is to create a state diagram. The machine is to be in the reset state (say R) when the first bit arrives. The partial state diagram after the first bit is shown in Figure 4a. There cannot be a 1 output until the parity bit arrives and completes a 4-bit string whose parity is even. At each input, the parity of the input bits up to that point is either even or odd, so transition is to one of two possible next states: an even-parity state or an odd-parity state.

The partial diagram after the second message bit is shown in Figure 4b. Why is it necessary for the next state at the second input bit to be a new state rather than one of the existing odd-parity or even-parity states, A or B? (Give it some thought before you look at the footnote.[9]) Confirm the details of the complete diagram shown in Figure 4c; for example, show all possible input sequences by which states E and F are reached. An output of 1 occurs only upon returning to reset from state E following one of the sequences 0000, 0110, and 1100.

The next step is to construct the flow table from the state diagram. It is shown in Figure 5. The fourth input bit is never 1; so what will the output be for

---

[9]If A can be reached both after a single 0 and after a string 00, for example, then the count of the number of input bits is lost. Hence, we can't tell when the third bit has arrived in order to decide whether or not to generate a 1-bit, nor can we tell when the fourth bit has arrived so as to go back to the reset state.

PS	NS,z $x = 0$	$x = 1$	y_1y_2	$(y_1y_2)^+$ $x = 0$	$x = 1$	y_1y_2	y_1^+ $x = 0$	$x = 1$	y_1y_2	y_2^+ $x = 0$	$x = 1$
A	A,0	B,0	A→00	00	01	00	0	0	00	0	1
B	A,0	C,0	B→01	00	10	01	0	1	01	0	0
C	A,1	D,0	D→11	00	11	11	0	1	11	0	1
D	A,0	D,0	C→10	00	11	10	0	1	10	0	1
	(a)			(b)			(c)			(d)	

Figure 6 State and transition tables for sequence detector.

$x = 1$ from states E and F? Confirm all the details of this table by reference to the state diagram. ∎

2 STATE ASSIGNMENTS

The preceding section described the initial stage in the design of a Mealy-model synchronous sequential machine. From a word description of the specifications of the problem, it consists of constructing a state/output table (possibly after constructing a state diagram). During this process, it is possible that redundant states are introduced; consequently, the table might have more states than necessary to perform the desired task. Procedures for eliminating redundant states would be useful, leading to a reduced table with fewer states that is equivalent, in some sense, to the original table. Fortunately, such procedures do exist, and we will discuss them later. For the present, assume that a reduced table is available.

The state of a sequential machine at a given time is the condition in which past inputs have left it. This information is stored in the flip-flops; the state is, thus, described collectively by the outputs of the flip-flops. The next step in the design, then, is to identify the states in the table with specific flip-flop outputs. That is the subject of concern in this section. We will develop the subject by reference to the examples in the preceding section.

EXAMPLE 3

The state table derived for the sequence detector in Example 1 was given in Figure 3 and is repeated in Figure 6a. The minimum number of state variables needed for a circuit implementation of this table is $\lceil \log_2 4 \rceil = 2$, where $\lceil k \rceil$ denotes the *ceiling* of k, the smallest integer not less than k. Let us designate the state variables as y_1 and y_2. There are four possible combinations of values of these two variables. How should these four combinations be assigned individually to each of the four states? Before considering a general answer to that question, let us arbitrarily make the assignment shown in Figure 6b. (Let's temporarily neglect the output z and concentrate on the states.) The result is a table that, for each present combination of state-variable values and each input value, specifies the next combination of state-variable values. This is a *state-transition table*.

Note that there are two different orders: the order of listing the states in the state table (alphabetical) and the order of the combinations of state-variable values. If the assignment is made so that both orders are maintained, there is no prob-

Transition Required

y	to y^+	J	K
0	0	0	×
0	1	1	×
1	0	×	1
1	1	×	0

(*a*)

y_1y_2 \ x	0	1
00		
01		1
11	×	×
10	×	×

(*b*) $J_1 = xy_2$

y_1y_2 \ x	0	1
00	×	×
01	×	×
11	1	
10	1	

(*c*) $K_1 = x'$

y_1y_2 \ x	0	1
00		1
01	×	×
11	×	×
10		1

(*d*) $J_2 = x$

y_1y_2 \ x	0	1
00	×	×
01	1	1
11	1	
10	×	×

(*e*) $K_2 = x' + y_1'$

Figure 7 Excitation maps for the sequence detector.

lem. For any other assignment, either one or the other order can be maintained, but not both. It is much more convenient to maintain the order of the value combinations than the order of the state names, since the former is directly transferable to a logical map. This is what has been done in Figure 6*b*; states C and D are out of alphabetical order, but the value combinations are in logic map order.

For simplicity, the transition table in Figure 6*b* is separated into two tables in Figures 6*c* and 6*d*, one for each state variable. If we were to implement the design with *D* flip-flops, each of these tables would represent the excitation map for one of the flip-flops; for a *D* flip-flop, the present input (excitation) is the same as the next state. Just for pedagogical reasons let's implement it with *JK* flip-flops instead.[10]

The first requirement is to determine logic maps for excitations *J* and *K* for each flip-flop. For this we use the excitation requirements for *JK* flip-flops given in Chapter 5, Figure 17, and repeated here in Figure 7*a*. For each flip-flop, this table gives the required values of *J* and *K* for each transition from a present-state value to a next-state value. In each case, *J* and *K* are to be obtained as the output of a combinational circuit whose inputs are the circuit input x and present states y_1 and y_2. This requires combining the transitions from present to next state in Figures 6*c* and 6*d* with the transition requirements table in Figure 7*a*. Thus, from state $y_1y_2 = 11$ and $x = 1$, transition is to $y_1^+ y_2^+ = 11$ from Figures 6*c* and 6*d*. That is, for both y_1 and y_2, transition is from 1 to 1 for $x = 1$. But from Figure 7*a*, the requirements for a 1 to 1 transition is

[10]Only one combinational logic circuit is needed for the excitation when implementation is with *D* flip-flops. With *JK* flop-flops, two circuits are needed, one each for *J* and *K*. It is, of course, possible that the circuit required with a *D* flip-flop is more complex; nevertheless, with present-day implementations with PLDs, it is usually preferable to use *D* flip-flops. Furthermore, *JK* flip-flops require more chip area in ASIC technology.

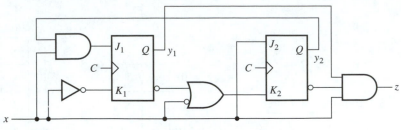

Figure 8 Implementation of the sequence detector.

$J = x$ and $K = 0$. These constitute the content of one square in the logic maps for each J and K. The completed logic maps are shown in Figures 7b to 7e. (Confirm each of them and confirm the expressions for J and K given under the maps.)

To complete the implementation, an expression for the output function must be obtained. From the flow table in Figure 6a the output is 1 for exactly one state (state C) and one input ($x = 0$). Since C has the assignment $y_1 y_2 = 10$, the expression for the output function is

$$z = x' y_1 y_2'$$

The complete implementation is shown in Figure 8. (Confirm it all.) ■

Analysis

Once a sequential circuit has been designed and a logic diagram constructed, how can we tell that errors have not been made and that the circuit outputs actually satisfy the original specifications? As discussed in relation to combinational circuits in Chapter 3, a process called *verification* is carried out. This involves making measurements (of voltage, say) at appropriate points in any circuit (not just logic circuits) to verify that the actual values are what they are supposed to be theoretically.

Again, as discussed in Chapter 3, there is no point in physically implementing the circuit before verification. Once a paper (or software-generated) sequential circuit has been obtained, one can *analyze* the circuit at various points to verify that the logic values at these points are indeed the values required by the design specs. Compared with the process of design, logic-circuit analysis is rather trivial. One starts at any point in the circuit (gate or circuit outputs, or MUX, flip-flop, or register inputs) and determines logic expressions for these variables. This is repeated until expressions for all outputs are obtained. Then one inserts all possible input values into these expressions. The values obtained are then compared with what they are supposed to be.

As an example, look back at the sequence-detector design in Figure 8. Carry out an analysis of the circuit and obtain expressions for the J's and K's and the output z. (Don't peek at Figure 9 until you have completed it.) The lines on the time axis in Figure 9 represent the rising edge of the clock signal. Using these expressions and the transition table of JK flip-flops, and assuming the input sequence shown on the first line, the values of the other variables are determined column by column. Note that, when $x = 0$, the values of the J's and K's do not depend on the states of the flip-flops; hence, the next states in the

Time

		x										
x		0	1	1	0	1	1	0	0	1	1	1
J_1	xy_2	0	0	1	0	0	1	0	0			
K_1	x'	1	0	0	1	0	0	1	1			
J_2	x	0	1	1	0	1	1	0	0			
K_2	$(xy_1)'$	1	1	1	1	1	1	1	1			
y_1^+		0	0	1	0	0	1	0	0			
y_2^+		0	1	0	0	1	0	0	0			
z	$x'y_1y_2'$	0	0	0	1	0	0	1	0			

Figure 9 Timing table for the implementation in Figure 8.

first column are based only on $x = 0$. When $x = 1$, the values of J_1 and K_2 do depend on the states (not the next states in the same column but the states in the previous column). Verify the remaining columns and complete the last three columns in Figure 9. Verify that the outputs satisfy the specifications.

Exercise 2   Using the information in Figure 9, choose an appropriate scale and draw a timing diagram that includes the circuit input, the flip-flop inputs, the next states, and the circuit output.   ◆

Rules of Thumb for Assigning States

A number of loose ends in the preceding development remain to be explored. The first is the simple observation that, when a transition table is obtained after a state assignment is made, as in Figure 6b, it is not essential to rewrite it in the form of the individual state variable transition tables, as was done in Figures 6c and 6d, before constructing the excitation maps. Instead, for each input value, concentrate on the column corresponding to one of the state variables, say y_1, mentally blocking the others from your perception, and construct the maps directly from the general transition table. Practice doing that for Figure 6b.

A more important consideration is the following. Given a state table having k states, the number of state variables needed to implement it is $n = \lceil \log_2 k \rceil$. An immediate decision is needed as to which of the 2^n combinations of state-variable values should be assigned to each of the k states. For a nontrivial number of states, many different possibilities exist for this assignment.

Exercise 3   The state table for the sequence detector in Example 1 was given in Figure 6a. The implementation of the circuit using the assignment in Figure 6b was given in Figure 8. Instead, use the following assignment and find an implementation for the circuit: A: 00, B: 01, C: 11, D: 10. Compare the number of gates with the number in Figure 8.
Answer[11]

---

[11]One more gate than the number in Figure 8.   ◆

1. Two present states should be assigned adjacent codes
   if they have the same next state for:
   a. Each input combination
   b. Different input combinations, if the next state can
      also be given adjacent assignments
   c. Some input combinations, but not necessarily all
2. For all inputs, codes assigned to the next states for
   each present state should be adjacent.
3. Assignments should simplify the output function.

Figure 10 State assignment rules.

In general, the choice of assignment will influence the implementation. Different assignments lead to different maps of flip-flop excitation and output and, hence, to different expressions for excitation functions and output functions. Unfortunately, there is no general theory on assignments—and so no algorithm—that will result in simplicity of implementation. Experience is the only guide to making a state assignment.

General models of sequential circuits were given in Figure 1. Although the actual circuit in Figure 8 is quite simple, it illustrates the Mealy model well. The nonsequential part of the circuit consists of two classes: the combinational circuit that implements the excitations and the one that implements the output, as expected. Once state reduction has been carried out (until you learn how, you'll have to subcontract it out), the extent of the memory (the number of flip-flops) is fixed. Economy of implementation, then, is a matter of reducing the number of IC packages (and gates) in either the state decoder, the output decoder, or both.

Recall that the number of prime implicants and the number of literals in a prime implicant can be reduced when there are many adjacent minterms. Hence, it comes down to this: How do we choose state assignments so as to achieve a large number of adjacencies? Not much in the way of generalities can be deduced from an examination of Example 3. However, on the basis of a great deal of experience, some heuristic "rules" have been formulated as guides in making a state assignment for the case of a single input. Figure 10 lists a number of such rules, in priority order.

For a given state table, it is unlikely that all the adjacencies specified by these rules can be achieved. When there is a conflict, the higher-priority rules take precedence. Even if the rules can be fully implemented, they do not guarantee an optimal assignment. That is, the rules do not constitute an optimal algorithm.

Furthermore, they do not necessarily lead to a unique assignment; it may be possible for the required adjacencies to be achieved by different assignments. Even so, the rules will reduce the number of alternatives that must be checked. Finally, even for the same assignment, the number of logic gates using *JK* flip-flops might be different from the number using *D* or other flip-flops. Notwithstanding all that, using these rules as a guide is reasonable.

EXAMPLE 4

An application of the assignment rules in the implementation of a sequential machine is illustrated in the state table shown in Figure 11*a*. Since there are seven states,

	NS,z	
PS	$x = 0$	$x = 1$
A	B,1	B,0
B	C,0	D,1
C	E,1	F,0
D	F,0	E,1
E	G,0	A,0
F	A,0	G,0
G	B,0	B,0

(a)

Rule	Adjacencies
1a	AG
1b	CD if EF
	EF if AG
2	CD, EF, AG
3	AC, BD

(b)

y_2y_3	y_1 0	1
00	A	E
01	G	F
11		B
10	C	D

(c)

Figure 11 Example state table, adjacencies, and assignment map.

the circuit will need $\lceil \log_2 7 \rceil = 3$ flip-flops. The adjacencies of states called for by the adjacency rules are shown in Figure 11b. (Don't fail to confirm these.) It is now a problem of determining a state assignment so that as many as possible of these adjacencies is achieved. To help in this process, an assignment map can be created, as shown in Figure 11c. This is a map whose coordinates are the three state variables. Each square in the map corresponds to a combination of state variable values.

The placement of the states in the map is initiated by deciding on the state that is to have the assignment 000. If there is a reset state in the problem, it is reasonable to give it this assignment; if not, the choice is arbitrary. Suppose the combination 000 is assigned to state A; then G must have an adjacent assignment. But each cell in the map has *three* others adjacent to it; which one is chosen for G depends on which other adjacencies are needed. In the present example, G is not required to be adjacent to any other state, so the placement is very flexible. One possibility is to assign 001 to G. The remaining assignments are made using the same approach, resulting in the assignment map in Figure 11c.

The next step in the design process is to construct transition and output tables. This is done using the assignment map and the given state table. The result is shown in Figure 12a. Let us again assume that *JK* flip-flops are to be used in the implementation; refer to Figure 7a for the transition requirements for this flip-flop. From Figures 7a and 12a, we construct the *J* and *K* excitation maps for each flip-flop. The result for the first flip-flop is shown in Figure 12b. The others are obtained similarly. Expressions for the excitation and output functions are given in Figure 12c   ■

Exercise 4   Construct the output map and the excitation maps for the other two flip-flops in Example 4 and confirm the expressions given in Figure 12c.   ◆

Exercise 5   In Example 4, suppose that the implementation is to use *D* flip-flops. Determine the maps for the *D* excitations and, from these, the state decoder. Compare the hardware with the case that uses *JK* flip-flops.   ◆

EXAMPLE 5

The objective of this example is to find a good assignment scheme for the state table in Figure 13a.

$y_1 y_2 y_3$	$y_1^+ y_2^+ y_3^+$ $x=0$	$x=1$	z $x=0$	$x=1$
A 000	111	111	1	0
G 001	111	111	0	0
011	—	—	—	—
C 010	100	101	1	0
E 100	001	000	0	0
F 101	000	001	0	0
B 111	010	110	0	1
D 110	101	100	0	1

(a)

xy_1

$y_2 y_3$	00	01	11	10
00	1	×	×	1
01	1	×	×	1
11	×	×	×	×
10	1	×	×	1

$J_1 = 1$

xy_1

$y_2 y_3$	00	01	11	10
00	×	1	1	×
01	×	1	1	×
11	×	1		×
10	×			×

$K_1 = y_2' + x'y_3$

(b)

$$J_2 = y_1'$$
$$K_2 = y_3'$$
$$J_3' = (x' + y_1')(x + y_1 + y_2')$$
$$K_3 = (x + y_1)(x' + y_3')$$
$$z = x'y_1'y_3' + xy_1y_2$$

(c)

Figure 12 Transition table and excitation maps.

PS	NS,z $x=0$	$x=1$
A	B,0	B,1
B	F,0	D,1
C	E,1	G,1
D	A,0	C,0
E	D,1	G,0
F	F,0	A,0
G	C,1	B,0

(a)

Rule	Adjacencies
1c	AG, BF, CE
2	AC, AF, BC
	DF, DG, EG
3	AB, AC, BC
	CE, CG, EG

(b)

y_1

$y_2 y_3$	0	1
00	A	C
01	G	E
11	D	B
10	F	

y_1

$y_2 y_3$	0	1
00	A	G
01	F	D
11	B	
10	C	E

(c)

Figure 13 State table, required adjacencies, and assignment maps for Example 5.

The first step is to determine the adjacencies using the adjacency rules; they are listed in Figure 13b (verify, please). With seven states, the number of state variables needed is three. Each cell in a three-variable map is adjacent to three other cells, so each state can be adjacent, at most, to three other states. In a full three-variable map, a total of 12 adjacencies are thus available. (Confirm this.) There are only seven states in this example; it turns out that the maximum number of adjacencies present is nine. From the list of required adjacencies, it is clear that A is required to be adjacent to four other states: G, C, F, and B. Since the lowest-priority adjacency is AB,

1. *State table:* Given the specifications of a problem in natural language, construct a state table satisfying the specifications, perhaps by first constructing a state diagram.
2. *Equivalent reduced table:* Use appropriate procedures to determine equivalent states and to remove redundant states, thus generating an equivalent reduced table. (Procedures will be considered in the following section.)
3. *State assignment:* Choose a state assignment.
4. *Transition and output tables:* Use the assignment to construct these.
5. *Excitation maps:* Choose a flip-flop type; using the transition table and the excitation requirements for the chosen flip-flop, construct excitation maps.
6. *Excitation functions:* Derive expressions for these from the maps.
7. *Output functions:* Derive expressions for these from the output table.
8. *Implementation:* Implement the state decoder from the excitation functions and the output decoder from the output functions.

Figure 14 Design procedure for Mealy machines.

that one should be the first to be abandoned. Similarly, the adjacency CG should be abandoned, since C (and G) cannot be made adjacent to four other states.

Whenever there is a choice, try to achieve those adjacencies at one priority level that are also required at a lower level. Aside from AB and CG (which we abandoned), the remaining adjacencies in rule 3 are also required by higher-order rules.

The number of achievable adjacencies in this example turns out to be nine, equal to the maximum possible. Figure 13c shows two assignment maps; each achieves all nine required adjacencies. Using *JK* flip-flops, the first one can be implemented with five AND and four OR gates. The second one needs one more OR gate. ∎

Exercise 6   Carry out implementations for the two assignments given in Example 5 and confirm the stated results. Convert to all NAND gates.   ◆

3 GENERAL DESIGN PROCEDURE

Each of the elements of a procedure for the design of synchronous sequential machines has been discussed in preceding sections of this chapter. We are now ready to consolidate these elements into a general design procedure. We will illustrate this general procedure by applying it to some specific examples.

Mealy Machine

The Mealy and Moore circuits are models that were shown in Figure 1. When a sequential circuit design problem is specified in terms of the outputs desired for specific sequences of inputs, no model is generally specified. For a given design requirement, it is conceivable to carry out the design based on either model. That means two different designs (state tables) can be obtained. It also means that one of the designs can be obtained from the other. In this book we will deal with both Mealy and Moore machines. The general design procedure for Mealy machines is given in Figure 14.

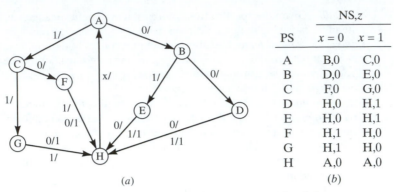

	NS,z	
PS	$x = 0$	$x = 1$
A	B,0	C,0
B	D,0	E,0
C	F,0	G,0
D	H,0	H,1
E	H,0	H,1
F	H,1	H,0
G	H,1	H,0
H	A,0	A,0

(a)   (b)

Figure 15 State diagram and state table of change-of-level detector.

EXAMPLE 6

A sequential circuit is to be designed having a single input line x and a single output line z. Starting in a reset state, the circuit receives input sequences consisting of 3-bit binary words. Each input word follows the preceding word with a delay of one clock period. The circuit must be in the reset state at the beginning of each word. The output is to be $z = 1$ upon receipt of the third bit of a word if the total number of level changes (from 0 to 1 or from 1 to 0) is odd (101, for example, has two level changes while 001 has only one). Such a circuit is called a *change-of-level detector.*

As the first step, we'll obtain the state diagram. Starting from the reset state, no matter what the third bit is, the circuit is to wait one clock period and return to the reset state at the fourth clock pulse. That means the third bit of the input word sends the circuit to a *waiting* state. Figure 15a shows the state diagram. The reset state is A and the waiting state is H. (To confirm this diagram, cover the waiting state and all the lines coming into it and out of it; then describe each of the states reached after 2 bits in terms of the number of level changes. Confirm the output resulting from each input that sends the circuit to the waiting state.) The diagram and the statement of the problem make it clear that each state is unique and there are no redundant states; no two states are equivalent to each other.

The next step is the state table; this is easily constructed from the state diagram and is shown in Figure 15b. (Confirm this, please.) Since each state is unique, the state table cannot be further reduced. The number of flip-flops needed is $\lceil \log_2 8 \rceil = 3$. The rules of thumb for adjacency, shown in Figure 16a, lead to the adjacency map in Figure 16b.

For variety, this time let's use D flip-flops, for which the present excitation is the same as the next state. Hence, the entries in the transition table also specify the D excitations, so logic maps for the D's can be constructed directly from the transition tables. Likewise for the output. We will supply the results here, but we expect you to confirm all this in an exercise. The following expressions result from the maps.

$$D_1 = y_2 + xy_1' + y_1'y_3 \qquad D_2 = y_2'y_3$$
$$D_3 = y_1'y_2'y_3' + xy_2'y_3 \qquad z = x'y_1 + xy_1y_2$$

Rule	Adjacencies
1a	DE, DF, DG
	EF, EG, FG
2	FG, BC, DE
3	DE, FG

Adjacencies not achieved:
DG, EF.

(a)

y_1

	0	1
00	A	H
01	B	C
11	G	E
10	F	D

$y_2 y_3$

(b)

$y_1 y_2 y_3$	$y_1^+ y_2^+ y_3^+$ $x=0$	$x=1$	z $x=0$	$x=1$
A → 000	001	101	0	0
B → 001	110	111	0	0
G → 011	100	100	1	0
F → 010	100	100	1	0
H → 100	000	000	0	0
C → 101	010	011	0	0
E → 111	100	100	0	1
D → 110	100	100	0	1

(c)

Figure 16 Adjacency rules, assignment map, and transition table for the change-of-level detector.

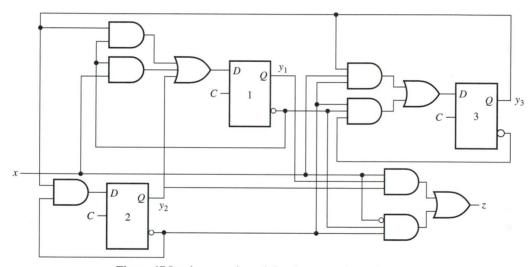

Figure 17 Implementation of the change-of-level detector.

The resulting circuit is shown in Figure 17. ■

Exercise 7 Using the transition table in Figure 16c, construct excitation and output maps. From these, confirm the preceding expressions for the D excitations and the output. Verify the implementation in Figure 17. ◆

> **EXAMPLE 7**

A synchronous sequential machine is to have a single input line and a single output line. The circuit is to receive messages of 5-bit words coded in 2-out-of-5 code. (See Chapter 1 for a description of codes.) The purpose of the circuit is to detect an error in any of the words. Thus, the output is to become 1 whenever a 5-bit word does not represent a valid code word. At the end of each word the machine is to return to the reset state.

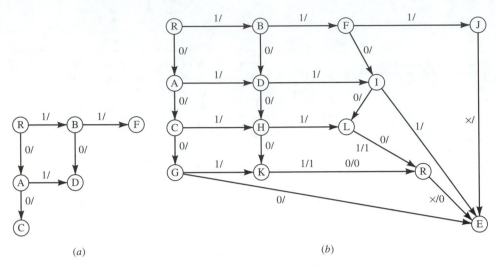

Figure 18 State diagram of error detector in 2-out-of-5 code words.

State Diagram Let the reset state be labeled R. One possibility for the structure of the state diagram is a tree, starting at state R. In such a diagram, when the machine is in any state, each of the two possible input values leads to a new state. Thus, the state diagram will have $2^5 = 32$ states. That not all of these are independent states is seen by examining the information that is needed upon receipt of the kth bit:

- How many bits have been received so far?
- How many of these are 1 bits?

Starting with the reset state, after the receipt of the second bit of a word, there are three possibilities: the number of 1 bits received is 0, 1, or 2. The partial state diagram is shown in Figure 18a. The three possibilities identify just three states after receipt of the second bit. A tree structure would require four states at this point. Upon receipt of the third bit of the word, there are four possibilities for the number of 1 bits received to that point—0, 1, 2, and 3—and thus four new states, labeled G, H, I, J in Figure 18b. Note that no matter what the fourth bit is while at present state J, the received word will never be in 2-out-of-5 code. Hence, from state J, the next state will be an error state (for which the letter E is reserved), but the output will not become 1 until the arrival of the fifth bit, whether a 1 or a 0.

The completed state diagram is shown in Figure 18b; to avoid clutter in the diagram, with lines running back to R from each of the states reached after 4 bits, a second copy of R is provided near these latter states. The two copies of R constitute the same state.

State Table The next step is to construct the state table from the state diagram. Do this and confirm the table given in Figure 19a.

Assignment Map The number of flip-flops needed is $\lceil \log_2 13 \rceil = 4$. Apply the rules of thumb for state adjacencies and confirm the list given in Figure 19. An assignment map that achieves all but one of the adjacencies is given in Figure 19b. It is impossible to achieve all three of the adjacencies required by rule 1. Since two of them are

PS	NS,z x=0	x=1
R	A,0	B,0
A	C,0	D,0
B	D,0	F,0
C	G,0	H,0
D	H,0	I,0
E	R,1	R,1
F	I,0	J,0
G	E,0	K,0
H	K,0	L,0
I	L,0	E,0
J	E,0	E,0
K	R,1	R,0
L	R,0	R,1

(a)

Rule	Adjacencies
1a	EK, EL, KL
1c	GJ, IJ
2	AB, CD, DF, GH, HI, IJ, EK, EL, KL
3	EK, EL

y_1y_2

y_3y_4	00	01	11	10
00	R	C	A	I
01	E	L		J
11	K	F		G
10		D	B	H

(b)

$y_1y_2y_3y_4$	$y_1'y_2'y_3'y_4'$ x=0	x=1
R → 0000	1100	1110
E → 0001	0000	0000
K → 0011	0000	0000
0010	xxxx	xxxx
C → 0100	1011	1010
L → 0101	0000	0000
F → 0111	1000	1001
D → 0110	1010	1000
I → 1000	0101	0001
J → 1001	0001	0001
G → 1011	0001	0011
H → 1010	0011	0101
A → 1100	0100	0110
1101	xxxx	xxxx
1111	xxxx	xxxx
B → 1110	0110	0111

(c)

Figure 19 State table, assignment map, and transition table for example.

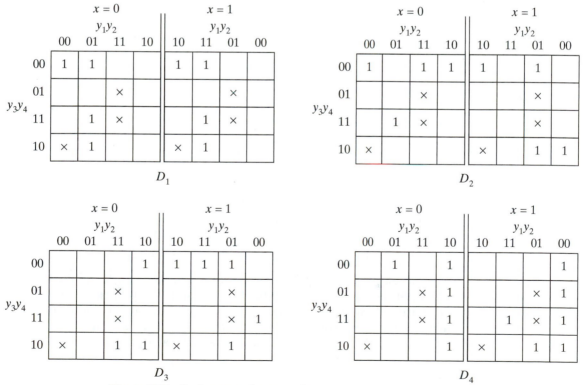

Figure 20 Excitation maps for error detector.

required by lower-priority rules, it is these two that are actually achieved. (Confirm everything as you go along.)

Transition Table   The transition table resulting from the adjacency map is shown in Figure 19*c*. Since four state variables imply 16 possible states, and there are actually only 13, three of the state variable combinations do not correspond to states of the machine. Hence, we don't care what the next states resulting from such nonexistent present states will be.

Flip-Flop Type, Excitation and Output Functions   Assume the use of *D* flip-flops in the implementation. The excitation and output maps require five variables. The excitation maps are shown in Figure 20. Confirm each of these maps and the resulting excitation and output expressions that follow. (The output map is simple enough that it is not shown.)

$$D_1 = y_1'y_4' + y_1'y_2y_3$$
$$D_2 = y_1y_2 + y_1'y_2'y_4' + x'y_1y_3'y_4' + xy_1y_3y_4'$$
$$D_3 = y_1y_2y_3 + x'y_1y_2'y_4' + xy_1y_2 + xy_1'y_3'y_4' + xy_1y_3y_4$$
$$D_4 = y_1y_2' + xy_1y_3 + xy_2y_3y_4 + xy_1'y_2y_3'y_4'$$
$$z = x'y_1'y_2'y_4 + xy_1'y_3'y_4$$

$\blacksquare$

Moore Machine

As shown in the model in Figure 1*b*, in a Moore machine the outputs do not depend directly on the inputs. Hence, in the state table of a Moore machine there is a single output column for each present state, independent of the input. An example will illustrate some of the features.

EXAMPLE 8

A synchronous state machine is to be designed to serve as an odd-parity checker. The inputs to be checked for odd parity arrive on an input line *x*, but the parity is checked only while the signal on another input line *y* (a synchronizing input) is 1. An output *z*, depending only on the state, is to become 1 when the parity fails to be odd. A possible sequence of inputs and output are

y: 0 0 0 1 1 1 1 1 1 1 1 1 1 1 0 0
x: × × × 1 0 1 0 1 1 1 0 1 0 0 × ×
z: - - - 0 0 0 0 1 1 1 1 1 0 - - -

Since the output is to depend only on the state, parity is determined from a memory of just the last two input *x* bits. This suggests two flip-flops, let's say *D* flip-flops. A register consisting of two *D* flip-flops, labeled 1 and 2, is shown in Figure 21*a*.[12] Note that, at any given clock tick, Q_1 will have whatever value *x* had on the preceding clock tick. Similarly, Q_2 will have whatever value Q_1 had at the preceding

---

[12]Registers were introduced in Chapter 5; review if necessary.

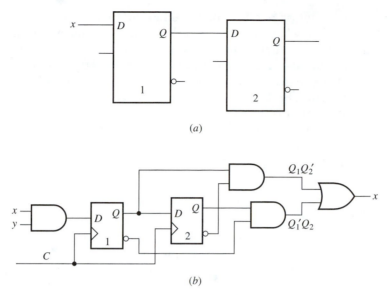

Figure 21 Moore circuit realizing Example 8.

clock tick. Using this information, construct three-variable logic maps (x, Q_1, and Q_2) for Q_1^+ and Q_2^+ and confirm that $Q_1^+ = x$ and $Q_2^+ = Q_1$. Since the output depends only on the state, a logic map for the output will be just a two-variable map (Q_1 and Q_2). Draw this map and determine an expression for the output.
Answer[13]

The requirement that the process will proceed only when $y = 1$ can be met by an AND gate with inputs x and y. The input to the first flip-flop will remain 0 when $y = 0$, but it will be x when $y = 1$. The full circuit, including the output decoder, is shown in Figure 21b. (Please verify.)   ∎

What was done in this example seems almost like cheating: no state diagram was drawn, no state tables created, no assignments made, and so on. We have seen that two flip-flops will do the trick and that, hence, there is a maximum of four states. As an exercise, construct a state table using the assignments (00, 01, 10, and 11) as state names and construct a separate column for the output, independent of the input. Using this table, confirm the logic maps you previously constructed.

4   STATE EQUIVALENCE AND MACHINE MINIMIZATION

Exercise 1 in section 1 (and also Example 7) demonstrated that more than one state diagram (or state table) can be constructed to satisfy the description of a design problem. If one such table has nine states and another has four, the first requires twice as many flip-flops in its implementation as the second. But no

---

[13]$z = Q_1 Q_2' + Q_1' Q_2 = Q_1 \oplus Q_2$

more information needs to be stored in the first machine than in the second. It must be that whatever task is performed by some of the nine states is performed also by other states, rendering some of the nine states superfluous.

It would be of great benefit to detect these superfluous states and remove them, thus leaving a reduced state table. We have already mentioned the possible reduction in the complexity of the circuit. Once a reduced machine is implemented, what is possibly of even more value is that the reduced complexity makes verification (the experimental determination of faults in a machine) considerably simpler. The purpose of this section is to develop procedures for reducing a given state table to one that carries out the same function with fewer states.

Distinguishability and Equivalence

A finite-state machine operates by receiving a sequence of input symbols, making transitions of state, and emitting output symbols. The input sequence becomes transformed into the output sequence. Look back at the state table in Figure 11 and suppose the circuit is in state B when an input 1 occurs. A transition is made to state D. We describe this by saying that state D *succeeds* state B under an input 1, or that D is the *1-successor* of B. Now suppose that the longer input string 011 arrives when the machine is initially in state A. The final state reached will be E. (Trace out the sequence of states encountered and verify.) So E might be called the 011-successor of A. In general,

> *If input sequence X is applied to a finite-state machine that is in state S_i and the machine makes a final transition to S_j, then S_j is the X-successor of S_i.*

If the same input sequence X is applied to a machine twice, once when the machine is in state S_i and once when it is in S_j, the two output sequences produced may or may not be the same. If they are the same, we will not be able to distinguish the two initial states by means of that particular input sequence. Now suppose that the output sequence is always the same, starting from each of the two states, no matter what input sequence is applied. Clearly, the two states could never be distinguished from each other. This leads to the following definition:

> *Two states S_i and S_j in a finite-state machine are* equivalent *if the same output sequence is produced in response to an input sequence, starting in either state, and this is true for every finite input sequence.*

If the potential equivalence of two states were to be checked using this definition, a whole career would be needed to check all possible input sequences. Clearly, a shorter test is needed. Suppose that, when the machine is started in two different states, the output sequences produced by the same input sequence are *not* the same. The two states can then be distinguished. We make the following statement:

> *Two states S_i and S_j of a machine are* distinguishable *if and only if there exists at least one finite input sequence that produces different output sequences starting first from state S_i and then from S_j. If the distinguishing sequence has length k, then the two states are k-distinguishable.*

PS	NS,z	
	$x = 0$	$x = 1$
A	B,0	D,1
B	C,1	D,1
C	C,0	A,0
D	A,0	C,1

(a)

PS	NS,z	
	$x = 0$	$x = 1$
A	B,0	D,1
B	C,1	D,1
C	E,0	A,1
D	A,0	C,1
E	B,0	D,1

(b)

Figure 22 Example machines.

To illustrate, consider states A and D in the state table given in Figure 22*a* as initial states. The outputs are the same for either input of length 1 (0 or 1). Hence, states A and D are not 1-distinguishable. Now take an input 00, of length 2. Starting from A, the output is 01, but starting from D, it is 00. Hence, states A and D are 2-distinguishable. This leads to the following definition:

Two states that are not k-distinguishable are k-equivalent.

The previous definition for equivalence can now be expressed as follows:

Two states of a machine are equivalent if they are k-equivalent for all k.

In the preceding discussion we have concentrated on output sequences in response to an input sequence. No attention has been paid to the next states that result along the way. In Figure 22*a*, for example, we found states A and D to be 1-equivalent; then we tested an input string of length 2. Suppose, instead, that we consider the next states after the first input. The next states are not the same starting from states A and D. Furthermore, it is evident that, using those next states as present states, the same output does not result for each input. Hence, the original states A and D are distinguishable.

Let's pursue this line, using Figure 22*b*, which is almost the same table as the one in Figure 22*a*. Now states A, D, and E are all 1-equivalent. Furthermore, the next states from A and E are the same for each input. Hence, after the first input bit, the transition from each of states A and E will be to the same next state; the outputs thereafter will be exactly the same. Hence, A and E are equivalent.

Machine Minimization

The preceding discussion gives a clue as to how to find the states of a machine that are equivalent to each other. Starting from an input sequence of length 1, we group those states that are 1-equivalent. These states are distinguishable from the others. Next we examine the next states to decide on their distinguishability, and so on. The details of the process are best described with an example.

EXAMPLE 9

A state table is given in Figure 23*a*. The objective is to find all groups of equivalent states and to reduce the table to one having a minimal number of states.

	NS,z	
PS	$x = 0$	$x = 1$
A	D,1	G,1
B	C,0	D,1
C	E,0	F,1
D	F,1	B,1
E	B,0	F,1
F	D,1	C,1
G	A,0	D,1

(a)

$P_1 = \{ADF; BCEG\}$

$P_2 = \{ADF; BCE; G\}$

$P_3 = \{A; DF; BCE; G\}$

$P_4 = P_3$

(b)

	NS,z	
PS	$x = 0$	$x = 1$
A $\rightarrow S_1$	S_3,1	S_4,1
BCE $\rightarrow S_2$	S_2,0	S_3,1
DF $\rightarrow S_3$	S_3,1	S_2,1
G $\rightarrow S_4$	S_1,0	S_3,1

(c)

Figure 23 Partitioning and machine minimization.

In accordance with the plan, we start by identifying those states that are distinguishable with an input sequence of length 1. We find that the group of states A, D, F have the same output for $x = 0$; they also have the same output for $x = 1$. Hence, they are not distinguishable with an input sequence of length 1. Similarly, confirm that the states B, C, E, G are indistinguishable with an input sequence of length 1. But these two groups of states are distinguishable from each other. Thus, the totality of all the states in the table can be *partitioned* into two blocks of states, written as follows: $P_1 = \{ADF; BCEG\}$. Within each block, the states are indistinguishable with an input sequence of length 1, but those in one block are distinguishable from those in the other.

Next we examine the successor states from all states in each block, one at a time. If, for each input symbol, the next states from all states in a block are not in the same block but fall in two distinct blocks, then the two sub-blocks are distinguishable. Hence, the original block must be subdivided. Thus, for $x = 1$, the next states from the block BCEG are DFFD; these are all in the block ADF. However, for $x = 0$, the next states from the block BCEG are CEBA; all except the next state from state G are in the same block. Hence, block BCEG must be subdivided into two blocks, BCE and G. The resulting partition is $P_2 = \{ADF; BCE; G\}$, as shown in Figure 23b; it is a *refined* version of P_1.

The next states from any one block in partition P_2 were found to be in the same block; hence, these next states will have the same outputs for each input bit. (That's because the outputs were the same, even for the larger blocks in partition P_1.) These next states in each block are, then, 1-equivalent. Hence, their predecessor states are 2-equivalent.

The process is now repeated with partition P_2. Again we take each block one at a time and, for each input, examine their next states to see if they fall in the same new block. (Clearly, blocks containing a single state need not be examined.) For each input, the next states from the block BCE fall in the same block. This is also true for $x = 0$ for block ADF; however, for $x = 1$, the next states for block ADF are GBC. These next states are not in the same block; hence, a further refinement of P_2 is needed, as shown in Figure 23b. The states in each new block are 3-equivalent.

The process must be repeated on the multistate blocks in partition P_3. Go through the process and confirm that no further refinements of the partition are needed. The states within each block cannot be distinguished; hence, they are equivalent. This final partition is called the *equivalence partition*. Each of the four blocks in the equivalence partition constitutes a state of the machine to which each of the original states in that block is equivalent. If these states are labeled S_i, a reduced state table can be constructed, as shown in Figure 23c. ■

The description of this process is far lengthier than the actual effort involved in carrying it out.[14] Note, in this example, that the reduced machine needs just two flip-flops in its implementation, whereas the original table required three.

The subject of this section constitutes the second step in the general design procedure described in the preceding section. As noted earlier, when you initially construct a state table to satisfy the specifications of a design problem, it isn't necessary to spend a lengthy amount of time to ensure that there are no redundant states. Any redundant states introduced earlier can always be removed in the machine minimization step.[15]

5 MACHINES WITH FINITE MEMORY SPANS[16]

What distinguishes sequential circuits from combinational circuits is memory. The information stored in memory, together with an input sequence, determines an output sequence. But how much past data is it necessary for the machine to remember? Is it necessary for the machine to remember past *inputs only,* or can its future behavior be determined from the present input and a memory of past *outputs*? Or is a memory of both inputs *and* outputs necessary?

We will examine three classes of machines in this section. In the three respective cases, the present output is determined by the present input plus

1. A limited number of the immediately preceding inputs
2. A limited number of the immediately preceding outputs
3. A limited number of the immediately preceding inputs *and outputs*

The machines are all said to have *finite memory spans.*

Not all finite-state machines have this characteristic; a machine with a waiting state, for example, does not. It cannot produce an output while in this state, no matter how long it stays there, even if it receives an acceptable input sequence. Each of the three classes of finite-memory machines will be discussed and implemented in certain specific structures.

---

[14]The process is algorithmic in nature. That means software can be produced to carry it out. Here we will concentrate on the principles.

[15]The treatment here has been limited to completely specified machines. Incompletely specified machines require the introduction of several additional concepts that require extensive development. These concepts apply also to asynchronous machines, and they will be treated in Chapter 7. If you skip Chapter 7, you will also be skipping coverage of incompletely specified synchronous machines.

[16]This section can be omitted without penalty in terms of preparation for material that follows.

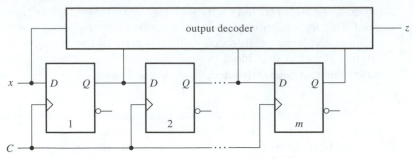

Figure 24 Canonic implementation of a finite-input-memory machine.

Machines with Finite Input Memory

One possible formal definition of the machines to be discussed in this section is the following.

> *A finite-state machine M is said to have* finite input memory *of memory span (or order) m if the present state of M can be determined uniquely from the preceding m input symbols but no fewer than m.*

It is clear from the definition that a circuit implementing the memory is an *m*-flip-flop shift register in which the last *m* inputs are stored. The present state consists of the outputs of the *m* flip-flops in the register. A *canonic* implementation consists of an *m*-flip-flop shift register and a combinational output decoder, as shown in Figure 24.

The shift register in Figure 24 is a serial-to-parallel converter. The input information arrives sequentially and is stored in the shift register. When the last bit of an input sequence of appropriate length arrives, both that input and the previously stored information are applied to the combinational logic at the same time, in parallel; the logic of the decoder then produces the desired output.

Since the state of a finite-input-memory machine of span *m* after an *m*-bit input sequence is known, the output will become known when the next bit arrives. Hence, such a machine can also be defined as one whose output is determined by the present input and the preceding input sequence of *m* bits.

The design procedure for a machine with finite memory span can be simplified if the specifications of the design problem allow us to recognize its nature; only the output decoder needs to be designed.

EXAMPLE 10

A synchronous sequential circuit with a single input line x and a single output line z is to be designed so as to produce an output $z = 1$ whenever an input symbol completes a sequence of 4 identical input bits; the output is to be 0 otherwise.

Just prior to receipt of each input symbol, the machine must remember only the preceding 3 bits. It then produces a 1 or a 0 on the basis of those 3 bits and the present input bit. This, then, is a machine having finite input memory of span 3. Labeling

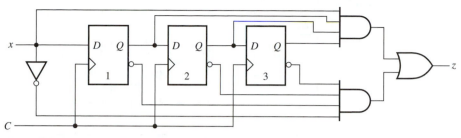

Figure 25 Canonic implementation of Example 10.

the states Q with subscripts $1, 2,$ and $3,$ from left to right, the minterms are 0000 and 1111. Hence, the output is easily written as

$$z = xQ_1Q_2Q_3 + x'Q_1'Q_2'Q_3'$$

The canonic implementation of the circuit is shown in Figure 25.   ■

Although the preceding example included just a single input line, the concept and the canonic implementation apply to any number of input lines. A separate shift register is needed for each input line.

In addition to the main inputs, some machines have one or more control inputs in order to change the instructions for the generation of an output. It is the present values of those inputs that do the controlling; past control inputs need not be stored. To illustrate, in the preceding example there might be a control input x_c that, under the previously given conditions of the problem, permits the output to become 1 only if $x_c = 1.$ For $x_c = 0$ the output could be specified as something else in terms of the main input x and its past values. In the implementation of the machine, the shift register in Figure 25 would not change; only the output decoder logic would be modified.

Machines with Finite Output Memory

In the second class of machines being considered, it is the preceding *outputs* that are to be remembered rather than the inputs. The definition of this class of machines is as follows:

> *A sequential machine M is said to have* finite output memory *of memory span (or order) m if the present output of M can be determined from the present input and the immediately preceding m (but no fewer) output symbols.*

Again the definition makes it clear that the memory can be implemented with a shift register (left shift) of m flip-flops in which the preceding m *outputs* are stored. Then, when the next input symbol arrives, the output is determined by both this input and the previously stored outputs. The canonical implementation is shown in Figure 26. The input to the shift register is the most recent output symbol. It is also clear that, if there is more than one output line, the canonic implementation will require more than one shift register.

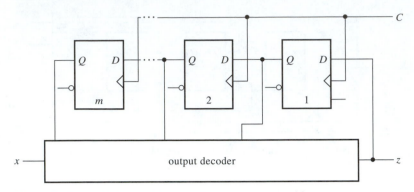

Figure 26 Canonic form of finite-output-memory machine.

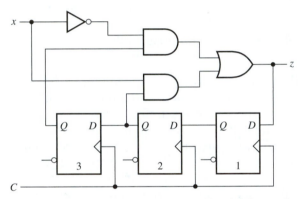

Figure 27 Finite-output-memory machine implementing Example 11.

EXAMPLE 11

A single-input, single-output sequential machine is to be designed. For an input $x = 1$, the output is to equal what the output was two clock periods earlier; and for $x = 0$, the output should equal what the output was three clock periods earlier.

The most that the machine must remember is three preceding output symbols. Hence, the desired machine has finite output memory of span 3. An expression for the output is easily written. If we label the states Q_1 to Q_3, then when $x = 1$, the output should be Q_2; and when $x = 0$ (which means $x' = 1$), the output should be Q_3. Hence,

$$z = xQ_2 + x'Q_3$$

(Verify the values of z when $x = 1$ and when $x = 0$.) The implementation is shown in Figure 27. ∎

Exercise 8 From the problem statement in Example 11, draw a logic map for the output, with the input and the three states as map variables. From the map, write a minimal sum-of-products expression and confirm the expression in the example. (This implementation does not depend on recognition of the nature of the circuit as finite-output memory.) ◆

Finite-Memory Machines

The third class of finite-state machines under discussion depends not just on a fixed number of past inputs or on a fixed number of past outputs, but on both. The formal definition follows:

> *A sequential machine M is a* finite-memory machine *of span m if the present state can be determined uniquely from the preceding m_i (but no fewer) input symbols and the preceding m_o (but no fewer) output symbols; the span is m =* max $\{m_i, m_o\}$.

You might conjecture that a canonic implementation of this machine would merge the two canonic implementations in Figures 24 and 26 by including two shift registers, one to store the m_i past inputs and one to store the m_o past outputs. While such an implementation is possible in some cases, it turns out that it is not universally possible.

Consider a finite-input-memory machine. From the definition just given, this machine is also a finite-memory machine. That is, if the present state is determined by the first m_i input symbols only, knowing also the first m_o output symbols will not detract from this. This finite-memory machine, however, is not a finite-output-memory machine; it is a finite-memory machine by virtue of having finite input memory. A similar argument can be made that a machine with finite output memory is, by virtue of this fact, a finite-memory machine, although not a finite-input-memory machine.

The converse is not true. That is, it is possible for a machine to be a finite-memory machine without having either finite input memory or finite output memory. To establish the validity of this claim requires the introduction of several additional concepts and algorithms that would take us too far afield. We will therefore abandon further consideration of the subject here but will provide some problems at the end of the chapter so you can explore it to some extent.

6 SYNCHRONOUS COUNTERS

We now turn to a class of sequential machines that perform a particular type of operation. A *counter* is a sequential machine that, starting at a particular state, cycles through a fixed sequence of states and then returns to its initial state; thereafter, it repeats this process. The number of distinct states in the counter is called its *modulo number.*

In some cases, the useful information from the counter may be simply the state it happens to be in. In this case, there is no output decoder circuit and no other output lines but its flip-flop outputs. In other cases, an output other than the state may be required. In synchronous counters, the signal that excites the counter is very often the clock. At other times, other inputs (called *control inputs*) are also provided. (It is also possible for counters to be asynchronous; such counters will be considered in Chapter 7.)

Counters can be used for a number of purposes. A common purpose is to extend the time scale—that is, to introduce delay in the inevitable march of the clock. This is done by producing an output (or control) signal for each *k* periods of the clock; this signal, rather than the clock, then controls the timing of a subsequent operation. Another purpose of a counter is to produce sequential words in some specific code. Of course, just plain counting is an important purpose — for example, counting how many times some process has been carried out. It would be very useful if

- The counter were cleared (set to 0) when first turned on or after it has gone through its count, or
- The counter were set at some specific value

This is done by external CLEAR and RESET inputs.

Single-Mode Counters

A counter is said to be *single mode* if the only external input is the clock and the only outputs are the states — the flip-flop outputs. The counter is described by specifying

- Its modulo number
- The code assigned to the states

The number of flip-flops needed in a counter is implicit in its modulo number. Thus, a modulo-k counter will require $\lceil \log_2 k \rceil$ flip-flops. Modulo-6 or modulo-8 counters, counting in binary or Gray code, will have three flip-flops; in common terminology we call them "3-bit counters."

If the code in which the counting takes place is specified, there is no point in giving the states arbitrary names (such as letters of the alphabet) and then later making a state assignment. Rather, each successive state is given an assignment on the basis of the code being used. It is convenient to assign the starting state of the counter the code word 00...0, unless there is some reason not to do so. (Return to Chapter 1 to review the subject of codes.) As a matter of fact, the state assignment problem, so important in the machines considered so far, disappears in a counter. The codes representing the states are specified beforehand.

The number of code words is fixed by the modulo number. Several codes are shown in Figure 28 for modulo numbers 6 and 8. Starting at code word 00...0, the counter is to cycle through each code word and return to 00...0 after the last word in each code.

Unit-Distance Counters

After the modulo number of a counter has been specified, the next question that comes up is, What code should be used in making the successive state assignments? Perhaps the simplest answer is: binary. The disadvantage of this code is that more than one bit value changes in a clock period. Thus, in going from 001 to 010, both the second and the third bits (counting left to right) must change value. This means that more than one flip-flop output must change simultaneously. If it should happen that one change occurs even slightly before the other(s), there may be a momentary transition to the wrong state — or even to an invalid state, one that is not among the states of the counter. For this reason, a code whose *distance* is 1 is preferable.[17]

---

[17]The distance between two code words is the number of bits that must be complemented in one word to transform it to the other word. A code in which the distance between each pair of consecutive code words is k is a *distance-k* code. See Chapter 1 for a discussion of codes.

Binary	Unit Distance	Creeping	One-hot
000	000	000	00000
001	001	100	10000
010	011	110	01000
011	010	111	00100
100	110	011	00010
101	100	001	00001

(a)

Binary	Gray	Creeping	One-hot
000	000	0000	0000000
001	001	1000	1000000
010	011	1100	0100000
011	010	1110	0010000
100	110	1111	0001000
101	111	0111	0000100
110	101	0011	0000010
111	100	0001	0000001

(b)

Figure 28 Codes for use in modulo-6 and modulo-8 counters. (*a*) Modulo-6. (*b*) Modulo-8.

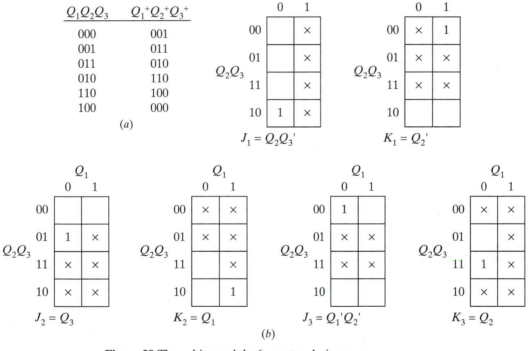

Figure 29 Three-bit, modulo-6 counter design.

The codes in the second and third columns in Figure 28*a* are unit-distance codes. Suppose the code in the second column is selected. Since there are no inputs (besides the clock) and no outputs, there will be no input/output information in the state diagram. Drawing the state diagram will be left to you (do it now). Since there is no output decoder, the only other hardware in the circuit besides the three flip-flops is the state decoder. Since the states are identified by their assignment in accordance with the code, the resulting state table is the transition table shown in Figure 29*a*. Assuming *JK* flip-flops, we use the excitation requirements tables in Chapter 5, Figure 17 to construct the logic maps for the *J* and *K* excitations. This is done line by line in the excitation table for each column. (These tables appear on the inside

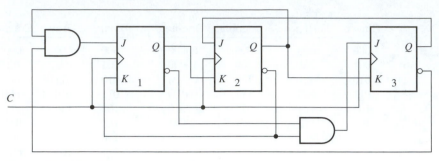

Figure 30 *JK* flip-flop implementation of modulo-6 counter.

cover of the book. You could make yourself a copy and have it handy so you can consult it without having to hunt in the book each time you want to use the tables.)

From the excitation requirements in Figure 17, Chapter 5, $J = 1$ only for the transition from 0 to 1. In the transition table of this example (for flip-flop 1), this occurs only when $Q_1Q_2Q_3 = 010$. Hence, a 1 is entered in the corresponding cell in the logic map for J_1 in Figure 29*b*. Also, J_1 is a don't-care for a present state $Q_1 = 1$, independent of the other states. In addition, all J's and K's are don't-cares for the present states that never occur (110, 111). All this confirms the logic map for J_1 in Figure 29*b*.

Exercise 9   Use the same approach to construct the logic maps for the other J's and K's. Confirm your results using Figure 29*b*.   ◆

Exercise 10   From the maps in Figure 29*b*, construct the combinational hardware of the state decoder. Confirm your circuit with the implementation in Figure 30.   ◆

Ring Counters

Another unit-distance code in Figure 28*a* is the creeping code. A counter designed to count in this code is called a *ring counter*. The state table, which is also the transition table (since the states are identified by their assignments), is shown in Figure 31*a*. This time, let's assume that *D* flip-flops are to be used. Then the excitations are the next states.

Exercise 11   Draw logic maps for the *D* inputs from the transition table. Confirm using Figure 31*b*, but only *after* drawing your own!   ◆

It is clear that the output of one flip-flop is the excitation to the next one, except that the output of the last flip-flop (the tail) is complemented before becoming the input to the first flip-flop. (For this reason it is called a *twisted-tail counter* or some colorful variation of this.) Hence, this counter is nothing but a serial shift register with its complemented output fed into its input. For practice, draw the implementation circuit.

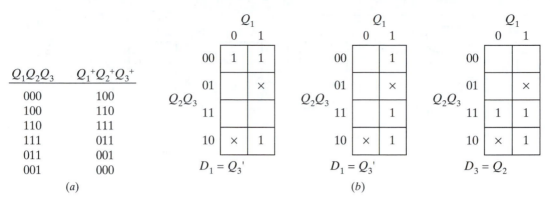

$Q_1Q_2Q_3$	$Q_1^+Q_2^+Q_3^+$
000	100
100	110
110	111
111	011
011	001
001	000

(a)

(b)

Figure 31 Transition table and excitation maps for twisted-tail ring counter.

Hang-up States

Ring counters have a problem that is not evident from Figure 31. What would happen if, for some reason, the counter enters one of the unused states (101 or 010; see Figure 28a)? This could happen when the power is first turned on, for example, or as a result of noise in the circuit. Nothing significant would happen if the next state were not an unused state. In this case, the count would resume on the next clock pulse. But, on the other hand, if each next state for subsequent clock pulses is an unused state, the counter will *hang up*. A state in which a counter hangs up so that the count cannot proceed is called (surprise!) a *hang-up state*.

Well, what is the situation in Figure 31? Because of the excitation equations, if the present state happens to be the unused state $Q_1Q_2Q_3 = 010$, then the next state will be $Q_1^+Q_2^+Q_3^+ = D_1D_2D_3 = Q_3'Q_1Q_2 = 0'01 = 101$; this is the other unused state. (By going through a similar process, show that the next state after that will be 010 again.) Once the circuit enters one of the two unused states, it hangs up and cycles between them; it will never return to the counting sequence. Hence, this counter would be defective and worthless.

Since the problem is caused by the excitation equations, resulting in a sequencing between the unused states, it can be solved by disrupting the excitation equations. Not all the equations need be modified. If we would like to retain the use of a shift register, we should concentrate on the excitation equation of only the first flip-flop. Considering the map for D_1, the hang-up problem arose because we used as a 1 the don't-care in the 010 position to form a 2-cube. Instead, let's reassign the value and take it as a 0. Then, the expression for D_1 becomes

$$D_1 = Q_2'Q_3' + Q_1Q_3'$$

For the unused present state $Q_1Q_2Q_3 = 010$, the next state of Q_1 will be $Q_1^+ = D_1 = 0$, and so the next state will be 001. We have escaped the hang-up state!

Exercise 12 Suppose the present state is the other unused state, 101. Using the new expression for D_1 and the old ones for D_2 and D_3, determine the next state and discuss whether the hang-up state is escaped. ◆

The ring counter design incorporating the preceding change is shown in Figure 32. It is a *self-correcting* design in that the hang-up states have been avoided. The

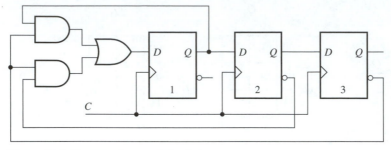

Figure 32 Self-correcting modulo-6 ring counter.

worst-case cost of this design is a two-clock-pulse delay in the count if the counter inadvertently enters one of the two hang-up states. Although individual flip-flops have been shown, an MSI shift register is appropriate for the implementation.

Multimode Counters

A counter is called a *multimode* counter if it has, in addition to the clock, external inputs and, possibly, external outputs besides the state outputs. It is "multimode" because the counting sequence might depend not only on the clock but also on some other control signals. Likewise, special output lines besides the flip-flop outputs may be provided.

Such a counter might be used, for example, in a system in which a number of consecutive operations are to be performed. Only when one operation is completed is the next one to start. So a control signal indicating the end of a particular operation will increase the count, thus causing a transition to the state in which the counter should remain while the next operation is performed. Completion of this next operation again generates a signal that increases the count. When the last operation in the system is completed, the state should revert to the initial, or reset, state. Along the way, while a specific operation is being performed and the machine is in some particular state, the occurrence of some circumstance before the operation is completed may require returning to an *earlier* count rather than advancing. The number of operations to be performed will determine the counter's modulo number.

Modulo-6 Up-Down Counter

We will now consider an example of a multimode counter. Besides the clock, a synchronous counter is to have one input line, x. The count is to increase by one when $x = 0$ and to decrease by one when $x = 1$. Assume that six operations are to be controlled, so the modulo number is 6. Suppose, also, that the creeping code is to be used. In this example, although there is an input besides the clock, there are no other outputs besides the flip-flop outputs.

Note from the creeping code in Figure 28*a* that, from any count (state), the count advances following $x = 0$ and regresses following $x = 1$. Draw the state (transition) diagram before peeking at Figure 33*a*; then confirm it. The transi-

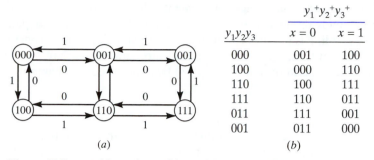

	$y_1^+y_2^+y_3^+$	
$y_1y_2y_3$	$x = 0$	$x = 1$
000	001	100
100	000	110
110	100	111
111	110	011
011	111	001
001	011	000

(a)   (b)

Figure 33 State table and transition table for a modulo-6 up-down counter.

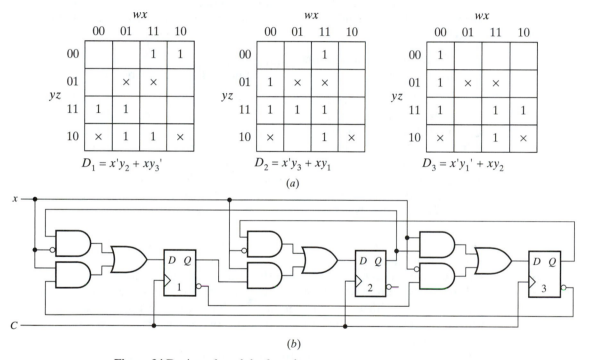

(a)

(b)

Figure 34 Design of modulo-6 up-down counter.

tion table is easily constructed from the transition diagram; do it yourself before confirming it in Figure 33b.

Assuming the use of D flip-flops, the excitations are the next states. The next step is to construct the logic maps for the excitations from the transition table; do this before you peek at Figure 34a. The final step is the implementation. Carry this out and then confirm it using Figure 34b.

7 ALGORITHMIC STATE MACHINES

Two tools were used early in this chapter for describing a sequential machine: state diagrams and state tables. Both of these are useful. In this section we will describe another tool of great utility. Review Example 1 (a single-input, single-output circuit) and

Figure 2, where state diagrams were introduced. Starting at any state, the arrival of an input sends the machine to one of the possible next states, depending on the input. This wording is reminiscent of the condition block in a flow chart representing an algorithm. Indeed, a diagram resembling a program flow chart can be created that contains the same information conveyed by either the state diagram or the state table.

Basic Principles

Review the statement of, and state diagram for, Example 1 early in the chapter. With the machine in any specific state, the arrival of an input bit requires a decision concerning the output to be emitted and the state to which the machine is to be directed. In the state diagram of Figure 2, the condition is shown as a directed line leaving each circle representing a state. A flow chart must include analogues of

- Circles representing states
- Directed lines leading to the next states with specified outputs

Remember that a sequential circuit is referred to as a *state machine* (short for finite-state machine). Since the problem statement for the design of a state machine is like an algorithm, the machine is also called an *algorithmic state machine,* or ASM. A flow chart, called an *ASM chart,* can be constructed that describes the operation of an algorithmic state machine. A flow chart describing *any* algorithm includes a condition box where a decision has to be made; this is the familiar diamond shape shown in Figure 35*a* or the variation of it in Figure 35*b*. It is called the *decision box*. The lines leading out from a decision box are the *exit paths*.

In an ASM flow chart, the circle enclosing a state in a state diagram is replaced by a rectangle called the *state box,* as shown in Figure 32*c*. Instead of writing the state name inside, as in the state diagram, both the state name and its binary code are written above the rectangle. (In a Moore machine, outputs are associated with each state; hence, in that case, the appropriate output is written inside the state box.) Note that the two boxes described so far exist in all ASM charts.

In Mealy machines the output depends on both the input and current state. In this case another box is used in the chart, called a *conditional output box*. To distinguish it from a state box, an oval shape is used, as shown in Figure 32*d*.

In a state machine, a tick of the clock initiates action, whether a new input is present or not. Starting at each state, decisions must be made as to the state transitions and the output. A basic unit in an ASM chart can be thought of as consisting of a single state box and all other decision and conditional boxes whose exit path leads to another state. Such a unit is called an ASM *block*. So an ASM chart is simply an interconnection of such ASM blocks. Each state to which a transition is made is the beginning point of another block. For clarity, it often helps to encircle the individual blocks (using dashed lines), but doing so adds nothing to the chart. The important entities are the state, condition, and conditional output boxes; whether or not these combinations of boxes are encircled to delineate an ASM block is secondary.

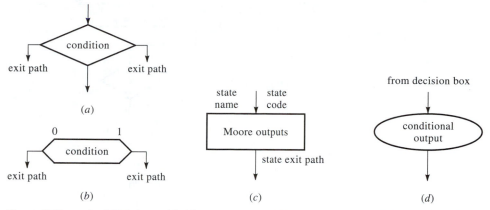

Figure 35 Boxes in ASM charts. (*a*), (*b*) Decision box. (*c*) State box. (*d*) Conditional output box.

EXAMPLE 12

An ASM chart is to be constructed for an automatic garage-door opener. Let x be the input signal resulting from a sensor actuated by a physical switch; the output is z. The value of x toggles when the switch is activated. The output signal z controls the mechanism that opens and closes the garage door; z also toggles. A low-frequency clock synchronizes the system. Assuming the garage door is closed, when the switch is activated the input becomes 1 and the machine changes state at the next clock edge. The next time the switch is activated, the input signal goes to 0. In response to this input the output goes to 0 at the next clock edge. We seek an ASM chart to describe this simple system.

The machine has two states: open and closed. An ASM block starting from the closed state is shown in Figure 36*a*. It does not indicate how this state is reached. After the "open" input signal is received, the output 1 is emitted. That means the garage door should open, so the next state is the "open" state. The next time the switch is actuated, the input toggles ($x = 0$), and so does the output ($z = 0$). The "closed" state is reached when $x = 0$ while in the open state. Figure 36*b* shows the completed chart.   ■

The utility of ASM charts is not evident from this simple example. To get on with something more substantial, let's return to Example 1 at the very beginning of this chapter, describing the sequence detector.

EXAMPLE 13

In Example 1, the output z is to become 1 only after receipt of the input sequence ...0110, independent of the sequence that precedes it. (Look over that example before going on.) There are two possibilities: Either the most recent 0 bit in the preceding sequence forms the first bit of an acceptable sequence (the

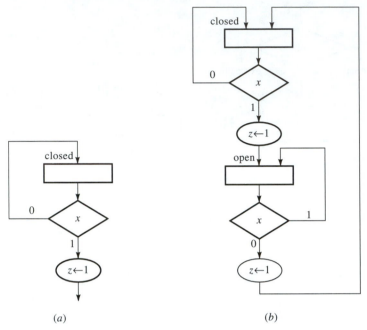

Figure 36 ASM chart for a garage-door opener.

overlapping possibility), or the machine returns to a reset state and we start over. Obviously the design will be different in the two cases. Example 1 dealt with the overlapping case.

Let state S_1 be the state arrived at by an $x = 0$ input. The first two blocks are shown in the partial ASM chart in Figure 37a. If the input is 0 while the machine is in either of the first two states, transition is back to the initial state, as shown. The completed ASM chart is shown in Figure 37b. While in state S_3, receipt of an input $x = 1$ will spoil the acceptable sequence and will send the machine to a state where it will stay for each consecutive $x = 1$. It escapes from this state to the initial state with an input $x = 0$. However, if $x = 0$ while the machine is in state S_3, it will emit an output of 1 and will also return to S_1. ■

EXAMPLE 14

ASM charts are particularly useful when there are a large number of inputs (state diagrams are unreasonable to use with three or more inputs). Consider the design of a traffic light controller for an intersection that is normally busy during the day and not very busy at night. There is a preferred direction to maintain the light green, but during the day the intersection is busy enough that the car detection sensors are not used. During the day the traffic light moves through a fixed sequence of green, yellow, and red for the same duration in both directions. During the night the signal for the preferred direction stays green unless a car is detected in the nonpreferred direction. The preferred direction is north/south.

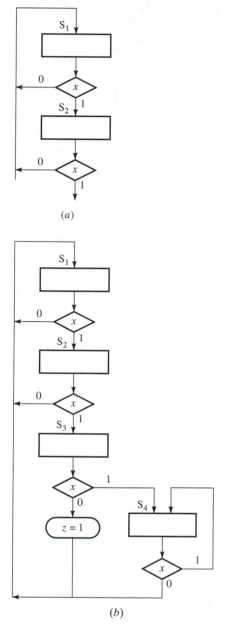

Figure 37 ASM chart for Example 13.

The traffic light controller has four inputs: one sensor input to detect the presence of a car on the nonpreferred route, one time-of-day input to signal day or night, and two timing inputs to determine the minimum duration of a green signal in a given direction and the duration of the yellow signal. The controller has six outputs, one for each color signal (green, yellow, and red) in each direction (north/south and east/west). The controller can be thought of as running two algorithms depending on the time of day, so the time-of-day input determines which algorithm is executed. The only difference between operation during the day and night is that the car presence sensor is used during the night to decide whether or not to maintain the light green

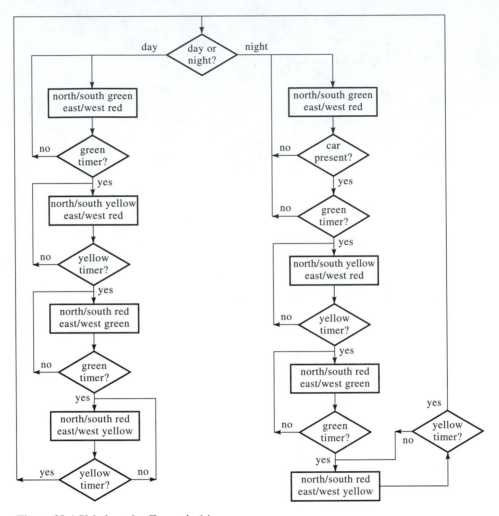

Figure 38 ASM chart for Example 14.

in the preferred direction. The ASM chart describing the system is shown in Figure 38. Before you look it over, try creating one yourself and confirm your version by comparison. ∎

8 ASYNCHRONOUS INPUTS

Recall from Chapter 5 that circuit implementations of state machines must satisfy the setup (t_{su}) and hold (t_h) time requirements of flip-flops. Excitation signals must be stable for a time period t_{su} before the clock transition and must be held for a time period t_h after the clock transition. Signals generated within the state machine are not a concern, since the delay of a flip-flop is typically longer than the hold time, and the clock-cycle time can be increased to provide sufficient margin to ensure that the setup time is honored. If a faster cycle time is

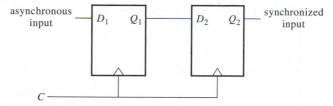

Figure 39 Synchronizing asynchronous inputs to a state machine.

required, then the next state and output decoders must be redesigned to decrease the delay.

Signals produced outside the state machine may or may not be synchronized by the same clock. They may be signals from a sensor, switch, keyboard, or communications channel. These signals switch at arbitrary times, so there is no guarantee that they will honor the setup and hold time requirements of the flip-flops. When these flip-flop requirements are violated, the state transition is unpredictable. The flip-flop can enter an undefined state (a voltage level between low and high) for a short period of time before switching (unpredictably) to a defined state (0 or 1). The undefined state is called a *metastable state*. A typical flip-flop remains in the metastable state for a short period of time, but the stable state it reaches after metastability cannot be predicted. This can cause a state machine to make an erroneous state transition.

To decrease the probability that an asynchronous input will cause an erroneous state transition, adding an extra flip-flop, as shown in Figure 39, can synchronize the signals. In this circuit the first flip-flop can enter metastability, but it will very likely reach a stable state before the next clock event. Thus, the probability that the output of the second flip-flop has a valid state is very high (much higher than if only one flip-flop were used to synchronize the signal).

Asynchronous Communication (Handshaking)

It is sometimes necessary for two synchronous sequential circuits with different clocks to communicate data to one another. The two systems could be two computers, a computer and an input or output device, or even a CPU and memory. This communication is asynchronous, so a simple protocol is required to ensure that the data is transmitted properly. The protocol used for asynchronous communication is commonly referred to as *handshaking*. Each of two independently clocked machines M_1 and M_2 that communicate with one another uses a sequence of signals to request data from, or send data to, the other machine.

Thus, M_1 and M_2 have control signals and data signals passing between them. The machine that initiates the communication (say, M_1) sends a request signal to the other machine, M_2; it also sends a control signal indicating the desire to read or write data to machine M_2. For a read request, M_2 responds with an acknowledge signal when the data is ready for M_1. For a write request, M_1 must make the data available before sending the request to M_2. M_1 holds that data on the data connection until M_2 responds with an acknowledge signal. Communication from

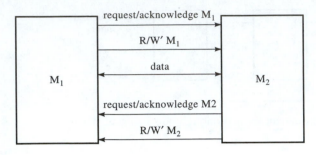

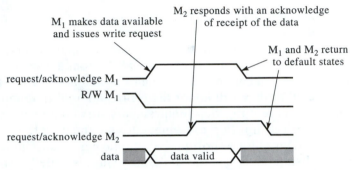

Figure 40 Connections required for asynchronous communication between two independently clocked state machines.

Figure 41 The timing diagram for a write operation from M_1 to M_2.

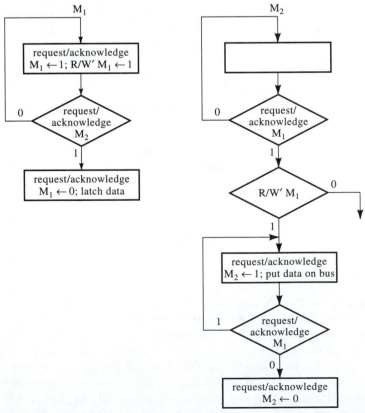

Figure 42 ASM charts for an asynchronous read request from machine M_1 to M_2.

M_2 to M_1 works in a similar manner. The connections required to implement this protocol are shown in Figure 40. The timing diagram for M_1 to write (or send) data to M_2 is shown in Figure 41.

Exercise 13   Draw the timing diagram for M_1 to read (or receive) data from M_2.   ◆

The sequence of control signals required for communication between machines M_1 and M_2 is produced by a sequence of state transitions within the machines themselves. ASM charts for a read request from M_1 to M_2 are shown in Figure 42.

CHAPTER SUMMARY AND REVIEW

This chapter introduced the design of sequential circuits of the synchronous variety. To describe how transitions from one state to another take place, we developed the tools of state diagrams, state tables, and algorithmic state machines. We introduced the class of sequential circuits called counters. Topics included the following:

- Mealy model and Moore model of a sequential machine
- State diagrams, translation of a performance specification into a state diagram, verifying the state diagram
- State table, construction of a state table from a state diagram or directly from a problem specification
- Difference between a reset state and an initial state
- Assignment of binary values to states
- Analysis of sequential circuits
- State transition and output tables
- Choice of flip-flop type in design
- Flip-flop excitation maps
- Implementation of sequential circuits from excitation maps
- Distinguishability and equivalence of states
- Minimization of machines
- Machines with finite memory spans:

  - Finite-input-memory machines
  - Finite-output-memory machines
  - Finite-memory machines

- Single-mode synchronous counters:

  - Unit-distance counters
  - Ring counters

- Hang-up states
- Multimode counters
- Algorithmic state machines
- ASM charts
- Condition box
- State box

- Conditional output
- ASM block
- Synchronizing machines with asynchronous inputs
- Asynchronous communications—handshaking

PROBLEMS

Many of the problems that follow require the design of a sequential machine. The design process involves several steps. As you study the chapter, you may wish to tackle early sections of each problem before you have studied everything necessary to complete the design. You can then return to each problem as you learn each successive step. Save the early parts of the solution as you go along.

1 A synchronous sequential circuit having a single input x and a single output z is to be designed. The output z is to become 1 upon completion of the input sequence 0101, whether it forms part of an overlapping string (such as 00010101) or not.

    a. Construct a state diagram and a state table.
    b. Construct an appropriate state assignment map.
    c. Assume the use of JK flip-flops and construct a transition table.
    d. Construct excitation and output maps.
    e. Draw the diagram of the resulting circuit.
    f. Repeat parts $c, d,$ and e but with D flip-flops. Compare the complexity of the circuits.

2 The output z of a single-input, single-output synchronous sequential circuit is to become 1 whenever the input sequence is either 1101 or 1001. (Overlapping sequences are to yield multiple outputs.) Carry out the six parts of the design specified in Problem 1.

3 Use the same conditions as Problem 2 except that the circuit is to return to a reset state upon emitting a 1 output. (What does that do to overlapping input sequences?)

4 Carry out all design parts specified in Problem 1 for each of the following specifications.

    a. The output of a single-input, single-output machine is to be $z = 1$ if the present input x is the XOR of its preceding two values.
    b. The output is $z = 0$ when consecutive input bits of 0 are of even length and consecutive input bits of 1 are of odd length. The output is to be $z = 1$ whenever there is a discrepancy in this pattern.
    c. The output becomes $z = 1$ whenever the input bit is the logical product of its previous two values.

5 The input sequence of a single-input, single-output sequential machine is made up of consecutive 4-bit words. Each word is an entity; words are not formed by overlapping strings. The output is to be 1 whenever the number of 1 bits in a word is odd. Carry out all parts of the design specified in Problem 1.

6 Repeat Problem 5 except that, besides having an odd number of 1 bits, the output becomes 1 only if the 4-bit word starts with a 1. Carry out all parts of the design specified in Problem 1.

7 Construct a state table for the parity-bit generator described in Example 2 directly from the problem description, without reference to the state diagram. Compare with Figure 4.

8 A parity-bit generator is to receive 4-bit coded messages followed by a blank space (a 0). A parity bit of 1 is to be generated and inserted into the blank space if and only if the parity (the number of 1's) of the preceding 4 bits is odd.

    a. Construct a state diagram and a state table for this parity-bit generator.
    b. Test your diagram to verify that, starting from the reset state, it gives the correct output for various 4-bit messages.
    c. Carry out the remaining parts of the design specified in Problem 1.

9 A sequential machine has been found to have three states, A, B, and C; there are, then, just two state variables. Specify three different assignments of 2-bit code words to the three states, such that any other assignment amounts to either interchanging the two state variables, inverting either variable, or both.

10 After assigning the combination 000 to state A in Example 4, the adjacency AG can be satisfied by assigning G to one of two other squares in the assignment map besides 001. Choose another square to satisfy this adjacency; then use the adjacency rules to obtain an assignment different from the one in Example 4.

    a. Complete the implementation, using JK flip-flops, and compare the complexity of the state and output decoders with that obtained in Example 4.

    b. Find an implementation using D flip-flops; again compare the complexity.

11 In Example 6 start from the transition table in Figure 16c and assume that implementation is to be carried out with JK flip-flops. Construct logic maps for the J and K excitations, determine the excitation functions, and draw the resulting sequential circuit diagram. Compare the complexity with that of the implementation using D flip-flops in Figure 17.

12 In Example 6 (the change-of-level detector), sets of adjacencies called for by the rules of thumb are {DE, DF, EF} and {DG, EG, DE}. Only two of the adjacencies in each set can be achieved. In Example 6, the unachieved adjacencies were chosen to be DG and EF. Suppose, instead, that the unachieved adjacencies are taken to be DF and EG, all others being achieved.

    a. Construct the resulting assignment map.

    b. Again assume implementation is with D flip-flops; construct the transition and output tables.

    c. Construct the excitation maps for the D flip-flops.

    d. Using the preceding results, draw the diagram of the resulting circuit.

    e. Compare the amount of hardware with the circuit in Example 6 (Figure 17).

13 A single-output sequential machine has a data input x and two control inputs c_2 and c_1. The output is to equal the input but delayed by one, two, three, or four clock pulses, as determined by the control-input code $c_2 c_1 = 00, 01, 10, 11$, respectively. Write an expression for the output function and design the circuit, explaining each step.

14 A 4-bit serial-in, parallel-out right-shift shift register with asynchronous preset has its initial state preset at $y_3 y_2 y_1 y_0 = 1101$, where y_0 is the state of the flip-flop at the input of the register. There is no external input except the clock, the register's excitation coming from the state decoder only. The desired output sequence is 110111001000; it repeats after these 12 bits. Design the combinational logic (state decoder) as a minimal circuit. An appropriate diagram is shown in Figure P14.

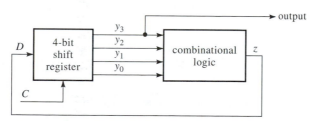

Figure P14

15 The same structure as the circuit in Figure P14 is to be used, but this time the excitation of the shift register is to be

$$D = y_1 y_3 + y_2 y_4 + y_3 y_4$$

Assume that the initial state of the shift register has been set at 0101. Find the output sequence from the shift register.

16 A combinational logic circuit is to be designed with a single output, D. This output is to become the input to a D flip-flop.

    a. Suppose that this circuit has two inputs, J and K. Design the circuit so that, together with the D flip-flop, it constitutes a JK flip-flop.

    b. Suppose, instead, that the combinational logic circuit has a single input labeled T. Design the circuit so that, together with the D flip-flop, it constitutes a toggle flip-flop.

17 A synchronous sequential circuit has two input lines, x_1 and x_2, and an output line, z. The data line is x_1 and x_2 is a reset line. Whenever $x_2 = 1$, the circuit is reset. When x_2 becomes 0, the first 4 bits on the message line constitute a message word. The output is to become 1 if the message received is 1010. At the end of the fourth bit of any word received when $x_2 = 0$, the circuit is to enter a waiting state, where it remains until it is reset and where the output is 0 for any input bits after the fourth bit.

    a. Construct a state diagram and a state table.

    b. Carry out the rest of the design and implement it using JK flip-flops.

    c. Construct a timing diagram showing the clock and the inputs and outputs.

18 Repeat Problem 17 if the message is to be 1100, using

    a. D flip-flops

    b. JK flip-flops

19 Repeat Problem 18 if the word length is 5 bits and the message is to be 11011.

20 Design a sequential machine that is to have all of the following characteristics:

    No 2 consecutive output bits can be 1.
    The output bit cannot be 1 upon receipt of the second of 2 consecutive input 0 bits.
    The output will be 1 if either of 2 consecutive input bits is 1, unless it conflicts with the first requirement.

A possible input/output sequence, for example, is the following:

    x: 0010101111010000
    z: -010101010101000

21 Design a single-input, single-output synchronous sequential circuit that is to have the following characteristics.

    No 3 consecutive output bits can be 1.
    The output will be 1 if, of 3 consecutive input bits, just two are 1, unless it conflicts with the first requirement.

A possible input/output sequence, for example, is the following:

    x: 0000110101110010101101110
    z: - -0001100110100010110110

22 Design a sequential machine for which the output becomes 1 if and only if just 1 of the following 3 bits is a 1: the present input bit and the last 2 output bits. A possible input/output sequence, for example, is the following:

    x: - - 001011011100101001001
    z: 11010100010000110110110

23 A sequential machine having the following properties is to be designed. It has two inputs: x and c, a control input. When $c = 0$, the machine returns to a reset state A, independent of x. With the machine in reset state A, whenever $c = 1$, the values of x at the next three clock pulses will constitute a binary word. An output of $z = 1$ is to occur at the third bit if and only if all 3 bits are the same: 000 or 111. Otherwise the output is to be 0. The state entered by the machine at the third bit of an input word is a waiting state. The machine does not leave this state until $c = 0$ occurs, at which time it is reset.

    a. Construct a state diagram and state table for this machine.
    b. If possible, reduce this table to one having the fewest states.
    c. Carry out the necessary steps to arrive at excitation equations, assuming the machine is to be implemented with D flip-flops. Then draw the corresponding circuit diagram.
    d. Repeat part c for the case where the machine is to be implemented with JK flip-flops.

24   **a.** Given the state table in Figure P24, assume the following input sequence with the machine initially in state A:

$$10011101011001$$

    Determine the resulting output sequence.
    b. Obtain a minimal reduced state table equivalent to the given one, letting new state A be the block in which the old state A appears.
    c. Again starting in state A of the reduced table, assume the same input sequence and find the resulting output sequence. Compare the result with the one in part b. Are you surprised?

PS	NS,z $x = 0$	$x = 1$	
A	B,0	G,0	
B	B,1	H,1	
C	F,1	D,0	
D	B,0	H,0	
E	F,1	D,0	
F	F,0	C,1	
G	E,0	A,0	
H	E,0	A,0	**Figure P24**

25   **a.** Design a single-input, single-output synchronous sequential circuit that produces an output of 1 whenever there is an odd number of 1's in the latest three input symbols.
    b. Draw several timing diagrams, assuming different combinations of inputs.

26 Design a single-input, single-output synchronous sequential circuit that generates an output of 1 whenever the latest 4 input symbols correspond to a binary number that is

    a. A multiple of 3
    b. A multiple of 5
    c. A multiple of either 3 or 5

27 Data appearing on a line synchronized with a clock should never have three or more consecutive 0's or four or more consecutive 1's.

    a. Design a sequential circuit that will detect such sequences and generate an output of 1 whenever they occur.
    b. Construct appropriate timing diagrams for different combinations of inputs.

28 The output of a synchronous sequential circuit is to be the same as the input but delayed by three or four clock periods under the control of a second input, c. The delay is to be three clock periods when $c = 0$ and four clock periods when $c = 1$.

    a. Design the circuit.
    b. Construct timing diagrams for the two cases.

29 A synchronous sequential circuit has two input lines, w and x, and a single output line, z. Let $W = w_2 w_1 w_0$ and $X = x_2 x_1 x_0$ be 3-bit sequences on the input lines representing binary numbers, the most recent bits being w_2 and x_2. The output is to be 1 whenever $W \geq X$.

    a. Design the circuit.
    b. Draw timing diagrams for several input word combinations.

30 A synchronous sequential circuit has two data inputs a and b, a control input c, and a single output z. The output is 0 except that $z = 1$ under either of two conditions:

    $c = 0$ and a and b had identical values two clock periods earlier
    $c = 1$ and $a = b'$ three clock periods earlier

    a. Design the circuit.
    b. Construct a timing diagram, showing the clock, input, and output signals for the two values of c.

31 A single-input, single-output synchronous sequential circuit is to generate an output of 1 whenever x has the same value it had three clock periods earlier; otherwise the output is to be 0.

    a. Design the circuit.
    b. Construct timing diagrams showing the clock, input, and output signals for the two cases.

32 A single-input, single-output synchronous sequential circuit is to generate an output of 1 whenever any of the following input sequences occurs: 011, 1001, 11011. The output is to be 0 otherwise.

    a. Design the circuit.
    b. Draw timing diagrams for the possible input sequences, showing the clock, input, and output signals.

33 A single-input, single-output machine is to have outputs as follows:

$$z(t) = z(t - 1)z(t - 2) \quad \text{when } x(t) = 0$$
$$z(t) = z(t - 3) \quad\quad\quad \text{when } x(t) = 1$$

If this machine has a finite memory span, specify its class and obtain a canonic implementation.

34 Modify Problem 33 by introducing a control input, c. The output specified in Problem 33 is to be obtained when $c = 1$. When $c = 0$, on the other hand, the output is to be 1 whenever the last 2 input bits are identical. Obtain a canonical implementation.

35 For each of the following machines, determine if it is a finite-memory machine. If it is, determine its type (finite-input-memory, finite-output-memory, or neither) and its order.

    a. *Serial parity generator:* The machine receives data bit-serially on its input and indicates on its output if the total number of 1's received so far is even or odd.
    b. *Serial adder:* The machine has two inputs and receives two binary numbers bit-serially on these input lines, least significant bit first. When bits of weight 2^i (for some i) are being received, the output is to be the sum bit of the same weight.

c. *Serial multiplier by a constant:* The machine receives a binary number bit-serially on its input, least significant bit first, and multiplies it by a fixed constant k (which need not be a power of 2). When an input bit of weight 2^i is being received, the output must be the product bit of the same weight.

d. *Divisible-by-k indicator:* The machine receives a binary number bit-serially on its input, least significant bit first, and indicates its divisibility by a fixed constant k that is not a power of 2. The output is 1 if and only if the binary number received up to that clock period (including the current input) is divisible by k.

e. Repeat part d assuming that k is a power of 2.

f. Repeat part d assuming that the binary number is received *most significant bit* first.

g. Repeat part e assuming that the binary number is received *most significant bit* first.

36 Sequential machines M_1 and M_2 are finite-input-memory of order m_1 and m_2, respectively. A new machine M is obtained by cascading M_1 with M_2. That is, the output of M_1 is the input of M_2. The input and output of M are, respectively, the input of M_1 and the output of M_2. Determine if M is a finite-input-memory machine and, if it is, determine its order.

37 M_1 and M_2 are finite-input-memory machines of order m_1 and m_2, respectively. A new machine M is obtained as follows. The inputs of M_1 and M_2 are tied together and constitute the input to M. The outputs of M_1 and M_2 are brought out separately and together form the output of M. Determine if M is a finite-input-memory machine and, if it is, determine its order.

38 Sequential machines M_1 and M_2 are finite-output-memory machines of order m_1 and m_2, respectively. A new machine M is constructed as in Problem 37. Determine if M is a finite-output-memory machine and, if it is, determine its order.

39 Odd-length counters do not have unit-distance codes, but almost do, having only one transition (usually the pivotal one in the middle of the count) in which more than 1 bit must change.

a. Design a self-correcting modulo-m ring counter, where $m = 5$.

b. Repeat for $m = 7$.

c. Repeat for $m = 9$.

40 The *creeping code* is a 5-bit code for decimal digits generated as follows. The code for digit 0 is 00000. The code for any digit d_i is obtained from the code for the preceding digit d_{i-1} by first setting the msb of d_i equal to the complement of the lsb of d_{i-1}, and then setting the lower 4 bits of d_i equal to the upper 4 bits of d_{i-1}, in the same order. (See the section in Chapter 1 on codes.)

a. Using D flip-flops, design a synchronous modulo-10 counter that counts in creeping code; draw the circuit.

b. Modify the design so that the circuit has an output $z = 1$ whenever the count is either 4 or 7.

41 A multimode counter has one pulse input line x that is synchronized with the clock and two level output lines f and g that respond to the rising edge of the clock. Level changes on the input line are separated by at least four clock periods. The operation of the counter is to be as follows:

f becomes 1 at each clock pulse.
g becomes 1 two clock pulses later.
f goes to 0 at the next clock pulse after g becomes 1.
g goes to 0 at the next clock pulse after f goes to 0.

a. Construct a timing diagram showing the clock waveform and the waveforms of $x, f,$ and g.

b. Design the counter using a distance-1 code and draw the circuit.

c. Design the counter using the creeping code, making sure that it is self-correcting.

42 The schematic diagram of a universal left/right shift register is shown in Chapter 5, Figure 24; this one is a 4-bit register.

    a. Draw a transition diagram (a state diagram whose states are already assigned codes) for a 3-bit universal register. Starting in any state, either a 0 or a 1 can be shifted either to the left or to the right. Hence, there will be four arcs leaving each node, indicated by 0L, 0R, 1L, and 1R.

    b. Notice how a standard ring counter sequence of length 4 and a twisted-tail ring counter sequence of length 6 are generated from this transition diagram.

    c. Design a decoder circuit for a 3-bit universal register so that the combined circuit with a schematic similar to Figure 24 in Chapter 5 is a modulo-m counter, starting with 000. Take

- $m = 4$ (two possibilities)
- $m = 5$ (four possibilities)
- $m = 6$ (six possibilities)
- $m = 7$ (two possibilities)
- $m = 8$ (four possibilities)

    d. Draw a transition diagram for a 4-bit universal register and notice how a standard ring counter sequence of length 5 and a twisted-tail ring counter sequence of length 8 are generated from this.

    e. Design a decoder circuit for a 4-bit universal register so that the entire circuit in Figure 24 in Chapter 5 is a modulo-m counter starting with 0000. Take $m = 8$; how many possible sequences of length 8 are there?

    f. Repeat e for $m = 9$.

43 Suppose that the counter implemented by JK flip-flops in Figure 30 is to be implemented by T flip-flops, obtained from JK flip-flops by setting $J = K$.

    a. Find the excitation maps (maps of T) of the three flip-flops.

    b. From these, determine the state decoder.

    c. Compare the hardware requirements with those using the JK flip-flops.

44 A state machine, with two inputs A and B and a single output C, is to be designed. The output is to become 1 only if the number of input 1's since the machine was reset is an exact multiple of 4. It doesn't matter on which input line a 1 occurs.

    a. Construct a state diagram. (As you go about this task, think of the following things. Is 0 a multiple of 4? With the machine in a particular state, what difference in next state and output would there be for inputs $AB = 01$ or 10? To what state would the machine go if, having already received three 1's, the next input is $AB = 11$? In such an event, what would the output be?)

    b. How many different assignments are possible? Select an appropriate assignment.

    c. Assume the use of D flip-flops and construct excitation maps.

    d. Write expressions for the excitations and the output.

    e. Draw a circuit implementing these expressions.

    f. If you want, try another assignment and repeat parts c, d, and e. Compare the two realizations. Was this fun?

45 For each table in Figure P45, the overall objective is to construct a minimal reduced table equivalent to the original one.

    a. Partition the states so that all states in a partition are 1-equivalent.

    b. Refine the partitions so that all states in the new partitions are 2-equivalent.

c. Continue refining the partitions until an equivalence partition is obtained.
d. Construct a reduced state table with each final partition as a state.
e. Compare the number of flip-flops needed in implementing both the original tables and the reduced tables.
f. Implement each reduced table using JK flip-flops.
g. Repeat f using D flip-flops.
h. Using D flip-flops, implement the table in Figure P45b before reduction. Compare the number of flip-flops and circuit complexity for the two implementations.

PS	NS,z $x=0$	$x=1$
A	A,0	C,0
B	D,1	A,0
C	F,0	F,0
D	E,1	B,0
E	G,1	G,0
F	C,0	C,0
G	B,1	H,0
H	H,0	C,0

(a)

PS	NS,z $x_1x_2 = 00$	$x_1x_2 = 01$	$x_1x_2 = 11$	$x_1x_2 = 10$
A	B,0	G,1	C,1	D,0
B	A,1	E,0	D,1	G,1
C	H,0	G,1	A,1	C,0
D	H,0	G,1	C,1	D,0
E	C,1	H,0	D,1	C,1
F	D,1	H,0	C,1	G,1
G	H,1	G,1	A,1	F,0
H	D,1	E,0	A,1	G,1

(b)

Figure P45

46 A state machine has a single input N and a single output D. Four-bit messages arrive at the input. The purpose of the circuit is to detect when a 4-bit message is not a BCD word. That is, $D = 1$ whenever the 4-bit word is not a decimal number in BCD code. Assume that the circuit returns to its initial (reset) state at the end of each 4-bit word.

a. Construct a state diagram and a state table. (Confirm that your diagram produces the correct outputs.)
b. By partitioning, reduce the table to a minimum.
c. Choose two 4-bit words, one that is and one that isn't a decimal number in BCD code. Draw timing diagrams for these two cases.

47 Modify Problem 46 as follows. The 4-bit words are not consecutive; when the last bit of a word is received, the machine enters a waiting state. While it is in this state, the signal that another 4-bit word is coming is the appearance of 3 consecutive 1 bits. Upon receipt of the third 1 bit, the machine enters the reset state, ready for the next 4-bit message.

a. Construct a new state diagram and a new state table.
b. Reduce the state table by partitioning the states.

48 a. Design a synchronous BCD counter. (It might be called a modulo-10 counter.) The only input is the clock. Draw a timing diagram that includes the clock and all flip-flop output waveforms.
b. Modify the design for a counter that is to be just one decade of a decimal BCD counter. That is, each decade is to represent a decimal digit in the 10^k position of a decimal number.

49 A certain binary signal consists of a periodic sequence of pulses having the same width as the clock pulse, synchronized with the clock. For a certain application, it is expected that the number of bits in a string of 0's is odd and the number of 1's is even. A state machine is to be designed to detect errors from this configuration. That is, $z = 1$ whenever an even string of 0's or an odd string of 1's is detected.

    a. Construct a state diagram and a state table. (Think about how the machine will know that a string of like bits has ended.)

    b. If your table is not minimal, reduce it.

    c. Make an "optimal" state assignment and construct a transition table.

    d. Assume the use of D flip-flops and write expressions for the excitations.

    e. Draw a circuit diagram implementing these expressions.

50 A synchronous sequential circuit has two input lines, x_1 and x_2, and two output lines, z_1 and z_2. At each clock tick, the combination $x_1 x_2$ constitutes a 2-bit binary number. If the present value of the input number is less than its immediately preceding value, then the outputs are $z_1 z_2 = 10$. If the present value is greater than the preceding value, then $z_1 z_2 = 01$. If it is the same, $z_1 z_2 = 00$.

    a. Design the circuit.

    b. Draw timing diagrams for the three cases.

51 Five-bit words arriving on a line are expected to be messages in 2-out-of-5 code. However, there may be errors. A synchronous machine is to be designed whose output is 1 only when the fifth bit is received and the completed word is not a valid word in 2-out-of-5 code. The 5-bit words are consecutive; as soon as one 5-bit word is completed, the circuit should be ready to receive the first bit of the next word.

    a. Construct a state diagram. (*Hint:* To how many distinct states can the circuit make a transition for each incoming bit after the first?)

    b. Construct a state table.

    c. Make an appropriate state assignment and construct a transition table.

    d. Assuming the use of D flip-flops, obtain expressions for the excitation and output functions.

    e. Construct timing diagrams for the clock, the input bits, and the resulting output.

52   **a.** A synchronous sequential machine has one input line x and one output line z. The machine is intended to receive a binary number of unknown length on the input line, with the *least* significant bit first, and to indicate on z its divisibility by 5. That is, for any time t, $z(t) = 1$ if and only if the binary number $x(t) \ldots x(0)$ is divisible by 5. Construct a state table for such a machine and minimize the number of states.

    b. Generalize part a: If divisibility by a number p is to be detected, where p is a known constant, determine a tight upper bound in terms of p on the number of states needed in the machine.

    c. Repeat part a assuming that the *most* significant bit is received first.

    d. Repeat part b assuming that the *most* significant bit is received first.

53 Figure P53 shows a schematic diagram of a synchronous modulo-10 counter whose states are 0–9. The Q_3–Q_0 outputs represent the count. If CE ("count enable") is 1, then the counter increments to the next state at the next clock pulse. Otherwise it retains the current count. The output TC is 1 if and only if the count is 9.

    The objective of this problem is to design a modulo-10^{16} counter by stringing together several modulo-10 counters. It is claimed that the CE inputs to the modulo-10 counters in the string are analogous to the carry inputs to the full adders in a multibit adder, and hence, the carry-lookahead principles can be used in designing the modulo-10^{16} counter.

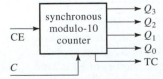

Figure P53

    a. Derive the generate and propagate expressions for a modulo-10 counter.

    b. Suppose 4-bit lookahead units of the type shown in Figure 8, Chapter 4, with inputs P_1–P_4, G_1–G_4, and C_1 and outputs C_2–C_5 are available. Using these and the modulo-10 counters (and no other logic) design a lookahead modulo-10^{16} counter.

    c. Obtain a modulo-10^{16} counter by replacing the lookahead units in the answer to *b* by only AND gates.

54   **a.** Design a modulo-2 (2-bit) binary counter using D flip-flops. The counter goes through the sequence 00 01 10 11 00 .... The machine is also to have an output line that emits a 1 at the count 11.

    b. Draw an appropriate timing diagram.

55 A state machine is to have a single input line and a single output line. The output is to remain 0 until the last bit of either of the sequences ...0000 or ...1111 occurs, at which time the output becomes 1.

    a. Construct a state diagram and a state table; then reduce the table to one having the fewest states.

    b. Construct an ASM chart.

    c. Assuming implementation with D flip-flops, construct the transition tables and, from these, construct a circuit implementation.

56 A certain state machine is to have the function of detecting when an incoming string of 7 bits is not the biquinary code for a decimal digit. The machine has two inputs: DATA and CONTROL. DATA consists of 7-bit words that are to represent the decimal digits in biquinary code. CONTROL is a signal that initiates an examination of DATA. When CONTROL = 0 for one or more clock ticks, the output remains 0. When CONTROL becomes 1 and stays 1, the machine is to examine the next 7 bits in DATA. Meanwhile the output remains 0; it becomes 1 only if the seventh bit completes a word that is *not* a decimal digit in biquinary code.

    a. Construct a state table for this machine.

    b. Assuming the use of D flip-flops, construct transition tables. Using these tables, design a circuit implementation.

57 A sequential comparator, with two input lines x and y and a single output line z, is to be designed. X $(x_n x_{n-1} x_{n-2})$ is a 3-bit word on line x and, similarly, Y $(y_n y_{n-1} y_{n-2})$ is a 3-bit word on line y. Taking X and Y as 3-bit binary numbers, the output is to be 1 only if $X \geq Y$.

    a. Construct a state diagram and a state table for this machine.

    b. Assume D flip-flops are to be used. Construct transition tables and, from these, a circuit implementation.

    c. Repeat *b* using JK flip-flops.

    d. Someone suggested implementing the circuit with two parallel-read shift registers and some combinational logic. Carry out this suggestion.

58 The objective of this problem is to design a Moore-model modulo-8 up-down counter. (Modulo-8 means that the machine counts from 0 to 7 in binary. "Up-down" means that when the count advances from 7 (111) it goes to 0 (000), and when it drops from 0 it goes to 7.) Besides the clock, the machine is to have a single input, x. When $x = 0$, the count will drop by 1 from its present value and, when $x = 1$, the count will increase by 1 from its present value, both occurring at the clock tick. Assume that D flip-flops are to be used and that there is no output decoder, the states being the outputs of the flip-flops taken as a binary number.

    a. Draw a diagram showing the three flip-flops and the state decoder as a rectangle. (Can you identify the nature of this machine?)

    b. Construct a transition table directly rather than using arbitrary names for the states and making a state assignment later; use the binary values of the count to identify the present and next states.

   c. Construct logic maps for each next state.
   d. Design the state decoder and complete the implementation.
   e. Using arbitrary times of input changes relative to the clock, draw timing diagrams showing the clock pulses, the input, and the flip-flop outputs.

59 The objective is to design a modulo-8 up-down counter with a single input x and three output lines. The binary number represented by the outputs $z_2 z_1 z_0$ is the count. It is to increase by 1 when $x = 1$ and decrease by 1 when $x = 0$. Design the circuit.

60 The purpose is to design a 3-bit binary up counter with no other inputs but the clock. At each clock tick, the counter cycles through the sequence 000, 001, 011, 111, 101, 100, after which it repeats the sequence. The other two possible states are not to occur.

   a. Using the state codes as state "names," construct a state table directly. Decide how to handle the entries corresponding to the rows of combinations that are not to occur.
   b. Assuming the use of D flip-flops, the next state for each position in any row is the same as the required value of D. Construct logic maps for the required value of D in terms of the present states. From these, write an expression for each D.
   c. Construct the circuit diagram to implement the counter.
   d. Draw a timing diagram, showing waveforms for the clock and for the outputs of the three flip-flops.
   e. Now assume toggle (T) flip-flops are to be used. From the excitation requirements for T flip-flops in Figure 17, Chapter 5, construct new logic maps for each T and construct a circuit diagram implementing the counter. Compare this with the implementation using D flip-flops. Show if there will be any changes in the timing diagrams.

61   a. Repeat Problem 60 if the counter sequence is to be the following: 000, 010, 001, 011, 101, 100, 000, ...
   b. Repeat Problem 60 if the counter sequence is to be the following: 000, 011, 111, 101, 001.

62 The state table in Figure P62a is to be implemented with two Lemon flip-flops.

PS	NS,z $x = 0$	$x = 1$		PS	NS,z $x = 0$	$x = 1$
A	C,0	B,1		A	C,0	B,0
B	A,0	A,0		B	A,1	C,1
C	A,0	D,0		C	B,0	D,0
D	C,0	A,1		D	C,1	C,0
	(a)				(b)	

Figure P62

   a. Using the results of Problem 18 in Chapter 5, specify all possible state assignments, justifying your response.
   b. Choose one of the possible assignments and carry out a circuit implementation.
   c. Repeat part b for a different assignment. Are there reasons for selecting one possible implementation over the other?
   d. Repeat each part for the state table shown in Figure P62b.

63 A counter is to have a single 1-bit control input C. When $C = 0$, the 3-bit counter is to sequence through the binary code. When $C = 1$, it is to sequence through the Gray code.

    Binary code: 000 001 010 011 100 101 110 111 $\rightarrow$ 000
    Gray code: 000 001 011 010 110 111 101 100 $\rightarrow$ 000

The only outputs are the flip-flop outputs representing the states.

a. Construct a state diagram, labeling the states by their 3-bit codes. Show the transitions to the appropriate next states for each C.
b. Draw an ASM chart for the counter.
c. Suppose the present state is 000 when the control input takes on the following sequence.

C: 1 1 0 0 0 1 0 0 0 1 1

Construct a table with the input as column 1, the present-state code as column 2 and the next-state code as column 3.
d. Draw a circuit implementing the counter.

64 A synchronous sequential circuit has a single output z and two inputs, x and r. The output is a delayed version of the x input under the control of r. When $r = 1$, the output equals what the x input was three clock periods earlier. When $r = 0$, the output equals what the x input was two clock periods earlier. Design the circuit using appropriate flip-flops. Show and explain all intermediate steps.

65 A state machine has a single output and—besides the clock—three inputs: a data input x and the outputs of a modulo-4 counter c_1 and c_0. The output of the machine is to equal the data input but delayed by a number of clock periods determined by the count $c_1 c_0$: the delay in output is one clock period at count 00, two clock periods at count 01, and so on.

a. Use whatever you need (state diagram, state table, logic maps, etc.) to arrive at an expression for the output. Explain your reasoning.
b. Find an implementation of the circuit.

66 A customer has placed an order from your engineering design shop for a single-input, single-output synchronous sequential machine. The output is to become 1 whenever, starting at some time, the number of input 1's exceeds the number of input 0's. An example is

x: 0 1 1 0 1 1 0 0 ...
z: 0 0 1 0 1 1 1 0 ...

Either (a) provide a statement as to why such a machine is impossible or (b) construct a state diagram of a machine that satisfies this requirement. In the latter case, construct a state table and implement it.

67 A synchronous sequential circuit is to have three inputs: A, B, and C. The single output z is to be 0 except for the following possible inputs:

$z = 1$ when $C = 0$ and A and B had identical values two clock periods earlier.
$z = 1$ when $C = 1$ and A and B had opposite values three clock periods earlier.

Obtain an implementation of the circuit and discuss its nature.

68 **a.** In Example 7, use the expressions for the flip-flop excitations and the output z on page 218 to construct a circuit realization.
b. Analyze the resulting circuit to verify the original expressions.
c. If these expressions cannot be verified by your circuit, repeat part a until verification is achieved.

69 Design a modulo-8 up-down counter. The count is to appear in BCD code on three output lines as $z_2 z_1 z_0$.

Chapter 7

Asynchronous Sequential Machines

The type of sequential circuit treated so far uses a clock to synchronize all flip-flop transitions. The speed of operations in a synchronous circuit depends on the frequency of the clock. This frequency must be set to accommodate the slowest subsystem in the overall digital system whose timing is to be controlled. Subsystems that could be operating at a higher speed—responding to changes in input as they occur, rather than waiting for the clock pulse—are thus unnecessarily slowed down. To increase overall speed, it is sometimes desirable to permit some subsystems to operate at their own speed—asynchronously, not in synchronism with the clock.

Furthermore, suppose there are two digital systems, each synchronized with its own clock, between which some communication is needed. At which clock frequency should such communication be carried out if done synchronously?[1] For such purposes also, an asynchronous system might prove useful.

In this chapter we will consider sequential circuits that don't use a timing source to synchronize state transitions. They fall under the generic class of *asynchronous* sequential circuits.[2] Indeed, such sequential circuits have already been considered. Latches and interconnections of them all fall in the class of asynchronous sequential circuits and are the most common such circuits. One might ask, Since the design of such a specialized class of circuits will occupy the attention of only a few people, why bother taking the time to study it? One reason is that many significant concepts in digital circuits, some of which are valid for combinational circuits as well, make their appearance in this study.

Furthermore, one never knows at what point in one's professional life some specific fundamental ideas will come in handy. Forgoing this study means narrowing one's vision. Under discussion are fundamental ideas—not the latest "how to's," things that would soon need changing anyway, since what is con-

---

[1]Indeed, in many large systems, using a single clock to synchronize all transitions becomes problematic. In such cases, it is becoming common to partition the overall system into subsystems, each synchronized by a local clock.

[2]They are sometimes called *feedback* sequential circuits.

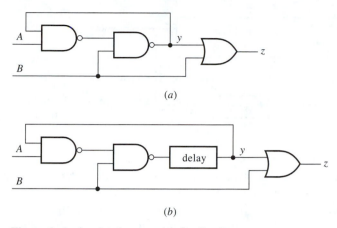

(a)

(b)

Figure 1 A circuit of gates with feedback.

temporary today will not be so tomorrow. On the other hand, because this topic can be considered specialized and not of central importance to the remainder of the book, the entire chapter can be treated as optional. If it is skipped, though, many important ideas will be missed.

1   THE FUNDAMENTAL-MODE MODEL

In one type of asynchronous circuit, the input and output signals are *levels*, and the level inputs can change at any time. Instead of using flip-flops to establish the states, the inevitable delay of signals propagating through gates is utilized by means of feedback. To see how sequential action is possible without flip-flops, consider the circuit shown in Figure 1a, which includes a feedback loop. Each of the gates in the loop introduces a certain delay. A possible model for representing the delays in the gates is to show the total delay from all gates lumped in one place in the loop, as in Figure 1b, with the gates considered ideal. The overall structure can then be described as a combinational circuit with a feedback loop containing delay.

The preceding discussion leads to the generalized model shown in Figure 2. The totality of all inputs to the combinational logic is called the *total state*. It is made up of two components:

$\{x_i\}$, the *primary* inputs, or the *input state*
$\{y_i\}$, the *secondary* inputs, or the *internal state*

The latter are the feedback variables following the delay. The combination of the primary and secondary inputs generates

$\{z_i\}$, the *outputs* to the outside world
$\{Y_i\}$, the *excitations* which, after a delay, become the secondary inputs

The primary inputs can change at any time from one level to another, but we assume that *no two inputs change simultaneously*. When inputs have remained stable for some time (longer than the propagation delay through any

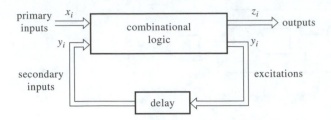

primary inputs x_i

combinational logic

y_i

z_i outputs

y_i

secondary inputs

excitations

delay

Figure 2 General model of fundamental-mode sequential circuits.

path in the circuit), no signal changes are taking place, and we say that the circuit is in a *stable state*. In this case $y_i = Y_i$ for all i. That is, the internal state and the excitations are the same.[3]

In response to an input-level change, the combinational logic generates a new set of excitations, and the circuit enters an *unstable state* in which $Y_i \neq y_i$. After a delay, the secondaries take on their new values. If another input change were to take place while the circuit is in an unstable state, the ultimate stable state would become uncertain. Since such uncertainty is intolerable, we assume that any changes in input level are separated in time by more than the propagation delay through the longest path in the circuit. Sequential circuits that function in the manner described are said to be operating in the *fundamental mode*.

2 THE FLOW TABLE

The design of fundamental-mode asynchronous circuits is analogous to the design of synchronous sequential circuits but without concern for flip-flops. In the synchronous case, an early step in the process is the translation of a problem specification into a state table. A similar step is carried out for fundamental-mode asynchronous circuits, except for the added complication of the unstable states. Nevertheless, as a result of input-level changes, transitions from one stable state to another, via an unstable state, are again shown in a table.

Primitive Flow Tables

The simplest way to describe the procedure for the construction of a table is in the context of a design problem. Suppose an asynchronous sequential circuit is to have two inputs and a single output. The output is to be 0 until the input becomes $x_1x_2 = 11$, but only if it follows an input of 10. (Why would it be impossible for the input to become $x_1x_2 = 11$ following a 00?) As example, the specifications are satisfied with the following output sequence as a result of the given input sequence:

x_1x_2: 00 01 11 10 11 10 00 10 11 01 11
  z: 0 0 0 0 1 0 0 0 1 0 0

Observe that z does not become 1 just because the input has become 11, but only if it reaches 11 from 10. A timing diagram illustrating these input output sequences is given in Figure 3.

---

[3]Both Mealy and Moore asynchronous models are possible. The difference from Figure 1 in Chapter 6 is that the state decoder and the clock are absent. No distinctions will be made here.

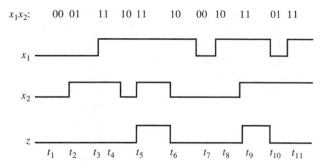

Figure 3 Possible sequences of inputs and output.

State, output
x_1x_2

00	01	11	10
ⓐ,0 ← initial output			

↑
initial stable state

(a)

S,z
x_1x_2

00	01	11	10
ⓐ,0			b
		ⓑ,0	

(b)

S,z
x_1x_2

00	01	11	10
ⓐ,0			b
		c	ⓑ,0
	d	ⓒ,1	
	ⓓ,0	e	
		ⓔ,0	

(c)

S,z
x_1x_2

00	01	11	10
ⓐ,0	d	—	b
a	—	c	ⓑ,0
—	d	ⓒ,1	b
a	ⓓ,0	e	—
—	d	ⓔ,0	b

(d)

Figure 4 Construction of a primitive flow table.

We need a mechanism to portray the "flow" of input changes that produce excitation changes, which then lead to state changes. This is done by means of a table, called a *flow table*, as shown in Figure 4 for this example.

Initially, with both inputs 0, the circuit is in a stable state labeled ⓐ. The circle around the state name indicates that the state is stable. An unstable state is indicated by simply naming the state without the circle. Now suppose that x_1 changes to 1, so that the input becomes $x_1x_2 = 10$. (That's a change in a single input.) Eventually, the circuit will enter a stable state, called ⓑ in Figure 4b; however, before it does so, this state is unstable. This sequence is shown in Figure 4b. The sequence is from a stable state at one input to an unstable state when the input changes. Then transition is to the corresponding stable state (after some delay but without any further change in input). This time the sequence is from one column to another in the same row, followed by a motion from one row to another in the same column. The latter occurs after some delay. The output corresponding to the specific input is also shown.

Suppose that x_2 next changes to 1, so that $x_1x_2 = 11$. This time, from stable state ⓑ we first move horizontally to the column corresponding to $x_1x_2 = 11$; the unstable state is marked c. After a delay, we move vertically to stable state ⓒ shown in Figure 4c, at which point an output of 1 is emitted, as required by the problem statement. Now an input change from $x_1x_2 = 11$ to 01 leads, after stabilization, to another state, ⓓ. When in state ⓓ (column $x_1x_2 = 01$), suppose x_1 changes to 1 so that x_1x_2 goes to 11. According to the problem specifications, this should not lead to $z = 1$. Hence, the stable state reached by this change of input should not be ⓒ, where the output is 1; it should be some other state, say ⓔ. These moves are also shown in Figure 4c.

The flow table can be completed in this manner. However, since there cannot be simultaneous changes in the two inputs, the input cannot change from 00 to 11 or vice versa. (How about from 01 to 10 or vice versa?) To complete the table, from each stable state any *allowable* input change is assumed and the appropriate transition is recorded, either as a state already entered in the table or as a new state. For each new state a new row is created and, at the column corresponding to the input symbol, the new stable state and the corresponding output are entered. When a particular input change is not permitted to occur, the state transition and the output are unspecified, indicated by a dash in the appropriate column and row. The completed flow table is shown in Figure 4d. This leads to the following definition:

> A primitive *flow table is a flow table having only one stable state per row, with outputs specified only for stable states.*

Note the features of the primitive flow table in Figure 4d and compare it with the state table for a synchronous machine. First, we see that the output has not been specified for the unstable states. This omission will be dealt with in the next section.[4] A second difference seems to be that, although there is a column for each input combination, there is no column listing the "present state" that identifies each row. Nevertheless, each row *is* identified with a particular stable state that can be considered the "present state" corresponding to this row. There is nothing to prevent using the names of the stable states as the row headings of the table.

Exercise 1   Reconstruct the flow table in Figure 4, providing a column on the left whose row entries are the names of the stable states corresponding to each row. For simplicity, there is no need to circle the names of the "present states" in this column.   ◆

---

[4]Since only one output is specified in each row of a primitive flow table, it is conceivable that the output can be detached from each state and placed in a separate output column. We will not do that, however, because there will be occasions when we will wish to assign different outputs to different unstable states in a given row, as we proceed to complete the table.

$$S, z$$
$$x_1 x_2$$

00	01	11	10
ⓐ, 0	b,	—	e,
a,	ⓑ, 0	c,	—
—	f,	©, 0	d,
a,	—	g,	ⓓ, 1
a,	—	g,	ⓔ, 0
a,	f,	g,	—
—	ⓕ, 0	ⓖ, 0	e,

$$S, z$$
$$x_1 x_2$$

00	01	11	10
ⓐ, 0	b,	—	
	ⓑ, 0	c	—
—		©, 0	d,
—			ⓓ, 1

(a) *(b)*

Figure 5 Construction of primitive flow table for Example 1.

Another feature different from anything in the state tables for synchronous machines is that some entries (both states and outputs) are unspecified. This, however, is not a fundamental difference. Incompletely specified synchronous machines also exist; it's just that we did not study them when discussing such machines in Chapter 6. We will correct that deficiency in the following section. But first, we will consider a slightly more complicated example of a fundamental-mode asynchronous circuit.

EXAMPLE 1

A fundamental-mode circuit is to have two inputs and a single output, which becomes 1 only upon the occurrence of the last in the following sequence of input combinations; otherwise $z = 0$:

$x_1 x_2$: 00 01 11 10

The objective is to construct a primitive flow table.

Different outputs are to result when the input symbol is 10, depending on the preceding input sequence. Hence, we would expect two different stable states in column 10; but what about column 11? Even if the output is 0 independent of the input sequence leading up to it, the output occurring when input 10 follows 11 will be different if input 11 is preceded by 00 01 than if it is preceded by 10. Hence, we should expect two distinct stable states in column 11 also. By the same reasoning, the same is true for column 01. (Reason it out for yourself.) Only for input 00 is it irrelevant how it is reached. Hence, even before constructing it, we should expect a total of seven distinct states in the primitive flow table.

Let us label as ⓐ the initial stable state in which the input is 00. If we label as ⓑ, ©, and ⓓ, respectively, the subsequent states that result from the input sequence 00, 01, 11, 10, then the partial primitive flow table takes the form shown in Figure 5a.

Next return to state ⓐ. If the input goes from 00 to 10, the eventual stable state cannot be ⓓ since the output should not be $z = 1$ with that input sequence; call it ⓔ,

| | State, z | | | |
| | $x_1 x_2$ | | | |
PS	00	01	11	10
a	ⓐ, 0	b, 0	—	e, 0
b	a, 0	ⓑ, 0	c, 0	—
c	—	f, 0	ⓒ, 0	d, –
d	a, –	—	g, –	ⓓ, 1
e	a, 0	—	g, 0	ⓔ, 0
f	a, 0	ⓕ, 0	g, 0	—
g	—	f, 0	ⓖ, 0	e, 0

Figure 6 Flow table for Example 1.

as in Figure 5b. Next, starting from stable state ⓓ, suppose the input goes to 11. The ensuing stable state should not be ⓒ because then an input change back to 11 would lead back to state ⓓ, with a resulting output of $z = 1$; but this is not the correct input sequence to produce an output of 1. Hence, starting from ⓓ with an input of 11, let the state reached be ⓖ (output 0). The rest of the flow table, shown in Figure 5b, can be completed using similar arguments. (Do it.) As expected, there are seven states. ∎

Exercise 2 Confirm the remaining two rows of the table in Figure 5b. ◆

Assigning Outputs to Unstable States

In the primitive flow tables in Figures 4 and 5, there are no output values associated with the unstable states. Two cases should be considered: a transition between two stable states: (a) with the same outputs or (b) with different outputs. As an example of the first kind, consider the transition from ⓐ to ⓑ in Figure 5, where the outputs are both 0. When the circuit is in unstable state b, would it make sense for the output to go to 1, even momentarily, on the way from state ⓐ to state ⓑ, both with output 0? The question makes the answer obvious:

> *When a transition is to take place between two stable states having the same output, the intermediate unstable state should be assigned that same output.*

In two cases in Figure 5 there is a transition between two states having different outputs: from ⓒ to ⓓ the output is to go from 0 to 1, and from ⓓ to ⓖ (also from ⓓ to ⓐ) the output is to go from 1 to 0. Since a change in output is to occur in any case, assigning an output to the intermediate unstable state will affect only how rapidly the output change takes place.

Exercise 3 Decide how to make the assignment of output to an unstable state in the preceding paragraph if there is a design requirement that the output change take place (*a*) as rapidly as possible, and (*b*) as slowly as possible. What would you do if there were no specification? ◆

The completed flow table for Example 1 is shown in Figure 6. Outputs are unspecified when a transition is between stable states with different outputs.

3   REDUCTION OF INCOMPLETELY SPECIFIED MACHINES

The fact that states and outputs in some columns of the flow table are unspecified does not constitute a basic difference between fundamental-mode asynchronous machines and synchronous sequential machines. The latter can also have unspecified next states or outputs. For this reason, the subject of this section applies equally to both the flow tables of asynchronous machines and the state tables of incompletely specified synchronous machines.

An important concept used in the reduction of (completely specified) state tables in Chapter 6 was equivalence. (Look up its meaning if you need to.) When there are unspecified next states and outputs in a state table or flow table, this definition becomes meaningless. As a result of some input, if transition is to be to an unspecified next state, what would happen at the arrival of the next input? Since the state of the machine when that input arrives is unknown, we won't know what the next state or output will be. Hence, equivalence is impossible in such cases.

The length of an input sequence must, therefore, be limited; if a particular input combination sends the machine to a next state that is unspecified, no further inputs are permitted. An input sequence is said to be *applicable to a state* if, when the machine is in that state, no unspecified next state is encountered, except possibly at the last input. In the asynchronous case, the restriction that no two input bits change value simultaneously was imposed. That automatically makes all permitted input sequences applicable. Thus, in the flow table in Figure 6, suppose the stable state is f at input 01. The only unspecified next state occurs for input 10. But this input is not applicable. For any other input, no unspecified next states will be encountered. Hence, the condition that for any input sequence no two input bits change simultaneously will automatically be applicable.

Now let's turn to unspecified outputs. Suppose an input sequence is applied to a machine that starts at either of two initial states. For any particular input combination, either the outputs from the two states are both specified, or only one is specified, or neither is specified. When the outputs are not both specified, we say that they *do not conflict* with each other. Now, in addition to this, suppose that when both outputs are specified, they are the same. This is cause for celebration; it leads to the following definition:

> *Suppose that for every input sequence applied to each of two initial states in a machine, the output sequences produced do not conflict. The two states are defined as* compatible.

Thus, compatibility in incompletely specified machines is the counterpart to equivalence in completely specified machines.

The idea of compatibility can be extended to more than two states. Thus, a set of states is called a *compatible set* if all members of the set are compatible with each other. In the set of three states A, B, and C, suppose that the pairs AB, BC, and AC are all compatible. Then the set ABC is compatible. Thus, to find the compatibility of sets larger than pairs, we need to examine only all compatible pairs. (What would have to be examined if one wanted to test the compatibility of sets larger than triplets?)

Applying these concepts permits the reduction of both incompletely specified state tables for synchronous sequential machines and flow tables for asynchronous machines. That's the task we turn to next.

The Merger Table

The concepts discussed in this section will be introduced in terms of the flow table in Figure 6. Compatibility of two states was defined in terms of the output sequences resulting from an input string. For each input bit applied to each of two states, there is a pair of next states.[5]

> *By an implied pair of states we mean the two next states resulting from the application of a specific input to each of two present states.*

Thus, for the present pair of states bf in Figure 6, the implied pair for input 11 is cg. Sometimes the implied "pair" is not a pair, either because the two next states are the same or because one or both next states are unspecified. In such cases, the "implied pair" is not listed.

Compatibility

In the process of deciding on the equivalence (indistinguishability) of two states for completely specified synchronous sequential circuits, we first found a condition for the two states to be distinguishable. Then we established a procedure so that at each step a set of states was not distinguishable. For the present problem we will do something analogous: we will seek a way of deciding if two states are not *incompatible*. This will eventually lead to a conclusion about compatibility.

It is easy to see that if two states S_1 and S_2 are compatible, then the implied pairs for all input symbols are compatible. If not, it would mean that, *starting from the implied pair,* output sequences would not be the same when both are specified. But those output sequences starting from the implied pair constitute all but the first symbol of the output sequences from S_1 and S_2; and those output sequences were known to be the same when they were specified, since the two states were specified as compatible.

From this, it also follows that, if a pair implied by the pair S_1S_2 is *not* a compatible pair, then S_1S_2 is not a compatible pair. This leads to the following conclusion:

> *In an incompletely specified machine, two states S_i and S_j are not incompatible if and only if (a) for each input their outputs don't conflict and (b) their implied pairs are not incompatible.*

This result forms the basis of a procedure for determining all the compatible pairs—and, from them, all larger groups of compatibles.

---

[5]For simplicity, we will use the terminology "applying an input to a state" when we mean "applying an input to a machine when the machine is in that state."

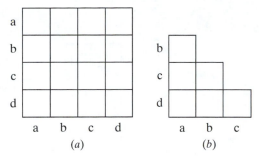

Figure 7 Merger table concept.

Construction of the Merger Table

We need a systematic approach for displaying *all* pairs of states in a machine. One method would be to create a matrix that lists all the states vertically and also horizontally. Such an arrangement is shown in Figure 7a. Each cell in the table represents a pair of states; on the main diagonal, the "pair" is just one state. Since the concept we are after is compatibility (or incompatibility), we don't need both the ab cells and the ba cells, for example. That is, if a is compatible with b, then b is compatible with a. So the squares either above the diagonal drawn from vertical a to horizontal d, or below that diagonal, can be deleted. Furthermore, the cells right on the diagonal (corresponding to aa, bb, etc.) are also unnecessary, since any state is compatible with itself. So a modified (staircase) table can be drawn, as shown in Figure 7b.

The top cell corresponding to aa is gone, so the vertical listing of states starts from b. Similarly, on the bottom right, the cell corresponding to dd is gone, so the horizontal listing of states starts from a but ends just before the last state. The table thus takes on a triangular staircase shape; the main diagonal and everything above it in the original matrix is removed.

To illustrate the use of a merger table, we will start with the flow table in Figure 6. The first step is the trivial one of setting up the structure of the merger table. The flow table in Figure 6 has seven states; hence the merger table will have six rows and and six columns. (Set up this blank table for yourself and carry out each step as it is described.) Using the flow table, the easiest thing to do next is to determine those states that are incompatible by virtue of having different outputs for the same input. For example, for input 10 the outputs for state a and d are different in Figure 6. So, the corresponding cell in Figure 8a is crossed out. From the flow table in Figure 6, verify the other two cells that are crossed out in Figure 8a.

Next we examine each pair of states whose corresponding cell has not yet been crossed out. For each pair of states that are not incompatible, what we intend to do first in Figure 6 is to find those whose next "pair" of states consists of

- A single state, because each next state is the same or one of them is unspecified
- No states at all, because both next states are unspecified

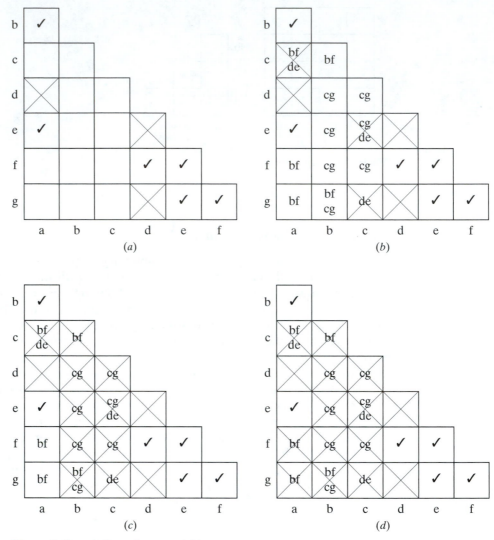

Figure 8 Completion of merger table.

In each of these cases the corresponding pair of states is compatible, so we enter a ✓ (check mark) in the corresponding cell. The result up to this point is shown in Figure 8a; verify that these are correct and there are no others.

So far, then, we have dealt with those pairs of states that are definitely incompatible and those that are definitely compatible. For each pair of remaining states, the next step using the flow table is to determine the implied pairs of next states—real pairs. For example, for input 01 and an initial pair of states ac, the implied pair in Figure 6 is bf. We write bf in the ac cell of the merger table. Carry out this plan for all remaining pairs of states in the flow table. Then confirm your work using Figure 8b.

Actually, we pulled a fast one on you by carrying out one extra step. Notice that Figure 8b has three more crossed-out cells than Figure 8a does. This results from an examination of the *implied pairs* inside each cell in Figure 8*b* to determine if any of them are pairs that are known to be incompatible from Figure 8*a*. For example, an implied pair in cell ac is de. But from Figure 8*a* we know that states d and e are incompatible. So, since pair ac implies pair de, and de is an incompatible pair, then ac must be incompatible also. That's why we crossed out square ac. Verify the other crossed-out cells in Figure 8*b*.

The process is now continued by noting if each newly discovered incompatible pair (ac, ce, and cg) is an implied pair written in any cell that is not yet crossed out. The incompatible pairs ac and ce, for example, are not implied pairs written in any cell. But cg *is* an implied pair appearing in six different cells; hence, since cg is an incompatible pair, all of these six cells must now be crossed out, as shown in Figure 8*c*; the corresponding pairs are incompatible.

After they are crossed out, we notice that the only implied pair remaining in any cell not crossed out is bf—and the pair bf is incompatible by virtue of its corresponding cell being crossed out in Figure 8*c*. So the two cells containing bf as an implied pair in Figure 8*c* are now crossed out. The final result is shown in Figure 8*d*.

Determination of Minimal, Closed Covers

Although the process just described appears to be lengthy, what is lengthy is *describing* it, not carrying it out. Indeed, the process is algorithmic; hence, software can be (and has been) written to carry it out.[6] The upshot is to discover all pairs of states that are compatible, by virtue of not being crossed out. In this example, the compatible pairs are

$$\{ab, ae, df, ef, eg, fg\} \rightarrow \{ab, ae, df, efg\}$$

Notice that the three states e, f, and g are compatible in pairs; hence, the triplet efg is a compatible set of states. No other pairs can combine to form a compatible triplet; nor can any others be combined with efg to form a larger compatible set. These observations lead to some definitions:

> *A compatible set of states C_1 covers another compatible set of states C_2 if every state in C_2 is contained in C_1.*

> *A compatible set of states is a maximal compatible if it is not covered by any other compatible set of states.*[7]

With this terminology, we see that each of the above compatible pairs covers each of the states making up the pair. The pairs ab, ae, and df are maximals, but ef, eg, and fg are not maximals since they are all covered by efg. Since state c is not covered by any of the compatible pairs, it itself is a maximal. We note that the individual states in the original flow table have been *merged* into combinations of states; hence, the table in Figure 8 is called a *merger table*.

---

[6]Three state assignment tools (Jedi, Nova, and Mustang) distributed with the Octtools integrated circuit design tools contain implementations of this procedure.

[7]When there is no possibility of confusion, "of states" will be omitted. "Minimal, closed cover" is the standard terminology.

A crucial difference is evident between sets of compatible states and sets of equivalent states in a synchronous machine: the maximals in the above list are not disjoint; they overlap. States a, e, and f each appear in two different compatibles.

The merger table is a tool; with the use of this tool the set of all maximal compatibles can be obtained for a given flow table. But the main problem of machine minimization (finding another machine with the fewest states whose performance is indistinguishable from that of the original machine) is still not completely solved. If such a new state table is to be formed by a set of compatibles, the compatibles must cover every original state. Furthermore, the "next state" resulting from any input must be covered by one of the present compatibles. This leads to the following idea:

A set of compatible states is closed if a compatible state implied by any one of them is covered by some compatible state in the set.

We are now ready to find a machine with the fewest states whose performance is indistinguishable from the original machine. What we need is a minimal set of compatibles that is closed and covers all the states of the original machine: a *minimal, closed, cover.*

In the preceding example, we found the set of all maximals to be

$$\{ab, ae, c, df, efg\}$$

Not all of these maximals are needed; the set {ab, c, df, efg} constitutes a minimal closed covering, but a minimal closed covering does not require that the sets of compatibles be maximals. We notice that state f is covered by both df and efg; we might consider removing f from either df or efg. The following minimal closed covering sets result if this is done: {ab, c, df, eg} or {ab, c, d, efg}.

Exercise 4 Confirm that each set of states in the preceding sentence includes all original states, is minimal, and is closed. ◆

Procedures exist for systematically seeking minimal, closed, covers similar to the Quine-McCluskey algorithm for completely specified machines. However, we shall not pursue such matters in general form here.

Let us suppose that the set {ab, c, d, efg} is chosen as a minimal, closed cover. The reduced flow table for this choice is shown in Figure 9. (The output table has been shown separately, for convenience.) A number of observations can be made. First, when an unstable state and a stable state are combined in a compatible, the resulting state must be stable.

Second, unlike the primitive flow table, the reduced table can have more than one stable state in each row. Since each row will be assigned specific values of the secondary variables, as we shall see, all stable states in a given row have the same value of secondaries; they are distinguished from each other only by the input states. This is a basic distinction from synchronous sequential circuits, where the inputs play no role in specifying the state. For example, the only way the two stable states in the first row in Figure 9—both labeled 1— differ from each other is in their input values: $x_1x_2 = 00$ in one case and 01 in the other.

	S x_1x_2				z x_1x_2			
	00	01	11	10	00	01	11	10
ab → 1	①	①	2	4	0	0	0	0
c → 2	—	4	②	3	—	0	0	—
d → 3	1	—	4	③	0	—	0	1
efg → 4	1	④	④	④	0	0	0	0

Figure 9 Reduced flow table.

y_1y_2	Y_1Y_2 x_1x_2			
	00	01	11	10
1 → 00	⑩0	⑩0		
2 → 01			⑪1	
3 → 11				⑪1
4 → 10	⑩0	⑩0	⑩0	

(a)

y_1y_2	Y_1Y_2 x_1x_2				z x_1x_2			
	00	01	11	10	00	01	11	10
00	⑩0	⑩0	01	10				
01	—	10	⑩1	11	—			—
11	00	—	10	⑪1	—			1
10	00	⑩0	⑩0	⑩0				

(b)

Figure 10 Transition and output tables for the reduced table in Figure 9.

Finally, some entries that were unspecified in a primitive flow table are now specified in the reduced table. The merging of ⓐ and ⓑ in column 11 requires the merging of unstable c with an unspecified entry. The result is unstable state 2, as shown in Figure 9, row 1 and column 11. This does not mean that a transition from ① for $x_1x_2 = 00$ to ② for $x_1x_2 = 11$ is now possible. The assumption that only one input bit changes at any one time still applies. A transition that is permitted is from ① in column 01 to ② in column 11. Similarly, the transition from ① to ④ is from column 00 to 10, and not from column 01 to 10.

Transition Tables

Once a reduced flow table is available, an assignment of secondary variable values must be made. For fundamental-mode asynchronous circuits, some major difficulties arise from the attempt to assign secondary values. Such difficulties will be temporarily disregarded here; they will be discussed in subsequent sections.

The four-state table in Figure 9 requires two secondary variables for its implementation. Let us arbitrarily make the assignment $y_1y_2 = 00, 01, 11, 10$ to the stable states in order. The resulting partial transition table that includes only the stable-state entries is shown in Figure 10a. Notice that the entries in any row are the excitations corresponding to that combination of secondary variables. For stable states, the excitations have the same values as the secondaries, as confirmed in the partial table. The question now is, What excitation values should the unstable states take on? Since, after some delay, the unstable states will become the corresponding stable states, the excitation value for an unstable state should be the same as the secondary values of the corresponding stable state.

> 1. Construct a primitive flow table satisfying problem specifications—one stable state per row with outputs specified for stable states only. Specify outputs of unstable states so that no momentary output flickers occur and possible fast or slow output changes in specifications are satisfied.
> 2. Obtain a minimal reduced table using a merger table to determine compatibles.
> 3. Assign secondary values to the states.
> 4. From the minimal table and the secondary-value assignment, construct transition and output tables. From these tables, derive expressions for the outputs and the excitations.
> 5. Implement these expressions in a circuit diagram.

Figure 11 Fundamental-mode circuit design procedure.

Using this concept, the resulting completed transition and output tables will be as shown in Figure 10*b*. Make sure that you confirm all entries in each table.

What has just been carried out by way of a specific example in the preceding discussion illustrates a general design procedure that is summarized in Figure 11.

EXAMPLE 2

An asynchronous sequential circuit is to have two inputs and a single output. The output is to become 1 (if it isn't already 1) when the input changes from $x_1 x_2 = 00$ to 10. It is to become 0 (if it isn't already 0) when the input changes from 00 to 01.

According to the design procedure, the first step is to construct the primitive flow table. We observe that, once the output has taken on the value 1, it will remain 1 until the input again reaches 00 and then 01. Similarly, when $z = 0$, it will remain 0 until the input reaches 00 and then 10. Hence, we would expect an output of both 0 and 1 for each input symbol, depending on the input sequence ending in that particular symbol. That means a total of eight stable states, two distinct stable states for each input symbol.

Starting at state ⓐ with input symbol $x_1 x_2 = 00$, the primitive flow table takes the form shown in Figure 12*a*. (Verify each of the rows.) As expected, there are eight stable states, two in each column. The next step is to construct the merger table; this is done in Figure 12*b*. (Verify the details.)

From the merger table the compatible pairs are determined and, from these, the maximal compatibles:

$$\text{set of maximal compatibles} = \{ab, bcd, efg, eh\}$$

There are two redundant states, b and e. Since the sets of compatibles implied by the maximal compatibles contain no pairs or larger sets, each redundant state can be deleted from either compatible set in which it appears, or each state can be retained in both compatibles. The considerations that guide this choice will be discussed in a later section. For now, let us arbitrarily delete state b from compatible ab and state e from compatible eh. The minimal reduced table can be constructed as shown in Figure 13*a*.

The next step is to assign secondary values to the states. (As noted earlier, the major problems connected with this step will be discussed in later sections.) For the present,

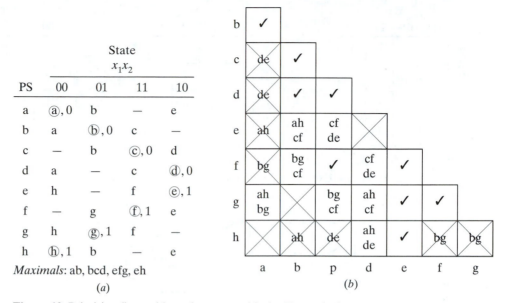

State x_1x_2

PS	00	01	11	10
a	ⓐ,0	b	—	e
b	a	ⓑ,0	c	—
c	—	b	ⓒ,0	d
d	a	—	c	ⓓ,0
e	h	—	f	ⓔ,1
f	—	g	ⓕ,1	e
g	h	ⓖ,1	f	—
h	ⓗ,1	b	—	e

Maximals: ab, bcd, efg, eh

(a)

(b)

Figure 12 Primitive flow table and merger table for Example 2.

	State x_1x_2				z x_1x_2					Y_1Y_2 x_1x_2			
PS	00	01	11	10	00	01	11	10	y_1y_2	00	01	11	10
a → 1	①	2	—	3	0	×	×	×	1 → 00	⓪⓪	01	—	10
bcd → 2	1	②	②	②	×	0	0	0	2 → 01	00	⓪①	⓪①	⓪①
efg → 3	4	③	③	③	×	1	1	1	4 → 11	①①	01	—	10
h → 4	④	2	—	3	1	×	×	×	3 → 10	11	①⓪	①⓪	①⓪

(a) (b)

Figure 13 Minimal flow table and transition table of example.

the assignment shown in Figure 13*b* is made, resulting in the transition table shown. Notice that stable state ②, assigned secondary value 01, appears three times in row 2. The *total* states represented by these three are all distinct, being distinguished from each other by their input values. Nevertheless, since they have the same secondary values, they are given the same name. The same is true for stable state ③ in the last row.

In a *primitive* flow table, the only way to reach a stable state is by a vertical transition through an unstable state. In a reduced table, on the other hand, a stable state can be reached (horizontally) simply by a change of input state alone. Thus, in Figure 13*a*, state ② in column 01 can be reached from state 1 in column 00. It can also be reached horizontally from state ② in column 11. However, state ② in column 10 can be reached only with a change of input values from state ② in column 11, without a change of secondary values.

As the last step, expressions for the excitation functions can be obtained directly from the transition table, or from excitation maps for the individual excitation functions obtained from this table. Maps for the excitations and the output obtained

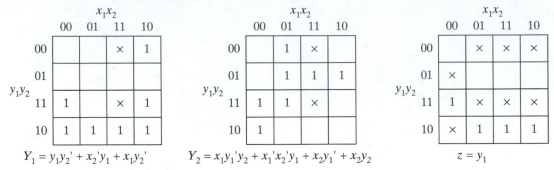

$$Y_1 = y_1 y_2' + x_2' y_1 + x_1 y_2'$$ $$Y_2 = x_1 y_1' y_2 + x_1' x_2' y_1 + x_2 y_1' + x_2 y_2$$ $$z = y_1$$

Figure 14 Excitation and output functions for Example 2.

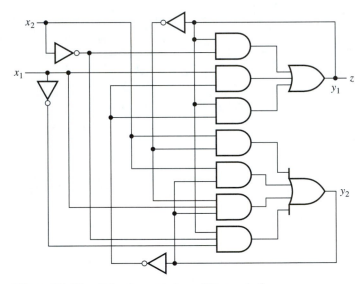

Figure 15 Circuit implementation of Example 2.

from Figure 13*b* are shown in Figure 14. The resulting expressions for the excitations and the output are given under the maps. (Confirm all of these.)

The circuit implementing these functions is shown in Figure 15. (Verify each of the above claims.) ∎

4 RACES AND CYCLES

Fundamental-mode circuits were defined as asynchronous circuits with level inputs where only one primary input change occurs at any one time. The reason for this restriction has to do with the practical difficulty of ensuring absolutely simultaneous changes. Since no two changes can be exactly simultaneous if the time scale is made fine enough, only nonsimultaneous changes in input are possible in practice. The real restriction under discussion, then, is that there should be sufficient separation in time between two input changes to ensure that the circuit has reached a stable state as a result of one change before the next input change is allowed to occur.

	Y_1Y_2 x_1x_2			
y_1y_2	00	01	11	10
$1 \rightarrow 00$	⓪⓪	01	—	11
$2 \rightarrow 01$	00	⑩①	⑩①	⑩①
$3 \rightarrow 11$	10	⑪①	⑪①	⑪①
$4 \rightarrow 10$	⑩⓪	01	—	11

Figure 16 Transition table for modified assignment.

A glance at the fundamental-mode circuit model in Figure 2 shows that, in this respect, the secondary inputs are not different from the primary inputs. For practical reasons, it is impossible to guarantee simultaneous changes for them as well. Hence, one secondary change must be sufficiently separated in time from another secondary change that a stable state has been reached before another secondary variable changes its value.

Glance back at the transition table in Figure 13b. Starting in any stable state, assume that a permissible input change occurs. What changes result in the secondary values on their way to the next stable state? Starting at $x_1x_2y_1y_2 = 0000$, let x_1x_2 change to 01. From the table, the required transition is to $y_1y_2 = 01$, through unstable 01. That is to say, only y_2 is to change its value, from 0 to 1. Similarly, from the same initial stable state, suppose that x_1x_2 changes to 10. The required secondary transition is to $y_1y_2 = 10$, through unstable 10. Again, only one secondary is to change its value: y_1 from 0 to 1. In this table, all other permissible input changes require only one secondary variable to change. (In other cases, permissible input changes may leave all secondaries unchanged, as illustrated in the preceding section.)

Critical and Noncritical Races

In order to prevent two secondaries from changing their values at the same time, we should determine how it is *possible* for two such changes to occur simultaneously. Such changes depend on the assignment of secondary values to the states. Return again to Figure 13b and suppose that the assignments of states 3 and 4 are interchanged. The result is shown in the transition table in Figure 16.

Now consider the permissible input change from total state $x_1x_2y_1y_2 = 0000$ to $x_1x_2 = 10$. Study the table carefully as you proceed. The required transition is to unstable state 11, from which the circuit should go to stable ⑪.

This requires a change of secondaries from 00 to 11, however. This required *simultaneous* change of two secondary variables is impossible in practice because it requires that the delays from inputs to each secondary variable be absolutely equal. In practice, either y_1 will change its value first or y_2 will. If y_1 changes first, the transition in column 10 is from $y_1y_2 = 00$ to 10. But the state to which the circuit is directed in column 10, row 10 remains unstable state 11, from which the required transition to $y_1y_2 = 11$ takes place. Thus, the originally required transition is achieved in two steps: from $y_1y_2 = 00$ to 10 to ⑪, but still it is achieved.

Now suppose that y_2 changes first, from the original value $y_1y_2 = 00$ to 01. The circuit then reaches stable state ⑩①, *an incorrect transition*. The machine is now in an incorrect stable state and its subsequent performance will be incorrect. The process can be viewed as a race between the secondaries to see which one can change first.

Y_1Y_2

y_1y_2 \ x_1x_2	00	01	11	10
00	(00)	(00)	01	10
01	—	11 ← (01) ↓	11	11
11	00	10 ↓	10	(11)
10	00	(10)	(10)	(10)

(a)

y_1y_2 \ x_1x_2	00	01	11	10
00			1	
01	×		1	1
11		×		1
10				

(b)

y_1y_2 \ x_1x_2	00	01	11	10
00			1	
01	×	1	1	1
11				1
10				

(c)

Figure 17 Formation of a cycle to avoid a critical race. (a) Transition table. (b) $Y_{2\text{old}}$. (c) $Y_{2\text{new}}$.

There exists a race in a fundamental-mode sequential circuit if, as a result of a permissible input change, more than one secondary value at a time is required to change. If the final stable state reached is dependent on the order in which the secondaries change their values, then the race is a critical race.

In the design of a circuit, a critical race cannot be tolerated; it must *always* be prevented. For the secondary assignments in Figure 16, there is a critical race. In the design of the same circuit, we found an assignment in Figure 13 that avoided this critical race. The solution is clear in this example: make the assignment of values to the states so as to avoid critical races. The situation is not always so easy, however, as we shall discover shortly.

In other cases a situation different from the one just illustrated is possible. In the table of Figure 10, assume a stable total state $x_1x_2y_1y_2 = 10\textcircled{11}$ when the input changes to a permissible $x_1x_2 = 00$. The requirement is for the secondary values to go to 00, a simultaneous change in two secondary variables. A race results, but this time, no matter which secondary changes first, the ultimate stable state is $y_1y_2 = \textcircled{00}$, the desired value. The race is said to be *noncritical,* and there is no need to lose sleep over it. (Verify that the stable state reached is $\textcircled{00}$.

The transition table in Figure 10 is really not *that* innocent; it contains another race occurring from state $11\textcircled{01}$ to the desired $01\textcircled{10}$. If y_2 changes before y_1, from 1 to 0, the circuit will go to stable state $\textcircled{00}$, which is not the correct state. This race is, therefore, critical. What to do? Notice that there is a don't-care in row 11, column 01. Suppose the don't-care is replaced by secondary values 10 in that position. Now, if we could manage to guide the changes in secondary values from the original $\textcircled{01}$ in column 11 to row 11, column 01, then we would be guided to the desired state $\textcircled{01}$ from there. Changing the assignment of secondaries in row 01, column 01, from 10 to 11 can do this. The changed transition table takes the form shown in Figure 17.

The arrows show the path taken by the change in total state:

$$x_1x_2y_1y_2 : 11\textcircled{01} \rightarrow 0101 \rightarrow 0111 \rightarrow 01\textcircled{10}$$

At each step, only one variable (primary or secondary) has changed, and the desired ultimate stable state is reached.

Cycles and Oscillations

The idea that a stable state can be reached from another stable state through a sequence of intermediate unstable states can be formalized by the following definition:

> *When a fundamental-mode sequential circuit makes a transition from one stable state to another through a unique sequence of unstable states, we say a* cycle *exists.*

Exercise 5   From Figure 16, determine a Boolean expression in sum-of-products form for excitation Y_1. Repeat for Y_1 from Figure 10b, and compare the two expressions.
Answer[8]

What would happen if, starting at a stable state, a sequence of unstable states results without ever reaching another stable state before the next input occurs? Because of the variability in gate delays, the final state of the circuit at the next input change will be uncertain. Hence, the subsequent performance will be uncertain. This leads to the following idea:

> *We say there is an* oscillation *if a permissible input change initiates a transition from some stable state through a sequence of unstable states without ever reaching another stable state before the next input change.*

An oscillation cannot be tolerated, and whatever measures are needed must be taken to prevent one.

By now some of you are getting anxious to find out what possible bearing this manipulation of flow tables has on actual circuit implementation. To answer the question, we will find the implementations of the circuit for the assignment in Figure 10 and the modified assignment in Figure 17. Part of the job will be left to you.

What follows will explicitly treat only excitation Y_2 using both the old assignment in Figure 10 and the new one in Figure 17. The expressions for excitation Y_2 are as follows:

$$Y_{2\text{old}} = x_1 x_2' y_2 + x_1 x_2 y_1' \tag{1}$$

$$Y_{2\text{new}} = x_1 x_2' y_2 + x_1 x_2 y_1' + y_1' y_2 = Y_{2\text{old}} + y_1' y_2 \tag{2}$$

Implementations of these two functions are shown in Figure 18. The new one is the same as the old one except for an added AND gate.

Let us now trace through each of these circuits to determine the consequences of a change in input $x_1 x_2$ from 11 to 01. For this purpose we will need the expressions for Y_1, both old and new; they happen to be the same as you discovered in Exercise 5. At each step in what follows, you should confirm the results.

---

[8]$Y_1 = x_1 x_2' + x_2 y_1 + x_1' x_2 y_2$. (The last term becomes $x_1' y_1' y_2$ if the don't-care is used to form a prime implicant.)   ◆

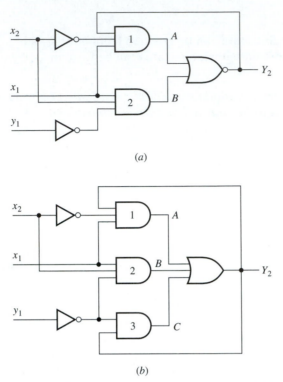

(a)

(b)

Figure 18 Implementations of Y_2 in example. (a) Old assignment. (b) New assignment.

When $x_1x_2 = 11$, the secondary values are $y_1y_2 = \textcircled{01}$. Now x_1 is to change from 1 to 0. In Figure 18a (the old circuit), when x_1 changes from 1 to 0, the outputs of both AND gates change to 0 (after a gate propagation delay). After one more gate delay through the OR gate, Y_2 changes from 1 to 0. As for Y_1 (whose value is initially 1), when x_1 goes from 1 to 0, $x_1{}'$ goes from 0 to 1, causing the third term in Y_1 (representing an AND gate) to go from 0 to 1. Since at least one input to the output OR gate in the implementation of Y_1 is 1, the value of Y_1 switches from 0 to 1. But this change occurs after three gate delays; if no special care is taken to put fast gates in the implementation of Y_1, the delay through this path will be somewhat greater than the delay through the Y_2 path. Hence, y_2 will change first, and so y_1y_2 will go from $\textcircled{01}$ to $\textcircled{00}$ — the wrong transition!

Now consider the new circuit for Y_2 in Figure 18b; it differs from the old one by having an additional AND gate. (Remember that Y_1 is the same.) The old Y_2 was 1 for $x_1x_2 = 11$ by virtue of the output of AND gate 2 being 1. In the new circuit, the outputs of both AND gates 2 and 3 are initially 1. When input x_1 changes to 0, the output of AND gate 2 goes to 0, but the output of AND gate 3 remains 1 until y_1 changes its value. Hence, y_1 changes before y_2; and so the state transition is from $y_1y_2 = \textcircled{01}$ to $Y_1Y_2 = 11$, after which Y_2 changes, making $Y_1Y_2 = 10$ and, finally, $y_1y_2 = \textcircled{10}$. So you see that manipulations in a transition table have effects in the real world.

By way of summarizing, we note that, one way or another, critical races must be avoided in a transition table. One way is simply to make an appropriate assignment of secondary values; that was the case in Figure 13. Another way is to utilize an unspecified entry to form a cycle in a column in which there is a critical race, as in Figure 17. Neither of these possibilities is a universal remedy because

- There may be no unspecified entries with which to form a cycle, and
- There may be no assignment that avoids a critical race—without additional measures.

An assignment that avoids critical races and oscillations deserves to be identified:

> A valid *assignment is any assignment of secondary values for which there are no critical races and no oscillations.*

A major problem in fundamental-mode circuit design is finding a valid assignment. An assignment is certainly valid if stable states between which transitions are to occur are given adjacent assignments. But, as previously discussed, an assignment will also be valid if cycles are formed from one stable state to another. There are other mechanisms for determining valid assignments; one is to give multiple assignments to some states so that all required adjacencies can be met. This might mean that the number of state variables—and so the complexity of the resulting circuit—is increased. Methods have been described in the literature that yield *universal* assignments for flow tables with certain numbers of states, sometimes without requiring extra state variables. We will not pursue the problem of finding valid assignments any further here.

5 HAZARDS

The design process described in the preceding sections proceeds through a series of steps to arrive at a transition table, such as the one in Figure 17*a*. But the transition table is simply another form of the truth table for the excitation functions. The only difference between the fundamental-mode circuit and a combinational circuit is the feedback in the former. There is no difference in the *process* of obtaining suitable expressions for the excitations from the transition table, on the one hand, and obtaining an expression for a combinational function from a logic map or minterm list, on the other.

There is, however, a difference in *meaning*. In the case of the transition table, some of the variables are not independent inputs but secondaries. Hence, the problem we will be discussing in this section applies to combinational circuits as well as to fundamental-mode sequential circuits. The consequences, however, are different.

Static Hazards

In the earlier discussion of the origins of a race in a fundamental-mode sequential circuit, we discovered that propagation delay through the gates plays a big role. The same factor—delay—also plays another role, as we will now discuss.

Suppose we have gone through a fundamental-mode design process and arrived at the map of an excitation function Y_1 shown in Figure 19. Also given are the minimum sum-of-products expression and its circuit implementation. The

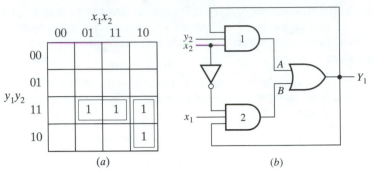

Figure 19 Excitation function and its circuit implementation. (a) $Y_1 = x_1x_2{}'y_1 + x_2y_1y_2$. (b) Circuit implementation.

implementation for excitation Y_2 is not shown, but let's assume that changes in Y_2 occur more slowly than changes in Y_1. Now suppose that $y_1y_2 = ⑪$ and x_1x_2 changes from 11 to 10. According to the *map*, the value of Y_1 is to remain unchanged ($Y_1 = 1$); but consider what can happen in the *circuit*.

Before the change in x_2 from 1 to 0, the AND gate outputs were $A = 1$ and $B = 0$, so Y_1 was 1. Now x_2 becomes 0; if AND gate 1 is faster than the path through the inverter and AND gate 2—which is not unlikely—then A becomes 0 *before* B has changed from 0 to 1. Hence, both inputs to the OR gate are 0 for some brief time interval, causing Y_1 to become 0. But Y_1 is fed back to the inputs of both AND gates. The result is that, depending on the differences in delay through various paths, a momentary change in Y_1 to 0 can become a permanent change—leading to an incorrect value of a secondary variable.

The same problem of differences in delay through various paths would exist *even if this were a combinational circuit*. In that case, however, a momentary incorrect value (blip) is not serious since it will eventually correct itself. It is the *feedback* that causes the trouble here.

Now, let us generalize the idea described in the preceding example. Suppose that the transition table of a fundamental-mode machine requires the value of one or more secondary variables to remain unchanged (at 0 or 1) when a permissible input change (only one variable) takes place. We make the following definition:

Suppose the value of one or more secondary variables in the transition table of an asynchronous sequential circuit is to remain unchanged (at 0 or 1). Suppose also that, as a result of unequal delays through different paths, it is possible for one or more secondary variables to have a blip—a momentary change of value—due to a permissible change of input. Then we say that a static hazard exists in the circuit.

(The term *static* refers to the fact that the secondaries are supposed to remain unchanging, that is, static.)

The existence of a static hazard means that the circuit is defective and the problem will have to be corrected. Both the cause of the problem and its remedy are evident from the logic map in Figure 19a. The expression for Y_1 is obtained from the two prime implicants shown on the map. Each prime implicant is im-

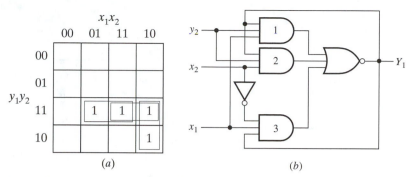

Figure 20 Adding a gate to avoid a hazard.

plemented by an AND gate. The hazard occurs because a change in a primary input (x) moves the circuit out of one prime implicant into the other. Over a short interval of time, neither prime implicant may be operative. In terms of the AND gates, one had an output of 1 and the other an output of 0. At the change of a primary input, both AND gates are supposed to change their outputs.

The problem arises when the one whose output is to change to 0 makes its transition first. One possible remedy you might think of is to introduce additional delay in this path in order to retard its transition. This can be done, for example, with two consecutive inverters or a buffer. The logic would remain unchanged; only the delay would increase. But speed is an important design parameter, and circuits with fast response are preferable to slow ones.

Suppose there had been a third AND gate feeding the OR gate such that its output would remain 1 for that particular change of primary input. The OR gate output would remain 1, no matter which of the other gates made its transition first and no matter how long the transitions took. This can be achieved by including another prime implicant in the expression, as illustrated in Figure 20a. Neither the expression ($Y_1 = x_1 x_2{}' y_1 + x_1 y_1 y_2 + x_2 y_1 y_2$) nor the circuit is now minimal. There is a redundant gate that serves no other purpose but to prevent a hazard.

We see here an illustration of two conflicting requirements: simplicity of circuit, and reliability of operation. The simplest circuit is the minimal one, but it is susceptible to hazards. To eliminate the hazards, we increase circuit complexity. It should not be assumed from the illustration that *all* minimal circuits will have hazards. For example, look at the map for $Y_{2\text{new}}$ in Figure 17 and its implementation in Figure 18. Its expression in (2) is the sum of three prime implicants. When $y_1 y_2 = 01$, the third prime implicant remains equal to 1 for any permissible changes in primary inputs. Hence, the circuit is hazard free even though it is minimal.

Considering only two-level sum-of-products implementations, the general principle in the design of hazard-free circuits is to ensure that any adjacent minterms for which the value of the function is 1 are covered by a prime implicant. (Note that this is a sufficient condition for only a two-level circuit to be hazard free. The situation is more complicated for higher-order circuits; it is possible for such circuits to be hazard free without satisfying the preceding principle.)

S	x_1x_2 = 00	01	11	10
1	①	2	①	3
2	3	②	②	②
3	③	4	2	③
4	1	④	④	2

(*a*)

$y_1y_2y_3$	$Y_1Y_2Y_3$, x_1x_2 = 00	01	11	10
1 → 000	(000)	001	(000)	100
2 → 001	011	(001)	(001)	(001)
3 → 011	(011)	010	001	(011)
4 → 010	000	(010)	(010)	110
3 → 100	(100)	101	110	(100)
4 → 101	111	(101)	(101)	001
1 → 111	(111)	110	(111)	011
2 → 110	100	(110)	(110)	(110)

(*b*)

Figure 21 Example transition table with assignments.

y_2y_3	$y_1 = 0$, x_1x_2 = 00	01	11	10	$y_1 = 1$, x_1x_2 = 10	11	01	00
00		1				1		
01	1	1	1	1	1	1	1	1
11	1		1	1	1		1	1
10								

Figure 22 Hazard-free design of Y_3 in Example 3.

EXAMPLE 3

The general principle is illustrated by the reduced transition table in Figure 21*a*. By examining the transitions required from each stable state for permissible input changes, we note that each state is required to be adjacent to all three other states. This is impossible to achieve with only two state variables. Hence, the number of state variables must be increased. A transition table in which each of the four states is given two different assignments is shown in Figure 21*b*. For this assignment, each permissible input change leads to a direct transition to the required stable state, without cycles. So the transition is made with the least possible transition time (TT).[9]

[9]Other assignments are also possible in which not all states are given multiple assignments. That leaves some rows with don't-care entries that can be utilized to form cycles. The resulting implementation will have a greater total transition time.

The map for Y_3 is shown in Figure 22. (Confirm everything.) The order of the variables for establishing the minterm numbers is $y_1 x_1 x_2 y_2 y_3$. There are three prime implicants having the following minterm lists:

$$x_1 y_3 = \Sigma(9, 11, 13, 15, 25, 27, 29, 31)$$
$$x_2{}' y_3 = \Sigma(1, 3, 9, 11, 17, 19, 25, 27) \tag{3}$$
$$x_1{}' x_2 y_2{}' = \Sigma(4, 5, 20, 21)$$

The corresponding minimal s-of-p expression for the excitation is

$$Y_3 = x_1 y_3 + x_2{}' y_3 + x_1{}' x_2 y_2{}' \tag{4}$$

Note: certain pairs of minterms are adjacent, namely, those for which excitation $Y_3 = 1$: 1 and 5, 17 and 21, 5 and 13, and 21 and 29. A permissible input change will change the secondaries from one member of each pair to the other. Hence, *each one represents a hazard.*

From the map, one can see that a set of eight minterms forms a 3-cube, but in the expression for Y_3, they are not all covered by a single prime implicant. The remedy is clear: to eliminate the hazards, a prime implicant covering all these minterms should be added to Y_3. The new expression for Y_3 will be

$$Y_3 = x_1 y_3 + x_2{}' y_3 + x_1{}' x_2 y_2{}' + y_2{}' y_3 \tag{5}$$

(Draw the circuit represented by this expression.) At the cost of one additional AND gate, Y_3 has been made hazard free, and speed has not been sacrificed in the process. ∎

A number of questions now arise. Suppose a hazard-free sum-of-products expression has been implemented. The circuit is drawn by looking at those cells in a logic map for which the excitation function is 1 and is to remain 1 during a permissible input change. Will the hazard-free circuit remain hazard free based on the 0's? Although the proof is beyond our scope, the answer has been provided in the literature: yes. This means that there is no need to worry separately about making sure the 0's of a function remain 0 when they are supposed to, provided that we have already ensured that the 1's remain 1 when *they* are supposed to.

The converse is also true. A hazard-free product-of-sums implementation can be found in a completely dual manner by ensuring that 0 cells in a logic map that are adjacent under permissible input changes are covered by a prime implicant. This ensures that when the value of the excitation function is to remain 0 for a permissible input change, no momentary flickers to a 1 will occur. But it also ensures that the excitation function will not have an incorrect transition from 1 to 0 when it is supposed to remain at 1. As is true in general for combinational circuits, one of these two circuits might be simpler than the other one.

A word of caution is needed at this point. What is hazard free is the *circuit* or the specific expression of which the circuit is an implementation. If this expression is manipulated into a different (though equivalent) form, there is no reason to believe that the result is also hazard free.

Dynamic Hazards

The preceding section discussed the case when a logic function is to remain unchanged (at 1 or 0) for a permissible input change. At other times, a given function is required to *change* value (from 1 to 0 or 0 to 1) for a permissible input change. An example is illustrated in the map in Figure 22. Suppose the total state is $y_1x_1x_2y_2y_3 = 00100$ and x_1 changes from 0 to 1. According to the map, the excitation is to go from 1 to 0. However, it is possible for the excitation to make the required transition from 1 to 0 but then have a glitch (return to 1 before settling into its desired value of 0). This might occur due to feedback, because of differences in delay through various paths in the implementing circuit. If this should happen, there is a danger that the second change (to the incorrect value of 1) could become permanent. Here again there would be a hazard. This time, because it occurs for a required *change* in the value of the function, it is called a *dynamic hazard*.

Fortunately, it has been proved in the literature that no special measures are needed to remove dynamic hazards; if a two-level circuit is free of static hazards, it will also be free of dynamic hazards. Hence, the remedy discussed for static hazards will also cure dynamic hazards.

Essential Hazards

Both static and dynamic hazards occur because of the way a given transition table is implemented. However, in some situations, it is possible for a hazard to exist from the very nature of the design specifications themselves, not because of klutziness in circuit design. That is, the possibility of making an incorrect transition due to differences in delay through various paths may result from the structure of the flow table itself.

Consider a stable state in a flow table. Suppose that, due to a single change in one input variable, the stable state reached in an adjacent column is different from the stable state reached starting from the same initial state after three consecutive permissible changes of the same input variable. It has been shown in the literature that, if there are at least two feedback paths in the circuit, this circumstance can cause an erroneous transition for certain combinations of delays through the feedback paths. Because this hazard is a consequence of the nature of the state table itself, it is called an *essential hazard*. Examples of flow table structures resulting in essential hazards are shown in Figure 23. The headings I_1 and I_2 refer to inputs.

Essential hazards are not uncommon. They occur in many practical circuits. They cannot be eliminated by adding gates, as static hazards can be. Eliminating an essential hazard requires careful adjustment of the delays in specific feedback loops in a circuit. Facility in doing this comes only with experience.

EXAMPLE 4

A fundamental-mode circuit is to be part of an electronic lock. It has two level inputs x_1x_2 and a level output z. The lock opens when $z = 1$. The input "combination" that opens the lock is

I_1	I_2
ⓐ	b
c	ⓑ
ⓒ	d
—	ⓓ

I_1	I_2
ⓐ	b
c	ⓑ
ⓒ	ⓒ

(a) (b) **Figure 23** Essential hazard possibilities.

S
x_1x_2

	00	01	11	10
a	ⓐ,0	b	—	g
b	c	ⓑ,0	h	—
c	ⓒ,0	d	—	g
d	e	ⓓ,0	h	—
e	ⓔ,1	f	—	g
f	a	ⓕ,0	h	—
g	a	—	h	ⓖ,0
h	—	f	ⓗ,0	g

(a)

S
x_1x_2

	00	01	11	10
ag → a	ⓐ,0	b,0	f,0	ⓐ,0
b	c,0	ⓑ,0	f,0	—
c	ⓒ,0	d,0	—	f,0
d	e,1	ⓓ,0	f,0	—
e	ⓔ,1	f,0	—	f,0
fgh → f	a,0	ⓕ,0	ⓕ,0	f,0

(b)

Figure 24 Flow tables illustrating essential hazards. (a) Primitive. (b) Reduced.

$$x_1x_2:\ 00\ 10\ 00\ 10\ 00$$
$$z:\ \ 0\ \ 0\ \ 0\ \ 0\ \ 1$$

The objective is to find the primitive flow table, reduce it, and then observe if any essential hazards exist.

The primitive flow table is shown in Figure 24a. One of the patterns in Figure 23 is visible here, so there is an essential hazard. (Verify that claim.) It might be argued that the pattern would disappear if the table were reduced. The minimal reduced table is shown in Figure 24b. Starting in stable state ⓐ for $x_1x_2 = 00$, if x_2 changes to 1, the stable state reached is ⓑ; but if x_2 changes from 0 to 1, then back to 0 and back to 1 again, the stable state reached is ⓓ—a different state. Hence, there is an essential hazard. ∎

Exercise 6 **a.** Confirm the reduced table in Figure 24b.
b. Find all other essential hazards in this flow table.
Answer[10]

---

[10]There are five other essential hazards.

◆

CHAPTER SUMMARY AND REVIEW

This chapter dealt with sequential circuits that have no clock to synchronize state transitions; they are asynchronous. (Basic latches and flip-flops themselves are asynchronous circuits.) The states are established by the unavoidable delays through logic gates. There are two general classes of state changes in such circuits: those brought about by changes of input levels and those caused by input pulses. Circuits in which state changes occur as a result of input-level changes are said to be operating in the fundamental mode. This chapter dealt only with such circuits. Among the topics included were

- The flow table versus the state table in synchronous machines
- Unstable states and stable states
- Primitive flow tables
- Assigning outputs to unstable states
- Rapid and slow transitions to a stable state
- Applicable input sequences in fundamental-mode circuits
- Minimizing flow tables in incompletely specified machines
- Implied pairs of states
- Compatibility and incompatibility of states
- The merger table
- Determination of compatible sets of states
- Determination of maximal compatibles
- Determination of closed sets of compatible states
- Assigning secondary-variable values and reducing merger tables
- State transition tables
- Design procedures for fundamental-mode circuits
- Races in fundamental-mode circuits:
  - Critical
  - Noncritical
- Cycles as transitions between two stable states
- Oscillations that never lead to a stable state
- Static hazards: momentary blips in secondary values
- Avoiding static hazards: hazard-free design
- Dynamic hazards
- Essential hazards

PROBLEMS

You will be able to carry out parts of many of these problems only after studying the later parts of the chapter. Save partial solutions; then complete the problems after you have studied the later material.

1 I. A fundamental-mode asynchronous machine has two inputs and a single output. The output becomes 1 only after the following input sequence:

$$x_1x_2: 00 \rightarrow 10 \rightarrow 11 \rightarrow 01$$

In addition, if x_1 and x_2 have the same value, then x_2 cannot change before x_1.

- **a.** Construct a primitive flow table.
- **b.** Assuming a flicker-free fast output, reduce the table to minimal form.

c. Make a valid assignment, and write expressions for the excitation functions.
d. Implement the circuit.
e. Discuss whether the circuit has essential hazards.

II. Repeat the preceding parts with the modification that, after the output becomes 1, it remains 1 until x_1 changes value.

2 I. An asynchronous machine has two inputs x_1 and x_2 and a single output z. The output becomes 1 only when the input combination goes from 01 to 11.

a. Construct a primitive flow table starting from the clear state. First follow any input sequences that result in $z = 1$. Then consider any other permissible input changes from existing stable states and add new states as needed to complete the table.
b. Assign unstable-state outputs so that the machine is flicker free.
c. Obtain a reduced, minimal table. If there is more than one possibility, select one that has advantages over the others.
d. Choose an assignment that is free of critical races. (If there is more than one possibility, try each one and note the difference in complexity.)
e. Write hazard-free expressions for the state variables and obtain a circuit implementation.

II. Repeat all of part I for the requirement that the output remains unchanged except for the following input changes:

$z = 1$ if the input sequence goes from 10 to 11.
$z = 0$ if the input sequence goes from 11 to 10.

3 An electronic lock has two level inputs x_1 and x_2 and a single level output z. The lock "opens" when $z = 1$. The "combination" that opens the lock is as follows. Starting at $x_1x_2 = 00$, x_1 turns *on* and *off* (goes to 1 and then 0) twice; then x_2 turns *on*. Any subsequent change in x_1 or x_2 closes the lock.

a. Construct a partial primitive flow table starting from the clear (00) state and following an input sequence that opens the lock.
b. Complete the primitive flow table, including all other possible states.
c. Assume flicker-free fast transitions and reduce the table to minimal form.
d. Make a race-free valid assignment and write expressions for the excitation functions.
e. Obtain an implementation for the circuit.

4 An electronic lock has three inputs A, B, and C and a single output z. The lock opens only after the following sequence starting from the clear (000) condition:

A turns *on* and *off* once; then *B* turns *on*; then *C* turns *on*.

a. Construct a flicker-free primitive flow table. (See if you can specify the number of states in the table before constructing it; were you right?)
b. Reduce the flow table to minimal form.
c. Specify any essential hazards present in this table.
d. Make a critical-race-free assignment.
e. Construct a transition table, specifying any cycles that are formed to prevent critical races.
f. Obtain hazard-free expressions for the state and output variables.
g. Implement the corresponding circuit.

5 A fundamental-mode sequential circuit has two inputs x_1x_2 and two outputs, z_1z_2. Whenever either input changes from 0 to 1 when the other input is 0, one of the two outputs changes its value; otherwise, there is no change in output. Assume that the outputs are initially 00 and that the first output to change is z_2.

a. Construct a fast flicker-free flow table.
b. Reduce this to a minimal table. If more than one is possible, pick one with the maximum number of unspecified entries.
c. Choose a valid assignment and construct a transition table.
d. Specify any cycles that may be formed, and indicate the number of transition times involved in each.
e. Write hazard-free expressions for the state variables and expressions for the output functions.
f. Determine an implementation of the expressions in part e.

6 A fundamental-mode sequential circuit has two inputs x_1, x_2 and a single output z. The output is to remain 0 until the last set in the following sequence of inputs, when it changes to 1:

$$x_1x_2: 00 \rightarrow 10 \rightarrow 11 \rightarrow 01$$

a. Construct a flicker-free flow table, first following a correct input sequence to produce $z = 1$.
b. Obtain a minimal reduced table. If there is more than one, choose one with the most unspecified entries.
c. Find a critical-race-free assignment and determine hazard-free expressions for the state variables. Determine an expression for the output also.
d. Construct an implementation that realizes the expressions in part c.
e. In part b, suppose that redundant states are not eliminated from compatibles in the minimal closed cover. What advantage or disadvantage can there be in this?

7 A fundamental-mode asynchronous circuit has two inputs x_1, x_2 and a single output z. The output is to become 1 if and only if the input sequence is either of the following:

$$x_1x_2: 00 \rightarrow 10 \rightarrow 11 \quad \text{or} \quad 11 \rightarrow 01 \rightarrow 11$$

a. Construct a primitive flow table starting from the clear state. First follow input sequences that produce $z = 1$; then follow all other permissible input changes from already established stable states and add new states, as needed, to complete the table.
b. Complete the table by assuming fast, flicker-free outputs.
c. Reduce to a minimal flow table.
d. Choose a valid assignment and construct a transition table. Verify that there are no critical races.
e. Implement the circuit.

8 A partially reduced flow table is given in Figure P8.

S

x_1x_2

	00	01	11	10
a	ⓐ, 0	b, 1	—	d, 0
b	ⓑ, 1	ⓑ, 1	c, 1	e, 1
c	—	b, 1	ⓒ, 1	e, 1
d	a, 0	b, –	—	ⓓ, 0
e	a, 0	ⓔ, 0	f, 0	ⓔ, 1
f	—	e, 0	ⓕ, 0	d, 0

Figure P8

a. Find a minimal reduced table.

b. If possible, choose a valid assignment and construct a transition table. If not possible, increase the number of state variables and choose a valid assignment that gives multiple assignments to some or all states.

c. In the latter case, the number of state variables will be the same as the number needed for the original table. Choose a valid assignment for the original table (before reducing) and construct a transition table. If there are any cycles, specify them. Compare the complexity of implementation.

S
$x_1 x_2$

00	01	11	10
ⓐ	ⓐ	ⓐ	e
a	c	ⓑ	ⓑ
ⓒ	ⓒ	ⓒ	d
ⓓ	a	b	ⓓ
d	ⓔ	c	ⓔ
a	ⓕ	ⓕ	b

(a)

S
$x_1 x_2$

00	01	11	10
ⓐ	c	f	ⓐ
d	ⓑ	e	a
ⓒ	ⓒ	e	ⓒ
ⓓ	f	ⓓ	c
a	b	ⓔ	f
c	ⓕ	ⓕ	ⓕ

(b)

S
$x_1 x_2$

00	01	11	10
ⓐ	d	b	ⓐ
ⓑ	c	ⓑ	c
a	ⓒ	d	ⓒ
b	ⓓ	ⓓ	a

(c)

S
$x_1 x_2$

00	01	11	10
①	4	①	①
②	②	②	1
5	4	③	6
2	④	3	5
⑤	2	1	⑤
1	⑥	2	⑥

(d)

Figure P9

9 For each table in Figure P9:

a. Find a valid assignment having no critical races and requiring the fewest secondary variables.

b. Construct a transition table and write expressions for the state variables. Find an implementing circuit for each case.

$$Y_1 Y_2$$
$$x_1 x_2$$

$y_1 y_2$	00	01	11	10
a → 00	⟨00⟩	01	⟨00⟩	⟨00⟩
b → 01	00	⟨01⟩	11	⟨01⟩
c → 11	10 ← ⟨11⟩	⟨11⟩ → 10		
d → 10	00	⟨10⟩	⟨10⟩	00

Figure P10

10 A race-free transition table is shown in Figure P10.

    a. Obtain a circuit implementation.

    b. Assume an initial total state of $x_1 x_2 y_1 y_2 = 00\langle00\rangle$ followed by an input change to $x_1 x_2 = 01$. Taking gate delays (assumed to be equal) into account, evaluate all initial gate outputs and changes in these outputs. Verify that the correct state transition is made.

    c. Repeat part *b* starting at total state $11\langle11\rangle$ followed by an input change to $x_1 x_2 = 10$.

11 Find a product-of-sums hazard-free implementation of the function in Figure 20 and compare its complexity with that of the sum-of-products hazard-free implementation in that figure.

12 Find a minimal product-of-sums implementation of the function whose map is given in Figure 22. Is it hazard free? Compare its complexity with that of the hazard-free sum-of-products expression in (5).

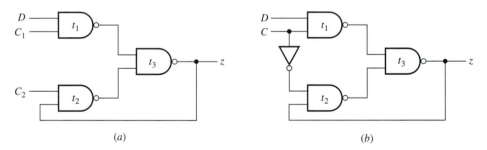

(a)   (b)

Figure P13

13 The circuit shown in Figure P13*a* is meant to be a latch. *D* is the data input; C_1 and C_2 are clock inputs; and t_1, t_2, and t_3 are the propagation delays of the gates. When $C_1 C_2 = 10$, the latch is *transparent,* or open; when $C_1 C_2 = 01$, the latch is *holding,* or closed. The 0-to-1 edge of C_1 and the 1-to-0 edge of C_2 will be referred to as the *opening* edges of the clocks (they open the latch; the other two edges of C_1 and C_2 will be called the *closing* edges of the clocks). The two clock inputs are meant to be complementary, but for "proper" operation of the latch, a small separation is needed between the two opening edges and also between the two closing edges. "Proper operation" means the following:

    The circuit must latch the data reliably. That is, if any value *x* is set up on *D* sufficiently before the latch closes, the value of output *y* after the latch has closed must be *D*.

    Y must not "glitch" if the value currently stored in the latch is reloaded into the latch. That is, if $y = z$ (where *z* may be either 0 or 1) and *z* is set up on *D* sufficiently before the latch opens, y must continue to be z without any momentary incorrect output when the latch opens.

a. Determine the proper separations between the opening edges of C_1 and C_2 and between their closing edges. In each case, indicate which clock should lead. Explain, with the help of waveforms, the improper operation that could occur if the required separation is not maintained.

b. Based on your answer to part a, explain why it is not advisable to obtain C_1 by inverting C_2, or vice versa.

c. Show that the combinational circuit in Figure P13b, not including the feedback, has a static hazard.

d. Show how this static hazard can be eliminated with the help of extra logic. With this extra logic, show that the sequential circuit formed by the feedback shown in the figure is a properly operating latch, according to the definition of proper operation given above.

14 Design an asynchronous circuit that has two inputs (x and y) and an output (z). If x and y agree ($x = y$), then $z = x = y$. Otherwise z retains the last value at which x and y agreed. (This circuit is called the Miller C-element.)

15 An asynchronous circuit has N inputs ($x_1 \ldots x_n$) and one output z. The inputs represent the opinions (yes/no) of N people on a controversial question. If two-thirds or more of the people have the same opinion, z should reflect it. Otherwise z should reflect the last opinion held by two-thirds or more of the people. Determine an implementation of this circuit.

16 A *control delay* circuit is an asynchronous circuit with two inputs, C and x, and a single output, z. C is a clock—a periodic pulse train—and x is a sequence of aperiodic pulses having the same width as the clock pulses but delayed from the clock pulse by a small fraction of a pulse width. The output z is a pulse of the same width initiated by an x pulse but delayed by exactly one clock period from the x pulse. (Indeed, an x pulse could be the output of another control delay circuit.) A timing diagram is shown in Figure P16. Take the clear state to be any time after an output pulse has occurred and before the next clock pulse.

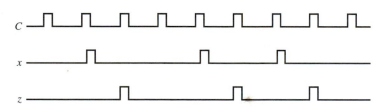

Figure P16

a. Construct a primitive flow table.
b. Find a minimal reduced table.
c. Make an assignment free of critical races
d. If there are noncritical races, different transition tables will result when one or the other secondary changes first. Choose the one yielding the simplest expressions for the state variables, and implement the circuit.

$$Y_1Y_2$$
$$x_1x_2$$

	00	01	11	10
a → 00	⓪⓪	11	⓪⓪	11
b → 01	11	⓪①	11	11
c → 11	⑪	⑪	00	⑪
d → 10	00	⑩	11	11

Figure P17

17 a. In the flow table of Figure P17 find all races, if any; specify those that are critical and those that are noncritical.
   b. Suppose the total input is $y_1y_2x_1x_2 = 1101$ and it changes to 1111. Describe the resulting changes in the excitations. Does this constitute a cycle?
   c. If possible, find another assignment that does not contain critical races. (Describe the process by which one gets such a new assignment or arrives at the conclusion that it's not possible.) If you find any noncritical races and cycles in the new assignment, specify them.

18 The state table of an incompletely specified synchronous machine is given in Figure P18.

NS,z
$$x_1x_2$$

PS	00	01	11	10
A	—	G, 1	C, 1	G, 1
B	F, 1	—	E, 0	D, 0
C	—	G, 1	E, 0	—
D	—	G, –	—	D, 0
E	B, 1	—	C, 1	—
F	B, 1	—	A, 0	A, 0
G	—	G, 1	—	A, 0

Figure P18

   a. Construct a merger table.
   b. Determine a minimal reduced table that covers the given table. Where there is a choice of states to include in a compatible, discuss the advantages and disadvantages of your choice.
   c. Make an optimal state assignment and construct a transition table.
   d. Assuming an implementation with D flip flops, obtain expressions for the flip-flop excitations.
   e. Implement the circuit.
   f. Repeat parts d and e if the flip-flops are to be JK.

State, z

x_1x_2

00	01	11	10
Ⓐ, 0	B, 0	—	D, 0
A, 0	Ⓑ, 0	C, 0	—
—	B, 0	Ⓒ, 0	D, 0
A, 0	—	E, 1	Ⓓ, 0
—	B, 0	Ⓔ, 1	F, 1
G, 1	—	E, 1	Ⓕ, 1
Ⓖ, 1	H, 1	—	F, 1
G, 1	Ⓗ, 1	E, 1	—

Figure P19

19 The primitive flow table for an asynchronous sequential circuit is given in Figure P19.

    a. Construct a merger table.

    b. Of all possible minimal closed covers, select one that is most likely to allow a race-free assignment and make a race-free assignment; construct a reduced flow table.

    c. Is it possible to make a race-free assignment without increasing the number of state variables? If so, make such an assignment and construct a transition table; specify the number of cycles you have formed to avoid critical races.

    d. If not, increase the number of state variables and repeat part *c*.

    e. Implement the appropriate table.

Chapter 8

Design Using Hardware Description Languages

The preceding chapters in this book presented the basic principles on which digital logic design is founded (switching algebra, logic maps, state diagrams, transition tables, and so on). Although some algorithmic procedures were described, and it was mentioned that software for carrying out these algorithms exists, such software tools were not utilized. The manual procedures described in the preceding chapters are inadequate to deal with machines needing more than, say, four or five flip-flops (16–32 states).

For larger machines the number of state assignments to consider would be huge; the derivation of the flip-flop excitation requirements would require logic maps of substantially more than six variables. Nevertheless, these principles form the indispensable basis for computer-aided design (CAD) tools used in large-scale logic system design.

Prior to the construction of a prototype, it is standard practice for engineers to create a computer-readable specification of a design that can be analyzed with simulation programs. This method permits verification of the functional correctness of a design in much less time and effort compared with the evaluation of a prototype. Until about 15 years before the end of the century a popular choice for the specification of a digital system was *schematic entry* (often called schematic capture). The process of schematic entry assumes that the designer has performed design synthesis, because the schematic that is being captured is indeed an implementation. Thus, schematic capture is limited to the design of systems with up to 100 or so gates and flip-flops (systems that can be manually synthesized).

The limits of schematic capture were realized quickly, and an alternative for the specification of digital systems was needed. A requirement for this alternative is that it allow the designer to specify behavior without performing manual synthesis. *Hardware description languages (HDLs)* circumvent problems with schematic capture and manual synthesis and are now universally used in design.

The major benefit of specifying a digital logic circuit using an HDL is that the designer need only capture the *specification* of the circuit in such a language, not the circuit itself. Following that, the implementation can be carried out in an automated way, through the use of CAD tools.

A hardware description language consists of a sequence of statements in a command language, very similar to a software program in a high-level language. However, hardware inherently operates in parallel, so tools that operate on the language, such as simulators, synthesizers, and translators, must interpret the language appropriately. Furthermore, in order to write correct specifications, the designer must understand how these tools interpret the language.

Use of an HDL has another benefit. One can conceive of two approaches in design:

- Carrying out a hardware implementation directly from the specifications, or
- If possible, simulating the design to verify its correctness before hardware is produced

Obviously, the latter is preferable. When a digital system specification is written in an HDL, it can be executed using a simulator to verify functional correctness and timing requirements. Only then is the hardware implemented. There are enormous savings in effort and cost this way.

Furthermore, once the HDL specification is verified for functional correctness, synthesis tools are available to carry out a gate-level design. Such a synthesis tool applies methods similar to those studied in Chapters 3 and 6.[1] Finally, the gate-level specification is translated into a code that can be used to program a specific integrated device.

Many hardware description languages are currently in use, some more popular than others. The two most popular in professional use are VHDL[2] and Verilog.[3] Each of these languages is comprehensive. However, the amount of coverage required for a book to present one of these languages—and for you to assimilate the details sufficiently to use it in design—is prohibitive at this level.

1   THE HARDWARE DESCRIPTION LANGUAGE ABEL

In this book we will present and utilize the hardware description language having the acronym ABEL[4] (Advanced Boolean Expression Language). Although not as comprehensive as VHDL or Verilog, ABEL is conceptually similar and is used extensively for the specification of systems implemented with programmable logic devices (PLDs).

We will introduce the ABEL language gradually, providing sufficient detail to enable you to use it in the design of circuits using PLDs.[5] Once you have

---

[1]The methods used in contemporary synthesis tools are more advanced than those we study in this book but are extensions of the same theory.

[2]VHDL was created through the VHSIC (Very-High-Speed Integrated Circuit) funding program of the U.S. Department of Defense; it is a public language. VHDL is a hierarchy of acronyms: it stands for VHSIC Hardware Description Language.

[3]Verilog is a hardware description language created by Cadence, Inc. in the early 1980s and has since become a public language.

[4]ABEL was created in 1984 by Data I/O Corporation.

[5]For additional language features and details, see David Pellerin and Michael Holley, *Digital Design Using ABEL,* Prentice Hall, 1994.

```
(1) module adder
(2) Title 'full adder cell'

(3) Declarations

(4) A PIN;
    B PIN;
    Cin PIN;
(5) S PIN istype 'com';    "combinational output
    Cout PIN istype 'com';

(6) Equations

(7) S = A $ B $ Cin;    "sum output
    Cout = A & B # A & Cin # B & Cin;    "carry output

(8) end adder
```

Figure 1 ABEL description of a full adder.

started practice as an engineer, your familiarity with ABEL will permit you to learn and easily apply one of the other hardware description languages.

ABEL uses constructs such as *pins* and *nodes* to represent points in a circuit where connections are made; it uses equations, truth tables, and state diagrams for the specification of circuit behavior. Rather than first describing the full details of the language, we will introduce it with an example, identifying general features of the language as we examine the specifics of this and other examples.

Adder Specification in ABEL

An ABEL description of a full adder is shown in Figure 1. (Review Chapter 4 on adders if you need to.) It looks similar to a program written in a high-level programming language, but it is conceptually different. The line numbering shown in parentheses is not a feature of the language but is given, *in this example only,* for ease of reference. (This numbering will be omitted in subsequent ABEL descriptions.)

The term *pin* is used with two interpretations. One is the physical location on a device to which an external connection is made; the other is a signal that is associated with that location. It is analogous to a variable in a programming language. The equations specify circuit behavior, where the output signal is on the left-hand side and an expression that produces the output signal is on the right. These equations are analogous to assignment statements in a programming language.

The main difference between an HDL and a programming language is that statements in a programming language are evaluated in sequence. On the other hand, statements in a hardware description language are interpreted as though execution is carried out in parallel. The combinational-logic circuit for a full adder, for example, does not produce the sum first and then the carry, but produces them in parallel. Hardware operates inherently in parallel; thus, an HDL

Operator Symbol	Operator Description
!	NOT
&	AND
#	OR
$	XOR
!$	XNOR

Figure 2 Some logic operations in ABEL.

must characterize this parallelism. In the ABEL description in Figure 1 it makes no difference which equation (S or Cout) is written first, since the evaluation by the simulator is as though they are executed in parallel.

Entries in an ABEL description of a logic task are called *statements*. An obvious need is for beginning and ending statements. The beginning one is a *module statement,* in which the module is given a name. Line (1) shows this identifier as module adder, without quotes or any other punctuation or capitals. The end statement has the form *end name;* for the adder, line (8) shows this as end adder, again without quotes or any other punctuation or capitals. A title statement is also necessary (for reasons to be described); line 2 shows this as Title 'full adder cell'. There can be any number of *comment statements,* which are ignored by language compilers. They begin with a double quote and are terminated by another double quote or simply by the end of a line. Their main purpose is to enlighten humans as to what that ABEL description is all about.

The main sections of an ABEL description are the *declarations* section, started by the word Declarations, as on line (3), and the *equations* section, started by the word Equations, as on line (6), both with no punctuation. The declarations section contains the specifications of

- The input and output pins of the circuit
- Internal node designators

In this example there are no internal nodes.

In Figure 1, the description specifies the input pins as A, B, and Cin and specifies the output pins as S and Cout. To specify a signal attribute, the terminology istype is used. Thus, the statement istype 'com' specifies the outputs as combinational. Declaration statements are terminated with a semicolon. (Confirm all of this by referring to Figure 1.) Note the words following the semicolon in line 5; what is their purpose? The equations line uses some symbols that will be described in what follows.

Exercise 1   Look over Figure 1 again. From this examination, can you figure out why we need both a name (given in the module statement) *and* a title? What role does each play?   ◆

The symbols used in ABEL for the logical operators are different from the usual ones. Those available in ABEL are summarized in the table in Figure 2. How many of these are used in Figure 1?

Exercise 2   Write ABEL assignment statements for the propagate and generate functions of a carry-lookahead adder. Rewrite the assignment statements

module 2to1mux
Title '2-to-1 multiplexer'
Declarations

D0 **PIN**;
D1 **PIN**;
S **PIN**;
Y **PIN istype** 'com';

Equations

When   S == 0 **then**
     Y = D0;
Else
     Y = D1;

end 2to1mux

(b)

module 2to1mux
Title '2-to-1 multiplexer'

Declarations

D0 **PIN**;
D1 **PIN**;
S **PIN**;
Y **PIN istype** 'com';

Equations

Y = D0 & !S # D1 & S;

end 2to1mux

(a)

Figure 3 Different ABEL descriptions of a 2-to-1 multiplexer.

for the sum and carry in Figure 1 using the propagate and generate functions. Does this new set of statements describe the same behavior as Figure 1?   ◆

Behavioral versus Operational Description

Look again at the ABEL description in Figure 1. The details make it clear that the designer has already synthesized the logic representation of the circuit. We could call this an *operational* description. In contrast, a *behavioral* description simply describes how a circuit is to behave, not what operations it is to carry out. The power of an HDL becomes evident when it can carry out a design from such a behavioral description.

We will illustrate the difference between an operational and a behavioral description by means of ABEL descriptions of a 2-to-1 multiplexer. This device will have two data inputs, one select input, and one data output. Two ABEL descriptions of such a circuit are shown in Figure 3.

The description in Figure 3a requires the designer to specify, in the equations section, the detailed logical operations of a particular implementation of a 2-to-1 multiplexer. The description in Figure 3b, on the other hand, requires only that the designer understand the multiplexer's behavior (or functional specification). This implies that one need not know exactly how to implement a system in order to create an initial, formal specification that can later be verified for correctness.

There is a tremendous advantage in verifying the correctness of a *behavioral* specification of a system over verifying the correctness of a *detailed operational* specification. Since the specification can be written without implementation details, the former generally requires less time. Also, once a behavioral specification has been verified for correctness, numerous different implementations can be explored to optimize such system characteristics as speed, power, size, or cost.

module adder
Title 'full adder cell'

Declarations

A **PIN**;
B **PIN**;
Cin **PIN**;
S **PIN** **istype** 'com';
Cout **PIN** **istype** 'com';

Equations

TRUTH_TABLE ([A, B, Cin] -> [S, Cout])
[0, 0, 0] -> [0, 0];
[0, 0, 1] -> [1, 0];
[0, 1, 0] -> [1, 0];
[0, 1, 1] -> [0, 1];
[1, 0, 0] -> [1, 0];
[1, 0, 1] -> [0, 1];
[1, 1, 0] -> [0, 1];
[1, 1, 1] -> [1, 1];

end adder

(a)

module 2to1mux
Title '2-to-1 multiplexer'

Declarations

D0 **PIN**;
D1 **PIN**;
S **PIN**;
Y **PIN** **istype** 'com';

Equations

TRUTH_TABLE ([S, D1, D0] -> [Y])
[0, 0, 0] -> [0];
[0, 0, 1] -> [1];
[0, 1, 0] -> [0];
[0, 1, 1] -> [1];
[1, 0, 0] -> [0];
[1, 0, 1] -> [0];
[1, 1, 0] -> [1];
[1, 1, 1] -> [1];

end 2to1mux

(b)

Figure 4 ABEL descriptions using truth table notation. (a) Adder. (b) Multiplexer.

A behavioral specification establishes a reference from which design synthesis can proceed. There are two requirements for the behavioral specification: It must be formal so that there is no ambiguity in its interpretation, and it must be functionally complete. A natural-language description cannot be interpreted in an automatic way, so a formal language is necessary. This is the motivation for hardware description languages.

In ABEL there are three mechanisms for specifying system behavior: equations, truth tables, and state diagrams. Equations and truth tables can be used for either combinational or sequential logic; state-diagram descriptions obviously apply only to the latter.

The ABEL descriptions shown in Figures 1 and 3a use equations to describe the circuits. The = symbol is used whenever the assignment is made to a combinational output signal. As already mentioned, it is also possible to describe a circuit by means of its truth table. Figure 4 illustrates this for the full adder and the 2-to-1 multiplexer.

Since a truth table can be translated into numerous implementations, a truth-table ABEL description is a behavioral description. Truth tables, however, are lengthy; from them, it is difficult to recognize what function a description is to carry out. For example, the function of the circuit description given in Figure 3b can easily be recognized as a multiplexer, but the function of the description in Figure 4b cannot. Thus, truth-table descriptions are suitable for modest-sized designs only.

module adder_be
Title 'adder_be'
"Behavioral description of 4-bit adder

Declarations
A3 .. A0 **PIN;**
B3 .. B0 **PIN;**
S3 .. S0 **PIN istype 'com';**
Cout **PIN istype 'com';**
Cin **PIN;**
@ carry 2
Equations

when Cin == 1 **then**
          [Cout, S3 .. S0] = [.x., A3 .. A0] + [.x., B3 .. B0] +1;
else
          [Cout, S3 .. S0] = [.x., A3 .. A0] + [.x., B3 .. B0];

end adder_be

Figure 5 ABEL description of a 4-bit adder.

Adder Specification in ABEL

The ABEL description of a single-bit adder cell was shown in Figure 1. For a multiple-bit adder there are numerous ways to write the description; the most abstract way is to use the addition operator available in the ABEL language. The description of a 4-bit adder is shown in Figure 5. We have not yet introduced some of the notation here; we will now correct that. When a group of signals forms a bus, the notation A3..A0 is used to designate the group.[6]

Any sub-sequence (e.g., A2A1) of the sequence A3..A0 can be referenced as a group, even individual signals (e.g., A2). The notation [Cout,S3..S0] is used to group signals into a bus; in this case the Cout signal is the most significant bit of the bus, but it doesn't have to be. The .x. syntax in [.x.,A3..A0] is used to "fill space" so that all arguments of the equation are of the same size. That syntax can be interpreted as though there is no connection on the corresponding bus signal for this argument. The equation

$$[Cout,S3..S0]=[.x.,A3..A0]+[.x.,B3..B0]$$

specifies a 4-bit addition that can produce a 5-bit result.

An ABEL *synthesis tool* translates an ABEL description into one that is suitable for implementation. By default, the synthesis tool generates a two-level maximally parallel implementation. Thus, the implementation generated for the description of an adder is a carry-lookahead adder.

---

[6]In ordinary language, a bus is a vehicle that carries many passengers. By analogy, in digital systems, rather than showing a separate line for each signal, a heavy line is used in diagrams to represent a *bus,* a hypothetical vehicle for carrying a collection of signals. The term *bus* is also used to refer to the collection of signals itself.

Operator Symbol	Operator Description	Type of Operator
==	Equality comparison	Relational
!=	Inequality comparison	
<	Less than	
>	Greater than	
<=	Less than or equal to	
>=	Greater than or equal to	
-	Negation (2's complement)	Arithmetic
-	Subtraction	
+	Addition	
*	Multiplication	
/	Division	
%	Modulus	
<<	Shift left	
>>	Shift right	

Figure 6 Additional ABEL operators.

It is possible to generate a carry ripple implementation of the adder without specifying the logic equations explicitly. This is done in the declarations portion of the description by giving the @carry directive. This directive has a numeric argument that specifies the number of bit stages between carry propagations. Thus, @carry 1 specifies a carry ripple adder, and @carry 2 specifies an adder where the carries of 2-bit groups are computed in parallel and rippled between the groups. The @carry directive is also useful for circuits other than adders. In general, it instructs the synthesis tool to generate a multiple-level circuit. There are numerous directives available in ABEL.

Only a few of the operators available in ABEL have been used so far. Additional operators are shown in Figure 6.

Exercise 3   Write the behavioral description of a magnitude comparator with two 4-bit inputs and three single-bit outputs. One output should be high if the inputs are equal, another one high if the first operand is larger than the second, and the third output high if the opposite is true.   ◆

Sequential Circuit Specification in ABEL

The next step in the study of ABEL is to turn attention to sequential circuits. What with state diagrams, state tables, don't-care conditions and the like, you might expect that producing an ABEL description for sequential circuits would be more difficult. As we go through the process, you be the judge as to whether this is the case.

Let's start by considering the sequence detector of Example 1 in Chapter 6. This sequence detector has one input and one output; it produces an output of 1 whenever the input sequence 0110 is detected. The state diagram for this sequence detector from Chapter 6 is shown in Figure 7.

An ABEL description of this finite-state machine is shown in Figure 8. The heart of the description is specification of the state diagram together with the state transitions. If necessary (in the case of multiple inputs), the state transition

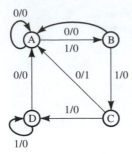

Figure 7 State diagram for 0110 sequence detector.

module detector
Title 'detector'
"0110 sequence detector

Declarations
clock, x **PIN;**
z **PIN istype 'com';**
Q1 .. Q0 **NODE istype 'reg';** "registered output
A = [0, 0]; "state assignments
B = [0, 1];
C = [1, 0];
D = [1, 1];

Equations
[Q1 .. Q0] . clk = clock; "definition of clock signal for registered outputs
z=Q1&!Q0&!x;

state_diagram [Q1, Q0]
state A: if x then B else A; "state transitions
state B: if x then C else A;
state C: if x then D else A;
state D: if x then D else A;

end detector

Figure 8 ABEL description of the 0110 sequence detector based on the state diagram.

can be a sequence of *if, else if, else if, ..., else* statements. For example, if there were two inputs to a state machine, one would write

If [x,y] == [0,0] then A
Else if [x,y] == [0,1] then B
Else if [x,y] == [1,0] then C
Else D

The state variables Q1..Q0 must be declared as istype 'reg' so that the synthesis tool will assign flip-flops for their implementation.

Two terms are used to designate a signal. One is NODE; this is an internal signal and not an output of the circuit. The other is PIN (previously discussed). As the name implies, a PIN is a signal that will be assigned to actual input/output pins of a device.

module detector
Title 'detector'
"0110 sequence detector

Declarations
clock, x **PIN**;
z **PIN istype 'com'**;
Q1 .. Q0 **NODE istype 'reg'**; "registered output
Qstate = [Q1, Q0];
A = [0, 0]; "state assignments
B = [0, 1];
C = [1, 0];
D = [1, 1];

Equations
[Q1 .. Q0] . clk = clock; "definition of clock signal for registered outputs
z = Q1 &! Q0 &! x;

truth_table ([x, Qstate.fb] :> [Qstate -> [z])
[0, A] :> [A] -> [0];
[1, A] :> [B] -> [0];
[0, B] :> [A] -> [0];
[1, B] :> [C] -> [0];
[0, C] :> [A] -> [1];
[1, C] :> [D] -> [0];
[0, D] :> [A] -> [0];
[1, D] :> [D] -> [0];

Figure 9 ABEL description of the 0110 sequence detector using a truth table.

There is no mechanism in ABEL for declaring symbolic state names; hence, state assignments must be specified from the start. An alternative is not to use symbolic state names at all but to use the numeric state assignments directly. The statement [Q1..Q0].clk = clock specifies that clk is the signal that is connected to the clock terminal of the flip-flops. The .clk expression just introduced is called a *signal dot extension*. Numerous other signal dot extensions are used besides .clk.

ABEL has no explicit way of distinguishing signals that are inputs from those that are outputs. (This can be observed in Figure 8 by the declarations clock and x; these do not specify whether or not they are inputs, and an ABEL compiler cannot determine it either.) To identify a signal explicitly as an output, the signal attributes 'com' and 'reg' are used.

Either truth tables or equations can be used to specify don't-care conditions in an ABEL description. Specification of don't-cares using truth tables is simple. Any input combinations that are omitted in a truth table are assumed to be don't-care conditions.

In addition to the description of the sequence detector based on the state diagram as in Figure 8, a description can also be written using either the truth table construct or equations. For a sequential circuit, the truth table corresponds to its transition table. (Review Chapter 5 if you need to.) The ABEL description of the sequence detector using the truth table construct is shown in Figure 9.

Exercise 4 Assume the present state is B and the input sequence 01001110110 is received. Verify that the descriptions in Figures 8 and 9 produce the same output sequence. ◆

Notice that the syntax for a sequential truth table uses :> instead of ->. The truth table in Figure 9 combines the transition and output tables into one truth table. The .fb extension instructs the synthesis tool to *feed back* the output of the Qstate register to the input of the next-state decoder. In this description, Qstate.fb is interpreted as the present state and Qstate is the next state. The flip-flop excitation expressions are synthesized automatically by the technology-mapping tool (described later) when a specific programmable device is chosen for implementation.

As already mentioned, the description of the 0110 sequence detector can also be written by specifying the next-state equations. The corresponding ABEL description is shown in Figure 10. The disadvantage of writing the description based on next-state equations is that synthesis must first be performed to derive the equations. As stated before, the most significant advantage of designing with a hardware description language is that a formal, unambiguous specification can be written and the synthesis can be performed later using software tools.

Don't-Care Conditions in ABEL

As you know from previous chapters, don't-care conditions can be used to minimize the size of the implementation of a system. ABEL can take advantage of don't-care conditions, provided it can determine from the specification that such conditions exist.

A specification using equations such as that shown in Figure 1 is a complete specification. The equations characterize input combinations for which the output is logic 1; those for which the output is logic 0 are implied.

module detector
Title 'detector'
"0110 sequence detector

Declarations
clock, x **PIN;**
z **PIN istype 'com';**
Q1 .. Q0 **NODE istype 'reg';** "registered output

Equations
[Q1 .. Q0] . clk = clock; "definitions of clock signal for registered outputs
z = Q1 & !Q0 & !x;
Q1 := Q0.fb & x # Q1.fb & x;
Q0 := !Q0.fb & x # Q1.fb & x;

end detector

Figure 10 ABEL description of the 0110 sequence detector using next-state equations.

```
module BCDto7seg
Title 'BCD to 7-segment code converter'
Declarations
[D3 .. D0] PIN;
a PIN istype 'com';
b PIN istype 'com';
c PIN istype 'com';
d PIN istype 'com';
e PIN istype 'com';
f PIN istype 'com';
g PIN istype 'com';
Equations
a=D3&!D2&!D1#!D2&!D1&!D0#!D3&D2&D0#!D3&D1;
a?=D3&D2#D3&D1; "expression covering the don't-care inputs
b=!D3&!D2#!D2&!D1#!D3&!D1&!D0#!D3&D1&D0;
b?=D3&D2#D3&D1;
c=!D3&D2#!D3&D0#!D2&!D1;
c?=D3&D2#D3&D1;
d=D3&!D2&!D1#!D2&!D1&!D0#!D3&D2&!D1&D0#!D3&!D2&D1#!D3&D1&!D0;
d?=D3&D2#D3&D1;
e=!D3&D1&!D0#!D2&!D1&!D0;
e?=D3&D2#D3&D1;
f=D3&!D2&!D1#!D3&!D1&!D0#!D3&D2&!D1#!D3&D2&!D0;
f?=D3&D2#D3&D1;
g=D3&!D2&!D1#!D3&D2&!D1#!D3&!D2&D1#!D3&D1&!D0;
g?=D3&D2#D3&D1;
end BCDto7seg
```

Figure 11 ABEL description of a BCD-to-seven-segment decoder using don't-care equations.

How does one specify don't-care conditions when describing a system using equations? For example, in the ABEL description of a BCD- to- seven-segment decoder using a truth table specification, the rows of the truth table corresponding to the 4-bit numbers 10 through 15 would be omitted. To specify don't-cares using equations, on the other hand, one must write separate equations using the assignment operator ?= or ?:=. These equations represent covers of the input combinations that are don't-cares. The specification of a BCD-to-seven-segment decoder using equations is shown in Figure 11.

Care must be exercised in specifying a don't-care equation so that it does not include the corresponding equation that specifies a logic 1 output. In other words, the equation describing the logic 1 cases for signal *a* should not cover any of the don't-care inputs. For this particular system, notice that the equations covering the don't-care inputs are the same for all of the output signals.

Hierarchical Specifications in ABEL

So far, the ABEL modules considered have been self-contained; no module has depended on another. In creating one ABEL module, however, it is possible to utilize other ABEL modules in its description. When ABEL module 1 is used in

module 2to1mux
interface (D1, D0, S->Y);
Title '2-to-1 multiplexer'
Declarations

D0 **PIN;**
D1 **PIN;**
S **PIN;**
Y **PIN istype** 'com';

Equations
When   S == 0 **then**
        Y = D0;
Else
        Y = D1;   **Figure 12** Interface statement in ABEL description of
end 2to1mux   Figure 3*b* permitting instantiation in other ABEL modules.

module 4to1mux
interface ([D3 .. D0], [S1 .. S0] -> Y);
Title '4-to-1 multiplexer'
"a 4-to-1 multiplexer built using three 2-to-1 multiplexers
Declarations

[D3 .. D0] **PIN;**
[S1 .. S0] **PIN;**
Y **PIN istype** 'com';

2to1mux **interface** (D1, D0, S -> Y); "declare the interface to the 2to1mux
mux2, mux1, mux0 **functional_block** 2to1mux; "declare 3 instances
tmp1, tmp0 **node istype** 'com'; "internal signals

Equations
mux2.D0 = D0; "S0 selects data from mux2 and mux1 for inputs to the
mux2.D1 = D1; "third multiplexer, mux0
mux2.S = S0;
tmp0 = mux2.Y;
mux1.D0 = D2;
mux1.D1 = D3;
mux1.S = S0;
tmp1 = mux1.Y;
mux0.D0 = tmp0;
mux0.D1 = tmp1;
mux0.S = S1;
Y = mux0.Y;
end 4to1mux

Figure 13 ABEL description of a 4-to-1 multiplexer using three 2-to-1 multiplexers.

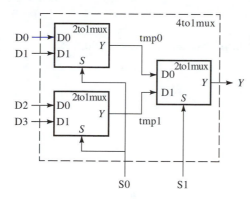

Figure 14 Schematic diagram corresponding to the ABEL description of the 4-to-1 multiplexer in Figure 13.

ABEL module 2, then 1 is said to be *instantiated* in 2. It can also be said that module 2 contains an *instance* of module 1. There is no limit to the number of times a module can be used (instantiated). This permits a hierarchical description in which an ABEL block for one function is utilized within another ABEL block. This feature of the language promotes the use of small-unit designs as blocks within larger blocks. The complexity of describing large systems is thereby decreased.

An ABEL module that is to be used within another ABEL module requires an *interface* statement, as shown in the second line of the multiplexer description in Figure 12. The interface statement specifies all inputs and outputs that are to be accessed from another module.

This 2-to-1 multiplexer can be used to build a 4-to-1 multiplexer, as shown in Figure 13. The 4-to-1 multiplexer uses three 2-to-1 multiplexers connected as shown in the schematic diagram in Figure 14. The functional-block statement in Figure 13 specifies the ABEL module to be used and the names of the instances. Each instance of the utilized module is required to have a unique name so that the connections to its terminals can be uniquely identified.

There is no limit to the number of levels of hierarchy in an ABEL description. Thus, an 8-to-1 multiplexer can be constructed using two 4-to-1 multiplexers and one 2-to-1 multiplexer. The ABEL description of such an 8-to-1 multiplexer is shown in Figure 15, and the corresponding schematic is shown in Figure 16. If the 8-to-1 multiplexer in Figure 15 is to be used in another ABEL specification, it too should have an interface statement.

The ABEL descriptions in Figures 13 and 15 describe the structure of a system because they instantiate functional blocks and specify the connections between them. In other words, there is a one-to-one correspondence between the ABEL description and the corresponding schematic (compare Figures 13 and 14 and Figures 15 and 16). As expected, these types of ABEL descriptions are called *structural descriptions*.

EXAMPLE 1

Write the ABEL description of a 4-bit serial adder. Assume one operand is stored in a shift register and the result is stored in the same shift register. Write the de-

module 8to1mux
Title '8-to-1 multiplexer'
"an 8-to-1 multiplexer built using 2 4-to-1 multiplexers and a 2-to-1 multiplexer
Declarations

[D7 .. D0] **PIN;**
[S2 .. S0] **PIN;**
Y **PIN** **istype** 'com';

2to1mux **interface** (D1, D0, S -> Y); "declare the interface to the 2to1mux
mux21 **functional_block** 2to1mux;
4to1mux **interface** ([D3 .. D0], [S1 .. S0] -> Y); "interface for the 4to1mux
mux411, mux410 **functional_block** 4to1mux;
tmp1, tmp0 **node** **istype** 'com'; "internal signals

Equations
mux410.D0 = D0;
mux410.D1 = D1;
mux410.D2 = D2;
mux410.D3 = D3;
mux410.S1 = S1;
mux410.S0 = S0;
tmp0 = mux410.Y;
mux411.D0 = D4;
mux411.D1 = D5;
mux411.D2 = D6;
mux411.D3 = D7;
mux411.S1 = S1;
mux411.S0 = S0;
tmp1 = mux411.Y
mux21.D0 = tmp0;
mux21.D1 = tmp1;
mux21.S = S2;
Y = mux21.Y;
end 8to1mux

Figure 15 Multiple levels of hierarchy in an ABEL description of an 8-to-1 multiplexer.

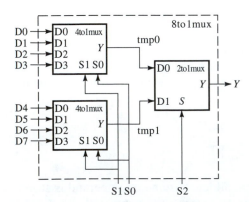

Figure 16 Schematic diagram corresponding
to the ABEL description in Figure 15.

```
module adder
interface (A, B, Cin -> S, Cout);
Title 'full adder cell'
Declarations
A PIN;
B PIN;
Cin PIN;
S PIN istype 'com';
Cout PIN istype 'com';
Equations
TRUTH_TABLE
([A, B, Cin] -> [S, Cout])
[0, 0, 0] -> [0, 0];
[0, 0, 1] -> [1, 0];
[0, 1, 0] -> [1, 0];
[0, 1, 1] -> [0, 1];
[1, 0, 0] -> [1, 0];
[1, 0, 1] -> [0, 1];
[1, 1, 0] -> [0, 1];
[1, 1, 1] -> [1, 1];
end adder
```

(a)

```
module sh_reg
interface (input, clock -> output);
Title 'shift register'

Declarations
clock, input PIN;
output PIN istype 'com'; 'reg';   "Q3 is MSB
Q3 .. Q0 NODE istype

Equations
[Q3 .. Q0].clk = clock;
output = Q0.fb;
Q0 := Q1.fb;
Q1 := Q2.fb;
Q2 := Q3.fb;
Q3 := input;
end sh_reg
```

(b)

```
module serial_add
Title '4-bit serial adder'
Declarations
clock, data PIN;
S PIN istype 'com';
carry NODE istype 'reg';
adder interface (A, B, Cin -> S, Cout);
sh_reg interface (input, clock -> output);
adder function_block adder;
sh_reg function_block sh_reg;
Equations
carry.clk = clock;
adder.A = data;
adder.B = sh_reg.output;
sh_reg.input = adder.S;
sh_reg.clock = clock;
carry := adder.Cout;
adder.Cin = carry.fb;
end serial_add
```

(c)

Figure 17 ABEL descriptions for Example 1.

scription hierarchically by creating a shift register module and an adder module and including them in the serial adder module.

The adder is a small circuit, so it is convenient to write a truth table description, as given in the ABEL description in Figure 17a. Confirm all details line by line.

The shift register in Figure 17b should shift right from the most to the least significant bit. It should have an external input to the most significant bit position, and the output should be the least significant bit position. The only input besides the data is the clock. Verify every line. ∎

The 4-bit serial adder can now be constructed by instantiating the adder and the shift register as in Figure 17c. The only other thing that is needed is a flip-flop to store the carry output of the adder between clock cycles. Draw a block diagram showing the flip-flop interconnection with each of the two units having their appropriate input and output lines.

EXAMPLE 2

A *barrel shifter* is a combinational logic block that shifts a data word a selected number of bit positions based on a control input. Write an ABEL description for an 8-bit barrel shifter with three control inputs. The input word should be shifted right (from most significant to least significant bit) with zeros shifted into empty positions.

We need a module statement and an end statement marking the beginning and end of this ABEL description. We can name the module anything, but it is better to choose a descriptive name such as *shifter*. The title can be more descriptive and should be sufficient to summarize the function of the module.

The desired shifter has eight data inputs, eight data outputs, and three control inputs. The data input signals are logically related, so they should be specified as a bus. The data outputs and control inputs should be specified similarly. What method of ABEL description should we choose, truth table or equations? Think of how many rows a truth table would have. Obviously, then, besides being inappropriate for that reason, a truth table ABEL description reproduced here would add 10 percent to the cost of the book! Should the description using equations be operational—that is, with written logic equations for the outputs—or behavioral? The former would be a huge job! The obvious thing to do is to write a behavioral description. When the control input has numeric value j $(0 \le j \le 7)$, the output is the input shifted j positions to the right. Although the ABEL description is given in Figure 18, write it yourself before looking at the solution. ∎

Before continuing with further development of ABEL, we will turn briefly to a few devices that can be used to implement circuits utilizing the results of an ABEL description.

2 PROGRAMMABLE LOGIC DEVICES (PLDs)

The architectures of programmable logic array (PLA) and programmable array logic (PAL) devices were introduced in Chapter 4. Review that section if necessary. Programmable logic devices are widely used in practice, and ABEL is especially useful in formulating designs utilizing them. This section is highly descriptive. Don't just read the description of each figure but examine the details of each one as it is discussed.

The configuration of the PAL16L8 is shown in Figure 19. (It is introduced here not because of its intrinsic value but as an introduction to the more generally useful PLDs to follow.) It is a conventional PAL with 64 AND gates having 32 inputs (16 variables and their complements). It has a maximum of 8 combinational out-

module shifter
Title '8-bit right barrel shifter'
Declarations
[in7 .. in0], [c2 .. c0] **PIN;** "in7 is the MSB
[out7 .. out0] **PIN istype 'com';** "out7 is the MSB
in = [in7 .. in0];
c = [c2 .. c0];
out = [out7 .. out0];
Equations
when c==0 then "no shift
        Out = in;
else when c==1 then
        [out7 .. out0] = [0, in7 .. in1];
else when c==2 then
        [out7 .. out0] = [0, 0, in7 .. in2];
else when c==3 then
        [out7 .. out0] = [0, 0, 0, in7 .. in3];
else when c==4 then
        [out7 .. out0] = [0, 0, 0, 0, in7 .. in4];
else when c==5 then
        [out7 .. out0] = [0, 0, 0, 0, 0, in7 .. in5];
else when c==6 then
        [out7 .. out0] = [0, 0, 0, 0, 0, 0, in7 .. in6];
else when c==7 then
        [out7 .. out0] = [0, 0, 0, 0, 0, 0, 0, in7];
end shifter

Figure 18 ABEL description of the barrel shifter.

puts, 10 dedicated inputs, and 6 bidirectional pins that can be programmed to be either inputs or outputs. (Examine the figure to confirm all this.) The buffers driving the bidirectional outputs are tristate buffers with programmable control. A bidirectional I/O pin can be programmed as an input by disabling the output buffer or as an output by enabling it. This device can be configured to implement multiple-level circuits, or even asynchronous sequential circuits by using one of the outputs as an input to the AND array. The I/O pins are fed back to the AND array; thus, they can be programmed as inputs to the same outputs or to different ones.

The PAL16L8 was introduced in circuit design when bipolar technology was in common use. Since larger scales of integration are readily available in CMOS technology, the industry transition to CMOS has reduced the utility of the PAL16L8 architecture. Its replacement in CMOS technology is the PALCE16V8; its logic schematic diagram is shown in Figure 20. The AND array in Figure 20 is exactly the same as the one in Figure 19. Note, however, the differences between the output circuitry of the two. The output of the PAL16L8 consists of an array of OR gates and tristate inverters.

A perusal of Figure 20 will make clear that the output of the PALCE16V8 can hardly be called an array of mere gates; instead, it is called an array of *macrocell*s (*macro* because the cells are rather large themselves). The output macrocell in this PAL device is versatile—hence the *V* in its name. A detailed schematic diagram of the output macrocell is shown in Figure 21. The transistor sizes in CMOS technology (MOSFET) decrease with the miniaturization of such integrated-circuit features as aluminum traces. Advances in CMOS technology allow increased system complexity on a single integrated circuit with little (if any) increase in cost.

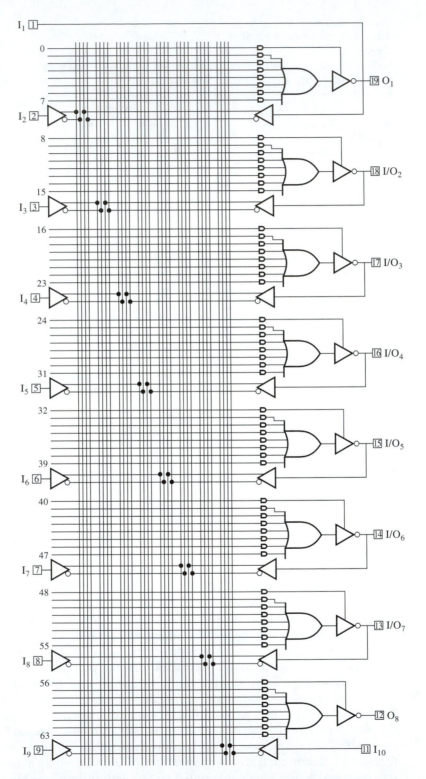

Figure 19 Logic schematic of the PAL16L8. (Courtesy of Texas Instruments)

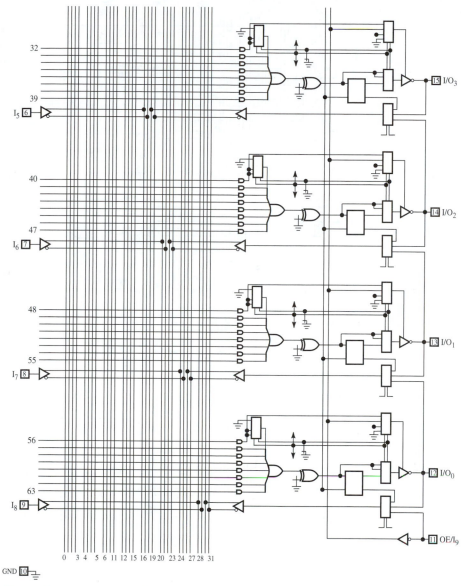

Figure 20 Logic schematic diagram of the PALCE16V8. (*continued on next page*) (Diagram provided by Lattice Semiconductor Corporation.)

Various configurations of the output macrocell are shown in Figure 22. If the configuration shown in Figure 22c is used, then the behavior of PALCE16V8 is identical to that of the PAL16L8.

The PALCE16V8 has significantly greater utility due to the existence of a flip-flop at the output of each OR gate. With the PAL16L8, on the other hand, it is necessary to use separate flip-flop chips (e.g., 7474s) to implement synchronous sequential circuits. Most of the synchronous circuits designed in Chapter 6 can be implemented with one PALCE16V8 but could require three or more chips if the PAL16L8 is used.

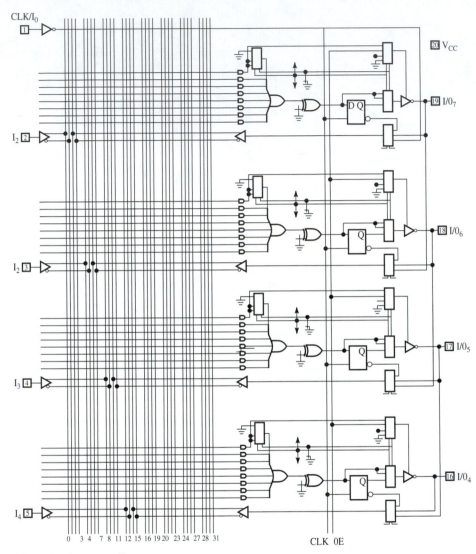

Figure 20 *(continued)*

Exercise 5 Describe how you would implement an *SR* latch with the PAL16L8. ◆

Exercise 6 Describe how to implement a circuit with four logic levels using the PALCE16V8. Is it possible to implement circuits with an odd number of levels also? If not, just how does one implement such circuits? ◆

Complex Programmable Logic Devices

The programmable logic devices described in the previous section seem intricate enough, but their complexity pales in comparison with that of interconnections of devices that are given the name *complex programmable logic devices*

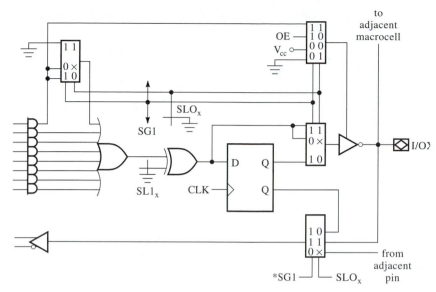

Figure 21 Logic schematic diagram of the PALCE16V8 output macrocell. (Diagram provided by Lattice Semiconductor Corporation.)

(CPLDs).[7] Such a device contains several conventional PLD architectures, such as the PALCE16V8, with an array of programmable interconnections often called an *interconnection matrix*. The architecture of one of these, the Xilinx XC9500 CPLD family, is shown in Figure 23. Each of its function blocks (containing 18 macrocells) has twice the capability of the entire PALCE16V8.

The architecture of the XC9500 function block is shown in Figure 24. The macrocell of a function block of the XC9500 is shown in Figure 25 and is similar in capability to a macrocell of the PALCE16V8 in Figure 21.

The XC9500 family includes several devices; details of the characteristics of six of them are summarized in the table in Figure 26. The smallest one, the XC9536, is equivalent in power to roughly 4.5 PALCE16V8 devices; the largest one, the XC95288, is approximately 36 times more powerful than the PALCE16V8.[8]

Notice in Figure 26 that the typical propagation delay increases with device complexity. The modular architecture suggests that the fan-outs of gates are similar for devices of all sizes. So why does the propagation delay increase with device complexity? The answer is that to maintain flexibility in device programming, the switch matrix complexity grows significantly with the number of functional blocks. And it is the delay through this switch matrix that causes the increase.

---

[7]At the time of publication, several vendors produce them.

[8]At the time of preparation of this manuscript, the cost of one PALCE16V8 (7 ns delay) in quantities of 100 or more was $2.00; that of the XC9536 was $3.70 per device. Thus, the XC9536 is 4.5 times more powerful and only 1.85 times more expensive. The power of the integrated circuit is rooted in the fact that device density (the amount of circuitry per unit area) has grown exponentially while device cost has remained roughly constant for more than a quarter century.

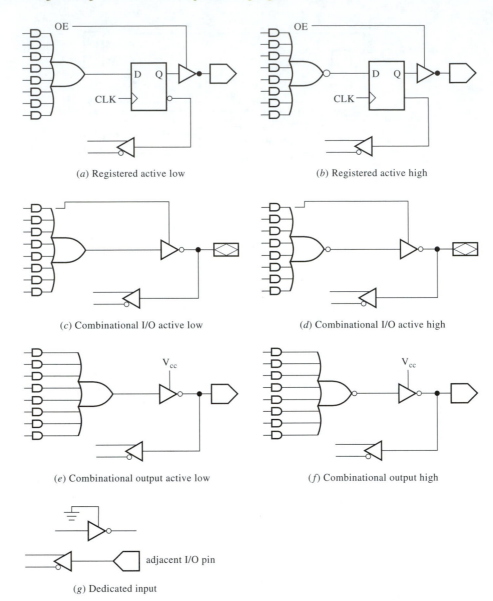

(a) Registered active low

(b) Registered active high

(c) Combinational I/O active low

(d) Combinational I/O active high

(e) Combinational output active low

(f) Combinational output high

(g) Dedicated input

Figure 22 Various configurations of the PALCE16V8 output macrocell. (Diagram provided by Lattice Semiconductor Corporation.)

Cost is not included in the specifications in Figure 26 because the cost changes rapidly over time. The cost per unit size of a device is roughly constant. So for the devices in Figure 26 the cost increases from left to right roughly in proportion to the number of usable gates. The cost difference between a PALCE16V8 and an XC9536 is not as substantial as the cost difference between an XC9536 and an XC95144 because the XC9500 family is fabricated in a more advanced technology than the PALCE16V8. As technology advances, more gates can be fabricated in an area of unit size.

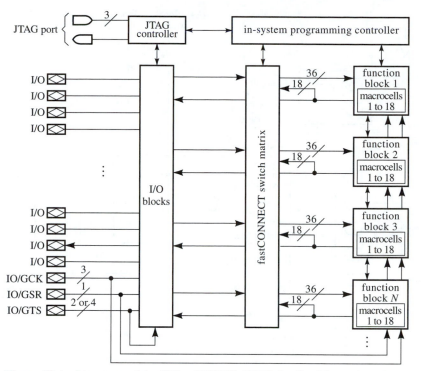

Figure 23 Architecture of the Xilinx XC9500 CPLD family. (Figure based on or adapted from figures and text owned by Xilinx, Inc., courtesy of Xilinx, Inc. © Xilinx, Inc. 1999. All rights reserved.)

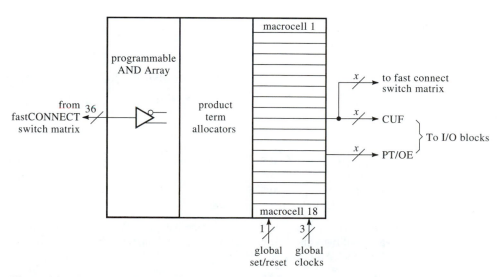

Figure 24 Function block architecture of the XC9500 PLD. (Figure based on or adapted from figures and text owned by Xilinx, Inc., courtesy of Xilinx, Inc. © Xilinx, Inc. 1999. All rights reserved.)

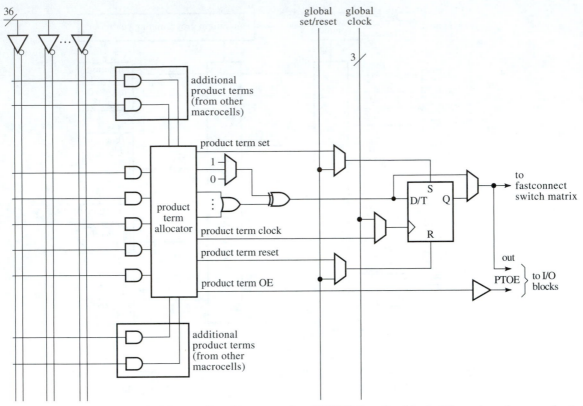

Figure 25 Macrocell architecture of an XC9500 function block. (Figure based on or adapted from figures and text owned by Xilinx, Inc., courtesy of Xilinx, Inc. © Xilinx, Inc. 1999. All rights reserved.)

	XC9536	XC9572	XC95108	XC95144	XC95216	XC95288
Macrocells	36	72	108	144	216	288
Usable Gates	800	1,600	2,400	3,200	4,800	6,400
Registers	36	72	108	144	216	288
t_{PD} (ns)	5	7.5	7.5	7.5	10	15
t_{SU} (ns)	4.5	5.5	5.5	5.5	6.5	8.0
t_{CO} (ns)	4.5	5.5	5.5	5.5	6.5	8.0
f_{CNT} (MHz)	100	125	125	125	111	95
f_{SYSTEM} (MHz)	100	83	83	83	67	56

Figure 26 Device specifications for the Xilinx XC9500 family. (Figure based on or adapted from figures and text owned by Xilinx, Inc., courtesy of Xilinx, Inc. © Xilinx, Inc. 1999. All rights reserved.)

Exercise 7   The XC95144 contains four times as many usable gates as the XC9536. If they were fabricated in the same technology, what would you expect the price difference to be between them? If the XC95144 were fabricated in a more advanced technology in which twice as many usable gates could be fabricated in the same space as before, then what do you expect the cost difference to be?   ◆

Field-Programmable Gate Arrays

A PLD not previously discussed is a field-programmable gate array (FPGA). FPGAs are programmable devices that differ from PLDs in their complexity and organization. FPGAs contain circuitry capable of implementing designs with up to tens of thousands of gates. They are organized as two-dimensional arrays of interconnected logic blocks.

The organization of a typical Xilinx FPGA is shown in Figure 27. It contains an array of configurable logic blocks (CLBs) with programmable interconnections between them. The structure of a CLB is shown in Figure 28. This is known as a RAM-based FPGA since, in each of the CLBs, the device is programmed by writing bits into RAM.

Xilinx (and other vendors) manufactures a series of FPGAs with varying complexity. A summary of their XC4000 series of devices is given in Figure 29. Since it is a RAM-based FPGA, it is possible to configure CLBs as randomly addressable read/write memory. This makes this particular technology efficient for the implementation of buffers and queues.

Each of the numerous FPGA vendors has its own architecture and logic block configuration. The vendors use various technologies for programming the devices. Xilinx devices can be reconfigured an arbitrary number of times, but a companion device, such as an EPROM, is necessary for permanent storage of the configuration.

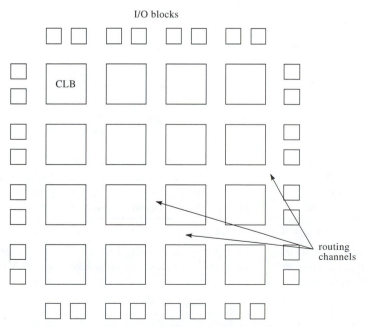

Figure 27 Architecture of the Xilinx FPGA.

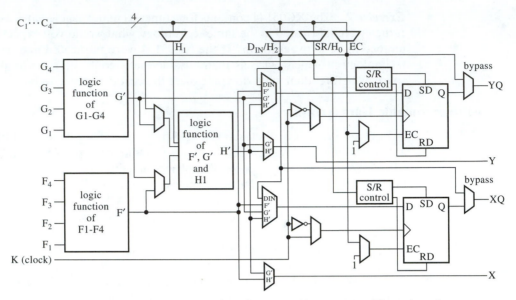

Figure 28 Configurable logic block (CLB) of the Xilinx FPGAs. (Figure based on or adapted from figures and text owned by Xilinx, Inc., courtesy of Xilinx, Inc. © Xilinx, Inc. 1999. All rights reserved.)

Device	Logic Cells	Max. Logic Gates (No. RAM)	Max. RAM Bits (No. Logic)	Typical Gate Range (Logic and RAM)*	CLB Matrix	Total CLBs	Number of Flip-Flops	Max. User I/O
XC4002XL	152	1,600	2,048	1,000–3,000	8×8	64	256	64
XC4003E	238	3,000	3,200	2,000–5,000	10×10	100	360	80
XC4005E/XL	466	5,000	6,272	3,000–9,000	14×14	196	616	112
XC4006E	608	6,000	8,192	4,000–12,000	16×16	256	768	128
XC4008E	770	8,000	10,368	6,000–15,000	18×18	324	936	144
XC4010E/XL	950	10,000	12,800	7,000–20,000	20×20	400	1,120	160
XC4013E/XL	1,368	13,000	18,432	10,000–30,000	24×24	576	1,536	192
XC4020E/XL	1,862	20,000	25,088	13,000–40,000	28×28	784	2,016	224
XC4025E	2,432	25,000	32,768	15,000–45,000	32×32	1,024	2,560	256
XC4028EX/XL	2,432	28,000	32,768	18,000–50,000	32×32	1,024	2,560	256
XC4036EX/XL	3,078	36,000	41,472	22,000–65,000	36×36	1,296	3,168	288
XC4044XL	3,800	44,000	51,200	27,000–80,000	40×40	1,600	3,840	320
XC4052XL	4,598	52,000	61,952	33,000–10,0000	44×44	1,936	4,576	352
XC4062XL	5,472	62,000	73,728	40,000–130,000	48×48	2,304	5,376	384
XC4085XL	7,448	85,000	100,352	55,000–180,000	56×56	3,136	7,168	448

*Maximum values of typical gate range include 20–30% of CLBs used as RAM.

Figure 29 Summary of the Xilinx XC4000 family of FPGAs. (Figure based on or adapted from figures and text owned by Xilinx, Inc., courtesy of Xilinx, Inc. © Xilinx, Inc. 1999. All rights reserved.)

The system is designed so that the EPROM loads the configuration into the FPGA upon start-up. Some manufacturers use technologies that are used in EPROMs or PLAs. These devices may be one-time programmable, or they may be erasable a limited number of times—on the order of hundreds or thousands.

At the time of this writing, there is some debate about the use of FPGAs versus ASICs (application-specific integrated circuits) in manufactured products.[9] FPGAs are mostly used for prototype development of products and are commonly replaced by ASICs when the product goes into production. Because the FPGA is generic, the gates are typically not well utilized. The ASIC is smaller because only those gates that are needed are included on the chip; they are arranged on the chip so as to minimize the total chip area. This makes the per-part cost of an ASIC less than that of an FPGA. Because the gates in an FPGA are placed in fixed locations and the structure is designed to accommodate arbitrary circuits, it is more difficult to use the gates efficiently. For this reason ASICs tend to have less delay than FPGAs and can be operated with a faster system clock.

However, as discussed in Chapter 3, ASICs have large development costs associated with them (called nonrecurring engineering, or NRE, charges). The large NRE charge for an ASIC comes from the engineering effort needed to design the chip and the tooling (in the form of photolithographic masks) needed to fabricate the chips. For this reason ASICs may not be used in products that are sold in low volume.

The decision of whether to use an FPGA or an ASIC in a product normally is an economic decision that weighs the cost of developing the ASIC against the per-part cost of the ASIC versus the FPGA. If the product is expected to sell in large volume (e.g., pagers or cell phones), then the investment required to develop an ASIC will be recovered several times over with the savings in per-part costs of the ASIC versus the FPGA. It still makes sense, however, to use FPGAs for designing a prototype for a product, since much is learned during the prototyping process and, hence, there may be a need for subsequent changes in the design. It would not make economical sense to use an ASIC before the product has been finalized.

3 THE DESIGN FLOW FOR HDL SPECIFICATIONS

Equipped with knowledge of ABEL as a hardware description language and a general understanding of certain CPLDs, we are now ready to undertake the design process utilizing these. Given a digital design specification, the design process consists of a sequence of several distinct steps. Because it flows from one step to the next in arriving at an implementation, this process is referred to as the *design flow*. The design flow uses a number of different computer-aided design tools, each of which operates on the specification in a way that makes it possible to arrive at a physical system at the end. In order to design effectively using any HDL, including ABEL, it is necessary to understand these computer-aided design tools and their methods of operation. This section describes some of these tools.

Before any hardware implementation is carried out, we need to be sure that the ABEL description—produced with so much blood, sweat, and tears—

---

[9]See Chapter 2 for a brief discussion of ASICs.

is correct. For verifying the ABEL description, a simulation would be most appropriate. Such simulation tools are commercially available.[10]

The synthesis tool translates the ABEL specification into one that is device independent and gate level. Then the technology mapper, or fitter, translates the gate-level description into a format that is used to program a specific physical device (e.g., a PALCE16V8). Why not go directly from the ABEL specification to the design using specific devices?[11]

Obviously, synthesis and technology mapping do not modify system behavior. After technology mapping, however, there is more information available about the system. For example, from the description obtained from the technology mapper, one can determine fan-in, fan-out, number of gate levels, and other parameters. Once such information has been absorbed, it might be useful to perform another simulation, paying close attention to timing requirements.

The design flow for arriving at hardware specifications consists of five steps:

1. Starting from a word description of the problem, write an ABEL description of the design.
2. Simulate the ABEL specification using an ABEL simulator.
3. Synthesize the ABEL specification into an intermediate, gate-level, technology-independent logic format.
4. Use technology mapping to translate the intermediate format into a physical specification (PLD, FPGA, or standard library).
5. Simulate the physical specification using a gate-level logic and timing simulator.

Synthesis and Technology Mapping of ABEL Specifications

The job of transforming an ABEL description into a physical specification is divided into two tasks: synthesis and technology mapping. The synthesis tool[12] produces a technology independent intermediate logic representation that corresponds to the logical structure of the physical implementation. A standard format for this representation is the *electronic data interchange format* (EDIF).

Depending on the technology used for implementation, the intermediate representation produced by the synthesis tool may have to be modified; this is the job of the technology mapping tool. For example, the synthesis tool may compute an implementation that uses a 10-input AND gate, but the maximum fan-in of gates in the target technology might be only six. The intermediate representation must be transformed into a physical specification that can be implemented using the devices available in the target technology (for example, an XC9500).

Synthesis tools cannot explore all possible implementations of a design to determine the best one. The implementation computed by a synthesis tool is

---

[10]An appropriate simulation tool is the one distributed with the Xilinx Webpack software and can be found at the Xilinx web site. At the time of this writing the site address is: http://www.xilinx.com/sxpresso/webpack.htm

[11]A device-independent specification would be carried out first so that a choice of specific hardware to use later is still available; the gate-level description is portable from one device to the next.

[12]See footnote 10.

Circuit	@ carry	Number of Product Terms	Number of Gates	Number of Gate Levels
A	0	135	455	5
B	1	77	152	10
C	2	97	266	8
D	3	84	220	7
E	4	145	517	5
F	5	135	455	5

Figure 30 Characteristics of synthesized adder implementations in Figure 5.

good with respect to some performance characteristics but not others. For example, a carry-lookahead adder has low propagation delay, but it has large power dissipation and requires more gates than other adder implementations. A ripple-carry adder is small and dissipates less power, but it has substantial delay. Numerous adder implementations lie between these two extremes, and the implementation best suited for a particular application may be any one of them.

To control the implementation computed by a synthesis tool, it is necessary for designers to understand how to use the tool and how the tool works (how it synthesizes specific implementations). Designers must direct the tools toward the implementation they are seeking.

By default, ABEL synthesis tools compute a two-level logic implementation of the specification. This corresponds to the implementation with maximum parallelism (lowest delay) and usually the largest number of gates. For most applications this may not be the desired implementation.

ABEL itself has procedures for varying the default implementation. ABEL contains instructions that are used specifically to direct the synthesis tool toward certain implementations. We saw one of these instructions (the @carry instruction) in the specification of an adder in Figure 5. These instructions are often called *directives*. The @carry directive controls the number of levels of logic in the implementation by specifying the number of levels between carry generations. For example, @carry 1 directs the synthesis tool to generate a carry between each level; it corresponds to a ripple-carry adder. @carry 4 directs the synthesis tool to generate a carry between each four stages of the adder; it corresponds to a carry-lookahead adder with lookahead in each 4-bit section and ripple carry between the sections.

We synthesized the 4-bit adder specification in Figure 5 with values of @carry between 0 and 5. A summary of the characteristics of the implementations is shown in Figure 30. @carry 0 means ignore the @carry directive. We expect the number of gates to increase and the number of gate levels to decrease as the value of @carry increases from 1, and the number of gate levels indeed decreases from 10 to 5. You might wonder why the number of gate levels for circuit F in Figure 30 hasn't been reduced further, even down to two levels. Indeed, the synthesis tool produces an intermediate representation that has two levels, but the target technology (in this case the Xilinx 9500 PLD) does not support gates with sufficient fan-in to physically implement the circuit with two levels of logic. Thus, the minimum number of gate levels achievable for a 4-bit carry-lookahead adder in the Xilinx 9500 is 5.

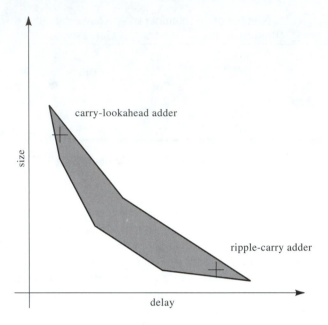

Figure 31 Typical implementation space for a digital circuit.

Consider the number of gates for each implementation in Figure 30. The reason the number of gates for circuit D is smaller than that for circuit C (and similarly for circuits E and F) is the way the carry is generated within the 4-bit adder. Implementation C has a ripple carry between bits 1 and 2 (bit 0 is the least significant bit) and between bit 3 and the carry output. Implementation D contains a ripple carry between bits 2 and 3 only (with carry lookahead from bits 0 to 2). Thus, circuit D is smaller (has fewer gates) and faster than circuit C. Circuit F is smaller than circuit E for a similar reason.

Generally, delay (number of gate levels) and size (number of gates) are inversely related. The design space can be characterized by the graph shown in Figure 31, where circuit implementations lie within the indicated boundaries in the two-dimensional space. Note that one can increase the number of gates to improve circuit speed, but after a point there is diminished return.

The job of the synthesis tool is to transform the HDL specification into an intermediate format that is optimized in some sense (e.g., smallest number of gates). The tool uses Boolean minimization techniques, state minimization techniques, and more advanced minimization techniques to optimize the circuit. In general, it is not possible for synthesis tools to produce optimal results overall, but they can produce results for complex designs that cannot be achieved by manual design methods. Furthermore, an implementation can be identified much more quickly with the assistance of synthesis tools.

Technology mapping tools transform the output of a synthesis tool into a form that specifies an implementation using a particular PLD or ASIC. These tools are sometimes called *fitters,* since they fit a design to a particular integrated circuit component. In the case of a PLD, the technology mapping tool decides which product terms and state variables to assign to each macrocell, and which inputs and outputs to assign to each pin. The technology mapping tool produces a binary computer file that is suitable for downloading to the PLD in order to configure it. Specialized software and hardware (known as a device programmer) are used to download the configuration to a device. This topic is not addressed further in this book but should be pursued in a laboratory.

Simulation Of ABEL Specifications

Once an initial hardware specification has been obtained (in our case, an ABEL description), it must be simulated to verify its correctness. This specification is technology independent, meaning that the simulator does not have knowledge about gate delays, setup and hold times, or interconnect parasitics.[13] The HDL simulation can be used to verify functional correctness prior to synthesizing an implementation, but it cannot be used to verify performance. A carefully crafted bench test must be used to verify that a specification is correct for arbitrary inputs. For large systems it is impossible to simulate all possible input patterns and states, but certainty of correctness must be established.

The device specification produced by the technology mapping tool contains relevant delay information such as gate delay, setup and hold times, and wire delay. A simulation model can be extracted from the device specification that is suitable for verifying the performance and timing of the design. Although an implementation may be functionally correct, it may not meet the performance requirement. Thus, a simulation is required after technology mapping to verify circuit performance.

If the performance requirement is not met by the implementation, the specification must be resynthesized, with more emphasis placed on circuit performance. This cycle is repeated until the requirement is met. If it cannot be met, a faster technology has to be chosen. It is also possible that the synthesized circuit is too fast. Does this have any consequence? Perhaps an implementation can be found that has fewer gates and still meets the timing requirement.

A summary of the design flow is depicted by the flow chart in Figure 32. Several iterations of the design flow with different initial specifications (functionally equivalent ones) and synthesis parameters are usually made to explore the implementation space (e.g., Figure 31) of a design. Subsystem requirements often change during the development of a system. It is necessary to understand the limits of various devices and technologies with respect to a specification.

Although ABEL is the hardware description language covered in this chapter, nearly all of the concepts apply to design with other HDLs, including VHDL and Verilog. The design flow depicted in Figure 32 is similar for all HDLs. All HDLs can be simulated, and numerous commercial simulators are available. Synthesis and technology mapping tools also are available from many CAD software companies, as is software for all tasks in the design flow.

---

[13]*Interconnect parasitics* are the capacitance and resistance of wires on an integrated circuit. Discussing them is beyond the scope of this book. Transistors are so fast in modern integrated circuits, however, that in estimating the delay of logic paths, it is necessary to take into account the delay of a signal on a wire. (The delay of a voltage signal on a wire is proportional to the product of the resistance and capacitance of the wire.)

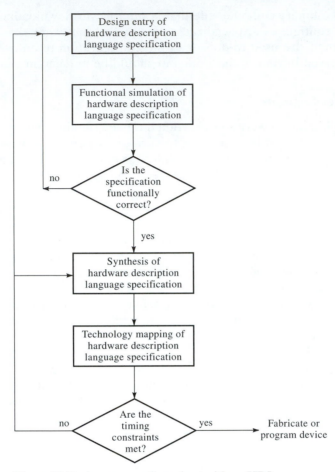

Figure 32 Design-process flow chart with an HDL.

CHAPTER SUMMARY AND REVIEW

This chapter utilized a hardware description language (HDL) in the design of logic circuits, including programmable logic devices (PLD). Programmable array logic (PAL) devices are featured, but complex PLDs (CPLDs) were also discussed. The specific language utilized is one called Advanced Boolean Expression Language (ABEL). Sufficient details of the language are provided to permit you to carry out such designs. Specific topics included the following:

- Differences between an HDL and a programming language
- Features of an ABEL description
- Module statements to identify a module name, title, and end
- Declarations section:
  - Specification of input and output pins
  - Specification of internal nodes
  - Equations section, including truth table listing

- Operational versus behavioral ABEL descriptions
- Specifying don't-care conditions in ABEL
- Hierarchical and structural specifications in ABEL
- Architectures of certain PLDs:

  - PAL16L8
  - PALCE16V8
  - Xilinx XC9500 CPLD

- Field-programmable gate arrays
- Using ABEL specifications in design
- Computer-aided design tools
- Simulation of ABEL specifications
- Synthesis into a logic format
- Mapping the logic format to a physical specification
- Simulation of the physical specification
- ABEL default implementation

PROBLEMS

Save the ABEL descriptions you write as called for in these problems so that you will have them available to carry out the further steps called for in later parts.

1 Write an ABEL description of a single-bit subtractor.

2 Write an ABEL description of a system that compares two 8-bit unsigned numbers.

3 Write an ABEL description of a BCD-to-seven-segment decoder using the truth table specification.

4 Synthesize the ABEL descriptions of Problem 3 and Figure 11.

5 Verify the ABEL description in Figure 11 by comparing the two implementations computed in Problem 4.

6 Draw the schematic diagram of the serial adder in Example 1.

7 Synthesize the ABEL description of Problem 2 using several different values for the @carry directive. Compile a table similar to the table in Figure 30.

8 Write an ABEL description based on state transitions for the specification given in Problem 1, Chapter 6, and synthesize it. Compare the result with your solution from Chapter 6.

9 Write an ABEL description based on a transition table for the specification given in Problem 2, Chapter 6, and synthesize it. Compare the result with your solution from Chapter 6.

10 Write an ABEL description based on next-state equations for the specification given in Problem 3, Chapter 6, and synthesize it. Compare the result with your solution from Chapter 6.

11 Write an ABEL description based on state transitions for the specification given in Problem 4, Chapter 6, and synthesize it. Compare the result with your solution from Chapter 6.

12 Write an ABEL description based on a transition table for the specification given in Problem 5, Chapter 6, and synthesize it. Compare the result with your solution from Chapter 6.

13 Write an ABEL description based on next-state equations for the specification given in Problem 6, Chapter 6, and synthesize it. Compare the result with your solution from Chapter 6.

14 Of the three methods for specifying sequential circuits in ABEL (state transition, state table, and next-state equations), which are behavioral descriptions? List the methods in order of abstraction from the most abstract to the least abstract. The most abstract method is the one that specifies the least detail of an implementation. Which method of specification do you find more efficient (which requires the least of your time)?

15 Write an ABEL description for the specification given in Problem 8, Chapter 6, and synthesize it. Compare the result with your solution from Chapter 6.

16 Write an ABEL description for the specification given in Problem 13, Chapter 6, and synthesize it. Compare the result with your solution from Chapter 6.

17 Write an ABEL description for the specification given in Problem 14, Chapter 6, and synthesize it. Compare the result with your solution from Chapter 6.

18 Write an ABEL description for the specification given in Problem 17, Chapter 6, and synthesize it. Compare the result with your solution from Chapter 6.

19 Write an ABEL description for the specification given in Problem 20, Chapter 6, and synthesize it. Compare the result with your solution from Chapter 6.

20 Write an ABEL description for the specification given in Problem 21, Chapter 6, and synthesize it. Compare the result with your solution from Chapter 6.

21 Write an ABEL description for the specification given in Problem 22, Chapter 6, and synthesize it. Compare the result with your solution from Chapter 6.

22 Write an ABEL description for the specification given in Problem 23, Chapter 6, and synthesize it. Compare the result with your solution from Chapter 6.

23 Write an ABEL description for the specification given in Problem 25, Chapter 6, and synthesize it. Compare the result with your solution from Chapter 6.

24 Write an ABEL description for the specification given in Problem 27, Chapter 6, and synthesize it. Compare the result with your solution from Chapter 6.

25 Write an ABEL description for the specification given in Problem 41, Chapter 6, and synthesize it. Compare the result with your solution from Chapter 6.

26 Write an ABEL description for the specification given in Problem 44, Chapter 6, and synthesize it. Compare the result with your solution from Chapter 6.

Chapter 9

Computer Organization

The coverage so far in this book has included certain components of digital systems: adders, multiplexers, flip-flops, registers, counters, programmable logic devices, sequential circuits (state machines), and the like. In this chapter we embark on what might be seen as a change in scale: the system organization of an entire digital computer. Principles of digital logic design and computer system organization might appear to be two different subjects. A digital computer, however, is simply a finite-state machine, although a very complicated one. It is therefore natural to present digital computer organization as an extension of the principles of digital logic design described in previous chapters.

Recall from Chapter 8 that two descriptions can be given for a digital system: a behavioral description and a structural description. A *register transfer level (RTL)* description is a behavioral description that specifies a machine's activities, clock cycle by clock cycle. If you think about it, you can conclude that the state table description, which is a behavioral description, is an example of an RTL description. In this chapter we will deal with an RTL description of a digital system, although we will not go into great detail.

1 CONTROL AND DATAPATH UNITS OF A PROCESSOR

One of the functions that a digital computer carries out is storage of information in memory. If you had to name the unit where this is done, wouldn't you call it a "memory unit" or a "memory section" or some other such name? Something else is also stored in the memory unit: instructions for carrying out the desired operations. What the computer does is to *process* both the stored information and any other information externally supplied, brought in by primary inputs. If you had to name the part of the machine that carries out this processing, wouldn't you call it a *processor* or a *processor unit?* Everybody else calls it a processor too. When the computer is in operation, data is exchanged back and forth between the processor and memory.[1]

For ease of description, it is useful to decompose the processor into two general blocks, called the *control unit* and the *dataflow* or *datapath unit,* as shown in

---

[1] A major component of any digital computer is the *central processing unit (CPU)*. More on this later.

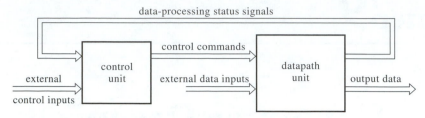

Figure 1 General structure of a digital system.

Figure 1. Some of the primary inputs to the processor are directed to the control unit and some to the datapath unit. The datapath unit contains circuitry that manipulates data to compute a desired arithmetic or logical result. The name of this unit comes from its organization as a set of paths through which the data flows. The control unit instructs the datapath unit to perform specific operations on the data, and it controls the sequence of these operations.

Besides primary inputs, the datapath unit receives inputs from the control unit also, as seen in Figure 1. The latter are viewed as control commands under which data processing is to be carried out. In turn, besides external (primary) inputs to control the details of the data-processing operations, the control unit also receives inputs from the datapath unit as to the status of the operations being carried out there. The only outputs from the composite machine come from the datapath unit.

Datapath Unit

A datapath unit includes both combinational and sequential circuits in a regular structure. Its combinational elements may include decoders, multiplexers, adders, and other circuits; its sequential elements include registers of all types, but not individual storage elements (flip-flops). Any system component that carries out some function, such as an adder or arithmetic logic unit (ALU), is referred to as a *functional unit.*

The complexity of a datapath unit depends on the number of registers, the number of functional units, their interconnections, and other factors, and these can vary substantially from one unit to another. The datapath is a generic circuit capable of being configured to perform any number of operations. It contains regular structures but not the circuitry required to select a particular operation. A sequence of datapath operations leads to the realization of some desired result.

Although all sequential circuits can be represented as finite-state machines described by state graphs or transition tables, those constituting more than a handful of states cannot efficiently be represented as such. A good example of such a circuit is a register. A simple 8-bit register has 256 possible states. Because of their regular structure, combinational logic circuits that perform arithmetic operations (e.g., multibit adders) are not candidates for logic minimization. In other words, we already know good implementations of adders, as discussed in Chapter 4; there is no need to perform logic minimization procedures on their specifications.

We can separate a processor into two parts: one part consists of units having regular structures, which can be synthesized easily. The second part has irregular structure and requires some effort to synthesize.

Control Unit

The control unit is a state machine that instructs the datapath unit to follow a series of operations culminating in a desired result. This state machine is synthesized using the techniques of Chapter 6. It is sometimes referred to as *random logic* because, unlike the datapath unit, it has no regular structure (no registers, only individual storage elements). Also unlike the datapath, the combinational-logic specification of the control unit calls for logic minimization; it is a good candidate for implementation in programmable logic, such as a PLA.

In the design of a processor there are fundamental trade-offs between the complexity of the datapath unit and that of the control unit. A simple datapath unit typically requires complex control. For example, if there is only one connection among all of the registers of the datapath, the control unit must provide the sequence of control signals to multiplex the data onto the connection. On the other hand, a complex datapath often requires simple control because separate data can be manipulated in parallel. Thus, the design processes of the datapath and control units cannot be independent.

Serial Multiplier Example

Let's make the preceding discussion concrete by considering a rather simple example: the design of a sequential circuit that computes the product of two 4-bit numbers. A numerical illustration follows.

0101 (5)	multiplicand
× 1101 (13)	multiplier
0101	partial product 0
0000	partial product 1
0101	partial product 2
0101	partial product 3
01000001 (65)	product

The system that carries out this operation is a processing unit. It can be designed in numerous ways, one of which performs a sequence of add and shift operations. The add-and-shift method is a direct implementation of the addition sequence specified by the preceding partial-product array. Even in this simple case, the implementation of the multiplier is separated into datapath and control units. The datapath contains registers and an adder; the control unit is a finite-state machine that sequences the datapath unit through each of the add and shift operations until all of the partial products are accumulated.

Because the control is dependent on the architecture of the datapath, the multiplier design should begin with the datapath. First, it is necessary to identify

- The kinds of operations to be performed in the datapath
- The necessary registers
- Interconnections (buses) among functional units (e.g., adders) and registers

Let's use two 4-bit registers and one 8-bit register in the datapath for this design.

A description of the sequence of operations performed by this multiplication circuit is needed; one possibility follows.

1. Initialize all registers (product←0, multiplier←input, multiplicand←input).
2. Add the first partial product to the product register.
3. Shift the product register.
4. Add the second partial product to the product register.
5. Shift the product register.
6. Add the third partial product to the product register.
7. Shift the product register.
8. Add the fourth partial product to the product register.
9. Shift the product register.

The notation A←0 is in register transfer language (RTL) notation; it specifies the transfer of the numerical value zero to register A. The preceding lengthy description of the operation of the multiplier can be written more concisely as follows.

1. Initialize all registers (product←0, multiplier←input, multiplicand←input).
2. For each bit j of the multiplier:

   - Add the jth partial product to the product register.
   - Shift the product register.

Numerous details of the implementation still need to be specified, but this description of the hardware helps in visualizing the structure of the datapath and the state machine needed to control its operation. A block diagram of the datapath is shown in Figure 2. It contains two 4-bit registers, one 8-bit register, a 4-bit adder, a functional unit to generate a partial product, and a flip-flop to store the output carry of the adder.

The states of the registers in the datapath at a few steps of the algorithm are shown in Figure 3. Notice that don't-care values are entered in locations where the value of the register bit is not important. This enables one to identify redundancy in the datapath. In particular, the rightmost bits of the product register can be used to store the multiplier, since the examined bits of the multiplier can be discarded. In addition, the content of the multiplicand register never changes. If we impose the requirement that these inputs be held constant during operation, then this register can also be eliminated.

Exercise 1   Fill in the missing steps in Figure 3.   ◆

A new block diagram containing only one 8-bit shift register is shown in Figure 4. Now another component is needed (a multiplexer) so that the least significant 4 bits of the product register can be initialized with the multiplier input. The partial-product generator contains four AND gates; any adder design (see Chapter 4) can be used for accumulation of the partial products.

Exercise 2   Determine the states of the datapath registers at each step of the multiplication procedure for a multiplicand of 0101 and a multiplier of 1101.   ◆

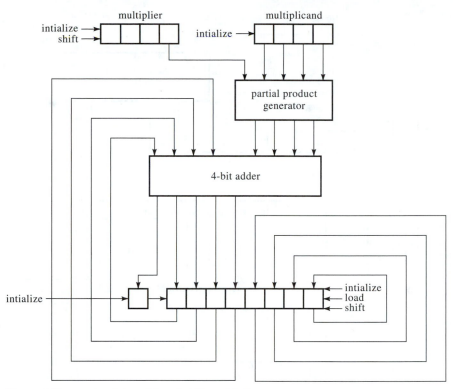

Figure 2 Datapath of the add-and-shift multiplier.

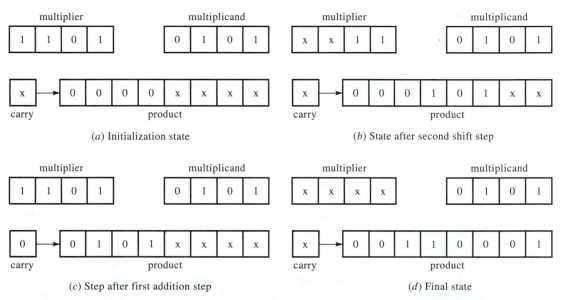

Figure 3 States of datapath registers of the add-and-shift multiplier for one set of inputs at selected steps of the algorithm.

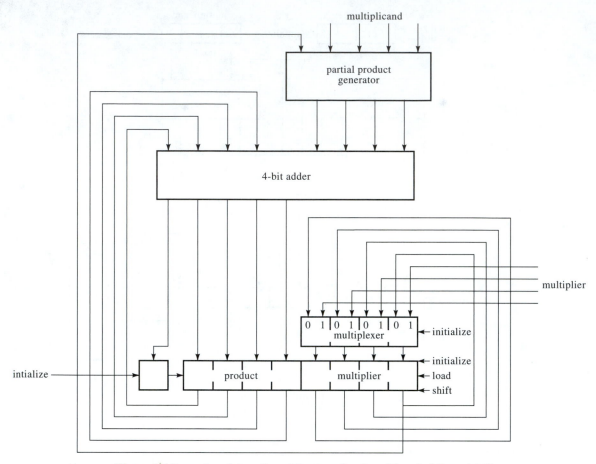

Figure 4 Alternative datapath architecture for the add-and-shift multiplier.

At the appropriate time in the sequence of the algorithm, the control unit for the multiplier under consideration must assert to the datapath unit certain control signals for each of the operations: initialize, load, and shift. The state diagram of the controller is shown in Figure 5; it is a direct implementation of the sequential control outlined in the preceding algorithm. The controller has a single input (the hardware reset) and three outputs.

Different implementation strategies can be chosen for the control-unit state machine; however, its structure suggests that a counter is a good choice. Is the number of states few enough that a BCD counter can be used? Work it out before glancing at the answer.

Answer[2]

Let's choose the 74LS90 for the implementation of this controller; it has the logic diagram and functional description shown in Figure 6. The state assign-

---

[2]There are nine states, so a BCD counter is adequate.

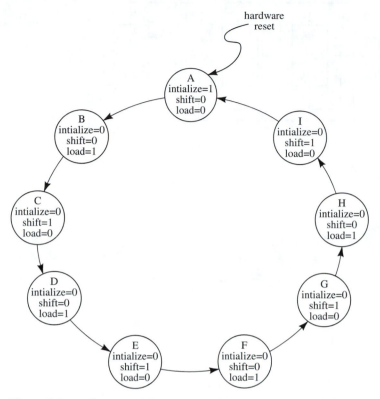

Figure 5 State diagram of the control unit for the add-and-shift multiplier.

ment should be selected to utilize the built-in functionality of the 74LS90. Thus, states A, B, C, ... are assigned the binary encodings 0000, 0001, 0010, ..., respectively. The state transition table of the controller and the excitation requirements for the 74LS90 are given in Figure 7.

Can you determine whether the finite-state machine described in Figure 7 is a Mealy or a Moore type? Does it necessarily have to be one or the other, or can it be redesigned to be either? What constraints does the use of the 74LS90 place on the type (Mealy or Moore) of implementation?
Answer[3]

The logic schematic of the implementation of the add-and-shift multiplier is shown in Figure 8.

Exercise 3 Verify the logic expressions for the 74LS90 excitations and the control-unit outputs in Figure 7. ◆

---

[3]Careful examination of the 74LS90 specifications (Figure 6) reveals that our machine is a Mealy machine and cannot be otherwise, because the reset inputs are asynchronous. That is, the state of the 74LS90 can be changed to 0000 without a clock event by applying the appropriate reset inputs.

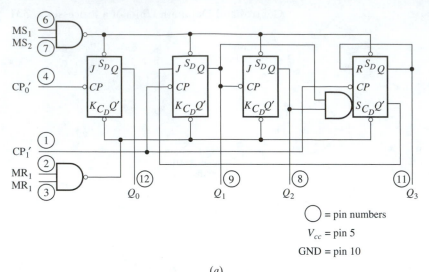

V_{cc} = pin 5

GND = pin 10

(a)

	Reset/Set Inputs				Outputs		
MR_1	MR_2	MS_1	MS_2	Q_0	Q_1	Q_2	Q_3
H	H	L	×	L	L	L	L
H	H	×	L	L	L	L	L
×	×	H	H	H	L	L	H
L	×	L	×		Count		
×	L	×	L		Count		
L	×	×	L		Count		
×	L	L	×		Count		

H = High voltage level
L = Low voltage level
× = Don't-care

(b)

		Output		
Count	Q_0	Q_1	Q_2	Q_3
0	L	L	L	L
1	H	L	L	L
2	L	H	L	L
3	H	H	L	L
4	L	L	H	L
5	H	L	H	L
6	L	H	H	L
7	H	H	H	L
8	L	L	L	H
9	H	L	L	H

Note: Output Q_0 is connected to input CP_1 for BCD count.

(c)

Figure 6 74LS90 block diagram and functional specifications. (*a*) Block diagram. (*b*) Mode selection. (*c*) BCD count sequence.

	Present State				Input (Reset)	Next State				Outputs			Excitations			
	Q_3	Q_2	Q_1	Q_0		Q_3^+	Q_2^+	Q_1^+	Q_0^+	Initialize	Load	Shift	MR_1	MR_2	MS_1	MS_2
	×	×	×	×	1	0	0	0	0	×	×	×	1	1	0	×
A	0	0	0	0	0	0	0	0	1	1	0	0	0	×	0	×
B	0	0	0	1	0	0	0	1	0	0	1	0	0	×	0	×
C	0	0	1	0	0	0	0	1	1	0	0	1	0	×	0	×
D	0	0	1	1	0	0	1	0	0	0	1	0	0	×	0	×
E	0	1	0	0	0	0	1	0	1	0	0	1	0	×	0	×
F	0	1	0	1	0	0	1	1	0	0	1	0	0	×	0	×
G	0	1	1	0	0	0	1	1	1	0	0	1	0	×	0	×
H	0	1	1	1	0	1	0	0	0	0	1	0	0	×	0	×
I	1	0	0	0	0	0	0	0	0	0	0	1	1	1	0	×

Figure 7 State transition table of the control unit and 74LS90 excitation requirements.

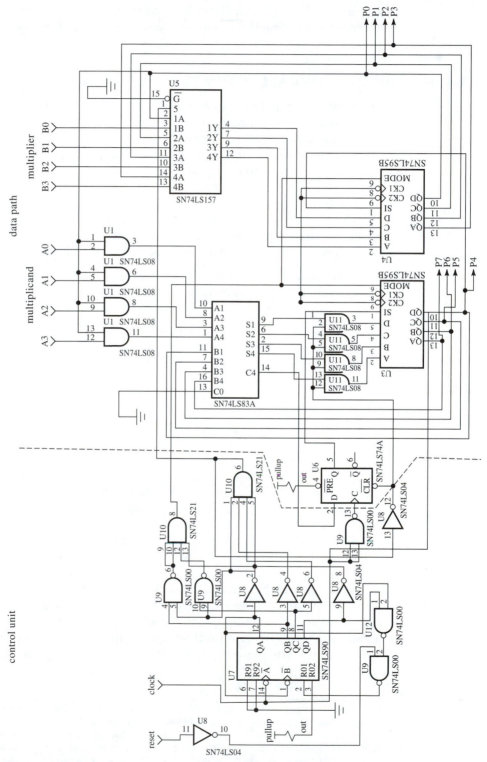

Figure 8 Schematic diagram of the entire add-and-shift multiplier.

Exercise 4 Using parameters obtained from the data sheets of the components in Figure 8 calculate the worst-case (critical) propagation delay in the datapath. Is the critical propagation delay in the control unit important? ◆

2 BASIC STORED-PROGRAM COMPUTER

The most common digital system on earth, as this book makes its appearance at the beginning of the twenty-first century, is the digital computer. No one reading this book can be unfamiliar with the ubiquitous desktop computer. The present-day stored-program digital computer was conceived by John von Neumann, who proposed that a machine be organized in such a way that a procedure could be implemented by the execution of a sequence of instructions stored in a central location (memory).[4]

The instructions for the tasks the computer is to carry out are stored in *memory*. Where the tasks are carried out (*executed*) in a number of steps is the *central processing unit* (CPU). A sequence of steps is carried out repeatedly:

- The machine *fetches* the instructions from memory to the CPU in sequence.
- In the CPU the instructions are first *decoded*.
- Finally, the instructions are carried out, or *executed,* again in the CPU.

This simple three-step process is known as the *fetch/decode/execute* cycle. While the computer is operating, this three-step process is performed repeatedly at a rate of millions of times per second.[5] It is the first thing a computer does when it starts up, and it never stops doing it. Since this is a fixed process, it can be implemented in hardware (i.e., as the control unit of a processing unit).

A computer performs different tasks by specifying the appropriate instructions and their sequence. Since this part of the control of the computer is programmable, it is specified in software. The computer must obviously have some hardware components to enable the execution of the software instructions: the datapath. The only instruction that should be hard-wired into the computer is the location of the very first instruction. Contemporary desktop personal computers have hardware that forces the machine to access the first instruction from the built-in operating system (BIOS).

The BIOS contains instructions that begin the start-up procedure of the computer. The BIOS program first instructs the computer to check the floppy disk drive for a disk that contains an operating-system start-up file. If there is no floppy disk in the drive, the BIOS starts checking the hard disks of the system for this file. If there is a floppy disk in the drive and it does not contain an operating-system start-up file, the machine gives an error message and asks for a valid start-up disk. If it finds an operating system start-up file on the floppy disk, then it starts the computer with that operating system. Every instruction

---

[4]Hungarian-born John von Neumann (1903–1957) received a Ph.D. in chemical engineering at the age of 23 but taught mathematics at the University of Berlin before joining Princeton and its Institute for Advanced Studies. He made significant contributions in many areas, but not in chemical engineering. His work on MANIAC-1 at Princeton laid the foundation for the design of all subsequent programmable computers.

[5]It is expected that by 2005, the rate of execution will reach billions of instructions per second.

that is executed during this process is done via the cycle: instruction fetch, decode, and execute.

Central Processing Unit (CPU)

The heart of a computer system is the central processing unit (CPU).[6] The microprocessor is only one type of CPU; examples of other types are *microcontrollers* and *digital-signal processors*. A digital system can be classified as a CPU if it contains the circuitry needed to execute a set of defined instructions that enables any desired, deterministic computation. But what is the minimum number of instructions necessary? Can a simple CPU be specified that can be constructed in a laboratory within a few hours or that can fit into a Xilinx 9500 series PLD? We will now explore such questions.

Simple Datapath

As discussed in section 1, the circuitry of a CPU is separated into a datapath unit and a control unit. Let's first consider the datapath of a CPU and ask, What minimum amount of hardware is needed to constitute a viable datapath? By definition, a CPU must read an instruction located in memory; this is referred to as an *instruction fetch*. For the CPU to do this, there must at least be some means of generating and storing the memory address from which to fetch the instruction. A register is good at storing things; we need one that can be *loaded* with what is to be stored and that can be *incremented* to go to the next task.

The instructions in the computer are normally executed in sequence. Hence, when the next instruction is to be fetched, the register can be incremented. It isn't really necessary to execute the very next instruction in a sequence. Consider *loops* in a program, that is, sequences of tasks that return to the initiating point and then repeat. To return to the beginning of a loop, it will not do to simply increment the instruction address; we need a means of loading an arbitrary address into the register in order to alter the consecutive sequence of execution. This register is often called the *program counter* (PC) because

- It is a counter.
- It points to the memory location where the program is executing.

An instruction that loads the PC with the address of the next instruction is called a *branch* instruction (or *jump* instruction) because the execution of the program branches to an instruction that is different from the next instruction in sequence.

The CPU can't be expected to fetch an instruction, decode it, and then execute it all in one clock cycle. Hence, we need a register for storage of the instruction, as the clock ticks on. Let's call it the *instruction register* (IR). Furthermore, there needs to be

- An arithmetic logic unit (ALU) to manipulate data
- A register to store data

---

[6]One CPU of a modern IBM-compatible personal computer is the Pentium microprocessor, most recently the Pentium III. For the Apple Macintosh computer the CPU is the PowerPC microprocessor.

Data to be stored can come from memory or it can be generated within the datapath. The two types of data can be stored in different registers. For simplicity, though, they can be stored in the same registers. Because a data register is used to accumulate arithmetic results, it is often called an *accumulator* (AC).

A memory address can be the address of an instruction or the address of data. Thus, a register is needed, called the *memory address register* (MAR), to serve as an interface between multiple address sources and the bus that carries the addresses and is connected to the memory.

What do you expect the sources for memory addresses in our datapath to be? Obviously, the PC is a source for the addresses of instructions. The datapath also needs data—for example, for an add operation. Data addresses are often contained in the instruction itself (we elaborate on this later). So in our system the MAR can take an address either from the PC (to fetch an instruction) or from the IR (to fetch data).

That's an outline of the minimum amount of hardware needed to execute instructions that are useful for the implementation of a program; with that, the architecture of the datapath can be sketched. One further thing needs to be determined: the width of the data and memory addresses in the system, and the kinds of instructions that the CPU is to execute. Generally, the complexity is constrained by time and cost. In this example, in order to minimize complexity and emphasize the concepts of CPU specification and design, we choose small data and address widths.

Although the minimum number of instruction types required for a functionally complete CPU is three, we choose four instructions for this example. The tasks that need to be performed are loading data from memory (*load* instruction), storing data in memory (*store* instruction), alter the consecutive execution sequence (*branch* instruction), and perform a logical operation between two operands. (We will choose the logical operation to be NAND, so this would be called the *NAND* instruction. Why choose NAND instead of, say, AND?[7])

To encode four instructions with binary data we need 2 binary digits. An instruction, therefore, will contain 2 bits of information that specify the operation to be performed by the CPU. These 2 bits are commonly referred to as the *operation code,* or *op-code,* of the instruction. By choosing the length of the op-code, we have begun to define the *instruction format* of the CPU. The instruction format specifies how we write each instruction (load, store, etc.) in binary format for storage in the computer memory. In other words, it defines precisely the language of the CPU or machine. This is the origin of the term *machine language,* which refers to the binary instructions executed by a CPU.

What other kind of information is needed in a machine language instruction? If we want to access data, then we need a means to address it. If the instructions themselves contain the address of the data, it makes sense to call the method of addressing *direct addressing.*[8] All of our instructions require a memory address.

The branch instruction requires an address that specifies the location in the memory from which the next instruction is to come. The branch instruction should

---

[7]Because NAND is logically complete; AND is not (refer to Chapter 2).

[8]There are numerous methods for addressing data in memory. Such methods can be found in a book on computer architecture but are beyond the scope of this book.

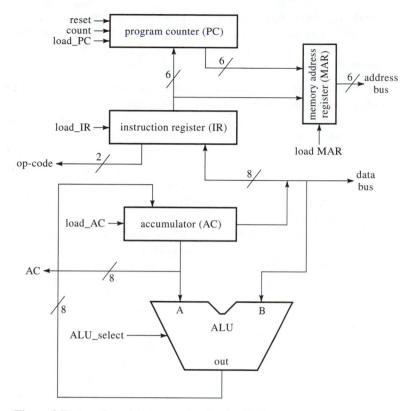

Figure 9 Datapath architecture of a simple CPU.

be conditional so that the branch occurs only under certain specific circumstances. In our machine we will branch if the content of the accumulator is zero.

The NAND instruction must contain the address of one of the operands; we will assume that the second operand is the accumulator. In this way the format of the NAND instruction is consistent with the other instructions. Where would be the most appropriate place to store the result of the NAND operation, in memory or in the accumulator? Something to avoid is the need to specify a second address in the instruction. Such a need would complicate the organization of our computer. It would require instructions that occupy two words in memory, and it would require more complicated datapath and control units. Hence, we will store the result of the NAND operation in the accumulator.

It would be useful if instructions and data were to have the same length. Assuming 8-bit data (and therefore 8-bit instructions), what would be the width of the address bus?

Answer[9]

The architecture of our datapath is shown in Figure 9. This should be viewed as a preliminary architecture, because during the design of the control unit we

---

[9]Six bits. In this case, all instructions fit into one memory location.

may decide to change some things. It is clear that in considering both the datapath and control units, trade-offs between speed and complexity have to be made.

Controlling the Simple Datapath

The instructions for the machine are shown in Figure 10. Each of these instructions may require several clock cycles for execution. The operations performed in one clock cycle are called *micro-operations*. The control unit is specified by identifying the micro-operations to be performed in each control state. For example, certain micro-operations are needed to perform an instruction fetch. One of these is to inform memory that we want to read. Another is to make sure that the data sent by the memory is stored. The control unit should assert load_IR and issue a memory-read request to accomplish this. (Remember that in positive logic, "asserting" means setting at logic 1.) As a result, that data is received by the datapath.

In order to confirm that the correct memory location is being addressed in any one operation, the control sequence must ensure that, in some previous micro-operation, the correct address was put in the MAR. Let's write the sequence of states and micro-operations to be performed in each state to fetch, decode, and execute the instructions of the machine.

Assume that a hardware reset leads the machine to state A. State A should reset the program counter to point to the first instruction in memory. As previously discussed, this can be written in RTL as PC←0. In state B of our control unit the content of the program counter must be transferred to the MAR so that the address can be used to read an instruction from memory. The RTL notation for this micro-operation is MAR←PC. At the same time, it may be possible, depending on the datapath architecture, to increment the PC to point to the next instruction. So far the state machine of our control unit looks like this:

State A: PC←0, go to state B.
State B: MAR←PC, PC←PC+1, go to state C.

The last line leads to state C; we need to figure out what is to be done in that state. The address of the instruction is in the MAR, so we are ready to read it. The memory will need a control signal to tell it that we are reading data from the address on the address bus as opposed to writing data to the address. This control signal can be called R/W' and asserted high (R/W' = 1) for a read operation and low (R/W' = 0) for a write operation. Write the register-transfer operations that should be performed next.

Instruction	Description	Op-Code
Load AC, memory	Load AC with content of memory location	00
Store memory, AC	Store AC in memory location	01
NAND AC, memory	NAND content of AC and memory location	10
Brz memory	Branch to memory location if AC is 0	11

Figure 10 Machine instructions for a simple CPU.

Answer[10]

When we reach state D, the instruction is in the IR, so we need to examine (decode) the op-code bits and execute a sequence of micro-operations based on their value. For example, if the op-code is 00, then the instruction is load, and we need to put the address of the data in the MAR and perform another memory read. The sequence of micro-operations for load and branch are as follows. The micro-operations for the remaining instructions are left for you to write.

State D: If op-code = 00, then go to state E.
            If op-code = 01, then go to state F.
            If op-code = 10, then go to state G.
            If op-code = 11, then go to state H.
State E: MAR←IR, go to state Ea.
State Ea: R/W'←1, ALU_B←data bus, ALU_select←pass, AC←ALU_out,
            go to state B.
State H: If AC = 0, then PC←IR<5:0>, go to state B.

There are a few new things in the above state sequence. The data bus is not a direct input to the AC, so it is necessary to pass the data through the ALU and into the AC. Thus, the ALU should have an operation that permits the data on its B input to pass directly through to the output without modification. When state Ea is complete, the execution of the instruction is complete, so the control unit sequences back to state B in order to prepare to fetch the next instruction. Finally, for the branch instruction, the least significant 6 bits of the IR are transferred to the PC if the content of the AC is zero. We use the notation IR<5:0> to identify bits 5 through 0 of the IR.

Exercise 5   Write the RTL notation of the micro-operations for the store and NAND operations of this simple CPU. (The answer follows, but work it out before you consult it.)

Answer   State F: MAR←IR, go to state Fa.
            State Fa: R/W'←0, data bus←AC, go to state B.
            State G: MAR←IR, go to state Ga.
            State Ga: R/W'←1, ALU_B←data bus, ALU_select←NAND,
                      AC ← ALU_out, go to state B.   ◆

It is good practice to divide the memory of the system into two parts: one for instructions and the other for data. The result is known as the *Harvard architecture*. The alternative is the *Princeton architecture*, where the data and instructions can coexist in the same part of the memory. Separating the instructions and data in memory gives a minimum level of protection to the instructions. A program should never be written such that it contains instructions designed to overwrite themselves. If the data and instructions are mixed in memory, it is more likely that instructions may be overwritten unintentionally.

---

[10]State C: R/W'←1, IR←data bus, go to state D.

3   CONTROL-UNIT IMPLEMENTATIONS

The control unit of the CPU sequences the datapath through the micro-operations. The control unit is a finite-state machine, where each state represents one micro-operation. The control unit can be implemented using any of the techniques introduced in previous chapters (e.g., gates, PLA, ROM). There are two common implementations, referred to as *hard-wired* and *micro-programmed*. Hard-wired control refers to implementation with gates and flip-flops. It is so called because once it is built, it can be modified only by changing the hardware.

Micro-programmed control, on the other hand, refers to implementation that uses a ROM and micro-program sequencer.

- The ROM is used to store control signals at various locations.
- The micro-program sequencer selects the location of the desired micro-operation within the ROM.

We will implement the control unit of our simple CPU using both of these techniques.

Hard-Wired Control Unit

The hard-wired control unit can be described using any of the specification tools studied earlier in the text: state diagram, ASM chart, or ABEL. Since use of a hardware description language is the most common method today, let's describe the hard-wired control unit in ABEL. For each state outlined in the previous section we need to assert the appropriate control signals given in Figure 9. When the PC is in state A, the reset signal has to be asserted and the next state is B. The ABEL code specifying this behavior is shown in Figure 11.

How do we code state B? Asserting the count_PC control signal is no problem, but how do we control what is clocked into the MAR? There are two inputs to the MAR: IR and PC. A register with two inputs does not generally exist, but we can put a multiplexer at the input and choose either PC or IR as the source. Obviously, we need to refine the architecture of Figure 9 a bit. The refinement may require additional hardware, such as the multiplexer at the input of the MAR, or it may require the addition of extra control steps to avoid conflict on a connection.

Exercise 6   Identify any ambiguities or conflicts in the architecture of Figure 9. Draw a block diagram of a new architecture. (The answer follows; carry out the work before checking.)
Answer   The AC has two destinations: the control unit and the data bus. The data bus is bidirectional, so when data is sent to the datapath, the output of the AC must not drive the bus. Thus, the AC must have tristate outputs, which requires an output-enable control signal. (See Chapter 2, section 10.) In addition, this means data cannot be sent on the data bus into the CPU while the AC output is being read by the control unit or the ALU. The control unit is interested only in the content of the AC in state H; the data bus is not utilized in this state. However, in states Ea and Ga data is sent to the ALU_B input on the data bus and from the AC to the ALU_A input. One way to alleviate the conflict on the

module CPU_control
Title 'Hard-wired control unit'
"Hard-wired control unit of simple CPU

Declarations
clock, reset, op_code1 .. op_code0, AC7 .. AC0 **PIN;**
reset_PC, count_PC, load_PC, load_IR, load_AC **PIN istype 'com';**
ALU_select, load_MAR, IR_PC, R_W **PIN istype 'com';**
Q3 .. Q0 **NODE istype 'reg';** "registered output
A = [0, 0, 0, 0]; "state assignments
B = [0, 0, 0, 1];
C = [0, 0, 1, 0];
D = [0, 0, 1, 1];

Equations
[Q3 .. Q0].clk = clock
[Q3 .. Q0].ar = reset; "asynchronous reset

state_diagram [Q3, Q2, Q1, Q0]
state A: reset_PC = 1; count_PC = 0; load_PC = 0;
goto B; "We don't care what happens to other registers

end CPU control
Figure 11 Beginning of ABEL specification of hard-wired control unit of a simple CPU.

data bus is to insert a tristate buffer between the output of the AC and the data bus. (The conflict we refer to here is often called *bus contention* because two sources of data contend for use of the bus.) When the AC is to be passed onto the bus, the tristate buffer must be enabled. A modified datapath architecture is shown in Figure 12. There may be several other ways to modify the architecture of Figure 9 to eliminate the conflicts. ◆

We now continue with the ABEL description of the control unit. A refined and detailed block diagram of the datapath unit is shown in Figure 12. Refer to it for all of the control signal names used in the following ABEL description. The code for state B must do two things:

- Select the PC to be transferred to the MAR.
- Increment the PC.

A question arises: Can these be done simultaneously? Our CPU is synchronous, so we assume that every memory element is clocked by the same signal. Thus, the control signals set up the present state of the PC (the next state of the MAR) at the input of the MAR, and the next state of the PC (PC + 1) at the input of the PC. The clock event transfers the next state into the present state, the propagation delays satisfy the hold times of the registers, and the next states are captured in the registers simultaneously with no problem. The ABEL code for state B can be written as follows:

State B: reset_PC = 0, load_PC = 0, count_PC = 1, IR/PC' = 0, load_MAR = 1, load_AC = 0, goto state C.

Why do we specify that nothing be loaded into the AC in state B?

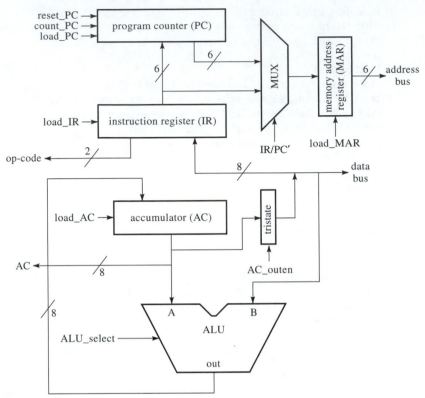

Figure 12 Refined architecture of Figure 9.

Memory and I/O Interface

The ABEL code for state C requires assertion of the R/W' control signal, which has not appeared in any of our block diagrams thus far. This is a signal sent directly from the control unit to the main memory of the system. Since the main memory is not considered part of the CPU, it is not shown in Figure 12. A top-level block diagram of the system is shown in Figure 13. In this figure, the input/output (I/O) devices are connected to the CPU, similar to the memory. A system configured in this way is said to have *memory-mapped I/O*. The instructions used to read and write to the I/O devices are the same as those used for memory; they are addressed in the same way as individual memory locations.

Some commercial CPUs (e.g., the Motorola 6800) restrict the designer to connecting I/O devices in this manner. Other commercial CPUs (e.g., the Intel 8085) have a separate I/O address space; even so, the designer is not barred from memory-mapping the I/O devices. Write the ABEL code for state C; then check it against what follows.

State C: R/W' = 1, load_IR = 1, load_AC = 0, load_PC = 0, goto state D.

In state D the control unit inspects the operation code bits of the IR and branches to one of four states for execution of one of the macro-instructions of

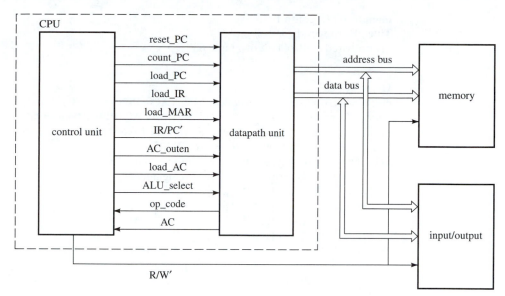

Figure 13 System-level block diagram of a simple computer illustrating the memory and I/O interface.

the machine. Label the states E through H and write the ABEL code before you proceed; then check your results against what follows.

State D: If op_code1..op_code0 == [0,0] then E
              Else if op_code1..op_code0 == [0,1] then F
              Else if op_code1..op_code0 == [1,0] then G
              Else H

States E–H specify the micro-instructions necessary to execute the corresponding instructions. The complete ABEL code for the control unit is shown in Figure 14.

Micro-Programmed Control Unit

A micro-programmed control unit uses a micro-program sequencer and a ROM to implement the control steps. Each location of the ROM stores the control signals for a single micro-operation, and the sequencer determines which location of the ROM to address during each clock cycle. For example, the first location of the ROM contains the control signals (micro-instruction) required to perform an instruction fetch. The sequencer can get its next address either from the micro-instruction itself or from an address translation table. If the next address is known, then it can be coded into the micro-instruction itself. This is the case for the instruction fetch (state C), since the next step is the instruction decode (state D). On the other hand, the next micro-instruction address after state D depends on the operation code of the macro-instruction. A look-up table (another small ROM or PLA) can be used to generate the next address. The advantage of the micro-programmed control unit is that the control can be changed without re-designing the hardware of the system. The firmware stored in the ROM of the control unit can be changed to implement a new control scheme.

module CPU_control
Title 'Hard-wired control unit'
"Hard-wired control unit of simple CPU

Declarations
clock, reset, op_code1 .. op_code0, AC7 .. AC0 **PIN;**
reset_PC, count_PC, load_PC, load_IR, load_AC **PIN istype 'com';**
ALU_select, load_MAR, IR_PC, R_W **PIN istype 'com';**
"ALU_select = 0 for pass and =1 for NAND
Q3 .. Q0 **NODE istype 'reg';** "registered output
A = [0, 0, 0, 0]; B = [0, 0, 0, 1]; "state assignments
C = [0, 0, 1, 0]; D = [0, 0, 1, 1];
E = [0, 1, 0, 0]; Ea = [0, 1, 0, 1];
F = [0, 1, 1, 0]; Fa = [0, 1, 1, 1];
G = [1, 0, 0, 0]; Ga = [1, 0, 0, 1];
H = [1, 0, 1, 0];

Equations
[Q3 .. Q0] . clk = clock;
[Q3 .. Q0] . ar = reset; "asynchronous reset

state_diagram [Q3, Q2, Q1, Q0]
state A: reset_PC = 1; count_PC = 0; load_PC = 0;
       goto B; "We don't care what happens to other registers
State B: reset_PC = 0; load_PC = 0; count_PC = 1; IR_PC = 0;
       load_MAR = 1; load_AC = 0; **goto** C;
State C: R_W=1; load_IR=1; load_AC=0; load_PC=0; goto D;
State D: **if** op_code1 .. op_code0 == [0, 0] **then** E
       Else if op_code1 .. op_code0 == [0, 1] **then** F
       Else if op_code1 .. op_code0 == [1, 0] **then** G
       Else H;
State E: load_MAR=1; IR_PC=1; **goto** Ea;
State Ea: R_W=1; ALU_outen=0; ALU_select=0; load_AC=1; **goto** B;
State F: load_MAR=1; IR_PC=1; **goto** Fa;
State Fa: R_W=0; AC_outen=1; **goto** B;
State G: load_MAR=1; IR_PC=1; **goto** Ga;
State Ga: R_W=1; ALU_outen=0; ALU_select=1; load_AC=1; **goto** B;
State H: if AC7 .. AC0 == [0, 0, 0, 0, 0, 0, 0, 0] **then** load_PC=1; **goto** B;
end CPU control

Figure 14 Complete ABEL description of the control unit for the datapath in Figure 12.

We can construct a simple block diagram of the micro-programmed control unit. We have a ROM with length sufficient to store all of the micro-instructions and width sufficient to store all of the control signals and micro-program sequence information. We know from the previous section that we have 11 control states; hence, the smallest ROM width we could use is 16 words. We also

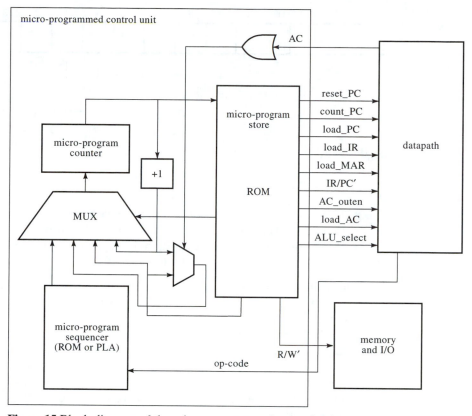

Figure 15 Block diagram of the micro-programmed control unit.

need a micro-program counter that is 4 bits wide.[11] The status inputs to the control unit (op_code and AC) must be processed by some hardware. We want to maintain flexibility in this hardware, so the translation of op_code to a micro-instruction address should be done in a ROM or PLA so that it can be changed without changing the hardware design. The AC input to the control unit is used to detect AC = 0. Hence, it is possible to create the status signal by ORing all of these bits together. The block diagram of the resulting micro-programmed control unit is shown in Figure 15.

The key to the design of the micro-programmed control unit is proper sequencing of the micro-program counter. There are four potential sources of the next micro-instruction address.

- If the next micro-instruction is in the next ROM address, then the micro-program counter can be incremented.
- If an unconditional branch is required, then the next micro-instruction address can be coded in the micro-instruction.

---

[11]We may need a few extra control steps compared with the hard-wired control unit implementation, but at this stage of design we can easily choose a larger ROM and micro-program counter.

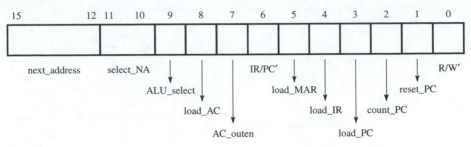

Figure 16 Micro-instruction format.

- If the next micro-instruction address is conditional on the operation code, then the next address can be looked up in a ROM or PLA.
- If the operation code is a conditional branch, then the next micro-instruction address depends on the content of the AC.

The following question now comes up: How many bits are needed in one micro-instruction? The micro-instruction format can be determined from the architecture of the micro-programmed controller in Figure 15. We need 1 bit each for the control signals to the datapath and to the memory. We need 2 bits for selection of the source of the next micro-instruction address, and we need 4 bits for the address. That makes a total of 16 bits in one word of the ROM. Let's format the micro-instruction as shown in Figure 16. The order of the bits is not important, but logically related bits (e.g., address bits) are usually placed next to each other.

We are now ready to specify the content of the micro-program store. Let's store the micro-instruction corresponding to state A in the first location; this will permit the micro-program counter to point to it after a hardware reset. What should the content of this ROM location be? In this clock cycle we want to reset the program counter (reset_PC = 1). None of the other control signals in bits 9 through 0 of the micro-instruction need to be set. The next micro-instruction corresponding to state B from the previous section can be located in the next ROM location. Thus, the source of the next micro-instruction address can be the present address plus 1.

The next task is to choose the encoding for the 2-bit control word select_NA, which selects the source of the next micro-instruction address. We make the following assignments:

select_NA = 00: Choose the present address plus 1.
select_NA = 01: Choose the next address field of the micro-instruction.
select_NA = 10: Choose the address selected by the AC = 0 detector.
select_NA = 11: Choose the address from the micro-program sequencer.

Thus, for the first micro-instruction, bits 11 and 10 should be 00. The next-address field of this micro-instruction doesn't matter, since it is not used. Let's arbitrarily make these bits 0's. Thus, our first micro-instruction (address 0 of the ROM) has a hexadecimal value of 0002. (Verify, please.)

Exercise 7 Determine the remaining contents of the micro-program store by placing states B, C, D, and so on in consecutive locations in ROM. If you need more micro-instructions than there are states, then insert the micro-instructions immediately following the states that require the additional micro-instructions. (Use hexadecimal code.)
Answer[12]

4   CONTEMPORARY MICROPROCESSOR ARCHITECTURES

The preceding section introduced the architecture of a very simple computer that resembles some of the very first electronic computers designed in the late 1940s. The ENIAC (Electronic Numerical Integrator and Calculator) and the IAS (by the Institute for Advanced Study), built at the University of Pennsylvania (1950) and Princeton (1946), respectively, are among the first electronic computers. These computers were built using vacuum tube technology, and each occupied a very large room and consumed a tremendous amount of power, requiring even more air-conditioning than computing equipment just to keep the temperature low enough for the equipment to continue operating.

The first commercial integrated-circuit microprocessor, the 4004, introduced by Intel Corporation in 1971, was more powerful than the ENIAC and occupied a space smaller than the tip of your finger. The 4004 microprocessor had 46 instructions, ran at a clock frequency of 108 kHz, computed 60,000 instructions per second, and was implemented with 2300 transistors. The 4004 marked the beginning of the microprocessor revolution. Since the 4004, the processing capability (speed and complexity) of microprocessors has grown exponentially. The Pentium II Xeon microprocessor, introduced in 1998 by Intel, runs at a clock frequency of 400 MHz, computes 1 billion instructions per second, and contains 5 million transistors. This frequency is 3700 times greater than that of the 4004, and the Pentium II computes 16,000 times as many instructions per second. The improvement in clock frequency is due largely to advances in integrated circuit manufacturing and transistor circuit design. Increasing the number of executed instructions beyond the clock frequency requires improvements in the architecture of the microprocessor. Contemporary microprocessors use parallel instruction execution to increase the number of instructions executed per second. On average, a contemporary microprocessor can execute more than one instruction per clock cycle.[13]

Instruction Pipelining

The most common form of parallel instruction execution is *pipelining*. Because instructions are executed in several small steps (e.g., fetch, decode, execute), the steps of different instructions can be overlapped. While an instruction is being executed, the next instruction is being decoded and the instruction after

---

[12]The ROM contents in consecutive locations beginning with address 1 are 0044, 0011, 0C00, 0060, 1501, 0060, 1480, 0060, 1701, 1800, 1408.   ◆

[13]By the year 2000, the Intel Pentium III (and comparable processors by other vendors) had far surpassed the capacity of Pentium II.

	Clock Cycle								
	1	2	3	4	5	6	7	8	9
Fetch	Inst1	Inst2	Inst3	Inst4	Inst5	Inst6	Inst7	Idle	Idle
Decode	Idle	Inst1	Inst2	Inst3	Inst4	Inst5	Inst6	Inst7	Idle
Execute	Idle	Idle	Inst1	Inst2	Inst3	Inst4	Inst5	Inst6	Inst7

Figure 17 Example reservation table for a pipelined architecture.

that is being fetched from memory. Once the pipeline is full, an instruction finishes execution every clock cycle.

Pipelined instruction execution can be analyzed using a *reservation table*. Such a table identifies the hardware resources occupied by an instruction at a given time. Consider an architecture where the fetch, decode, and execute stages are separate hardware units that can operate on different instructions at the same time. If instructions 1 through 7 require one clock cycle each for fetch, decode, and execute, then their execution can be represented by the reservation table shown in Figure 17.

When execution begins, the decode and execute units do not have any work to do (they are idle). Once the pipeline is "filled" (clock cycle 3), an instruction finishes execution every clock cycle. With the pipelined architecture, seven instructions are executed in 9 clock cycles. If the architecture is not pipelined, then 21 clock cycles are required for the same number of instructions.

It is likely that one or more instructions will require more than one clock cycle for the execution stage. In this stage alone, for example, a multiplication or division may take 10 or more clock cycles; an add may take only 1 clock cycle. What does the reservation table look like when such an instruction, say a multiplication, is executed? The next instruction must wait for the multiplication to finish. If instruction 3 in Figure 17 is a multiplication and requires 2 clock cycles for execution, then the reservation table will look like the one in Figure 18.

Instruction 4 can not use the execution unit until instruction 3 is finished. Thus, the fetch and decode units are idle during clock cycle 6, and the total execution time for the seven instructions requires one additional clock cycle.

Exercise 8   Redraw the reservation table of Figure 17 assuming that instruction 2 needs to fetch data from memory to complete its execution.
Answer[14]

The regularity of a pipeline can also be disrupted by conditional branch instructions. When a conditional branch (e.g., branch if the result of an operation is zero) is fetched from memory, the next instruction to be executed is not known until after the branch is executed. To keep the pipeline full, the instruc-

---

[14]Since there is only one hardware unit for fetching from memory, the fetch of instruction 4 must be delayed for the data fetch for instruction 2. Note that instruction 3 is fetched in cycle 3 because there is no way to know that instruction 2 requires a data fetch until after it is decoded.   ◆

Clock Cycle

	1	2	3	4	5	6	7	8	9	10
Fetch	Inst1	Inst2	Inst3	Inst4	Inst5	Idle	Inst6	Inst7	Idle	Idle
Decode	Idle	Inst1	Inst2	Inst3	Inst4	Idle	Inst5	Inst6	Inst7	Idle
Execute	Idle	Idle	Inst1	Inst2	Inst3	Inst3	Inst4	Inst5	Inst6	Inst7

Figure 18 Reservation table where an instruction in the execution stage requires two clock cycles.

tion should be fetched from memory while the branch is being decoded. How does the processor know which instruction to fetch? There are two possibilities for the next instruction, and most contemporary microprocessors try to guess which one is correct and then choose it. If the guess is wrong, then the pipeline must be purged and the correct instruction fetched from memory.

Parallel Hardware Units

In the preceding section, we introduced three examples that reduce the effectiveness of pipelining: multiple cycles for execution, multiple fetches per instruction, and branches. Also, the fastest rate of execution using pipelining cannot exceed one instruction per clock cycle.[15] How can a microprocessor execute more than one instruction per clock cycle? It has to fetch multiple instructions at the same time, and decode and execute them in parallel as well. Each parallel path can be pipelined to maximize the throughput in each.

Most modern microprocessors have two or more instruction pipelines so they can execute more than one instruction per clock cycle on average. Most often multiple-instruction pipelines are used to execute instructions in parallel. From what you know about programming, programming languages are sequential and statements in a program are executed sequentially. However, many statements in a typical program can be executed in parallel.

Consider two program statements—one that sums the values stored in variables A and B and stores the result in X, and another that adds the values stored in variables C and D and stores the result in Y. The order of execution of these two statements does not make any difference to the result of the computation. Can the statements be executed in parallel? Of course they can, why not? Suppose the second statement adds C and X rather than C and D. Can the statements be executed in parallel now? Since the first statement computes the value of X, the execution of the first statement must be completed before the second one can be executed (the second requires the result of the first). This is known as a *data dependency* between statements (or instructions). Contemporary compilers are able to identify statements that can be executed in parallel. Sometimes it is necessary to reorder statements in a program to maximize the amount of parallel computation.

---

[15]A pipeline must always be filled before it can produce useful results. Thus, the number of clock cycles for executing n instructions can be at most n plus however many clock cycles are required to fill the pipeline.

EXAMPLE 1

Suppose there is a loop that sums the elements of two vectors, *A* and *B,* and stores the result in a third vector, *C.* The program statements look something like this:

```
For i from 0 to 99 loop
    C[i] = A[i] + B[i];
End loop;
```

Now suppose there is a processor with two instruction pipelines (instead of one) for executing the code. Since all of the additions are independent, this processor would be twice as fast as the original one.   ■

Another use of parallel instruction pipelines is the execution of both possible paths for a conditional branch instruction. Since both paths are being executed, no time is lost due to a wrong guess.

Memory Hierarchy

A simplified block diagram of a computer system is shown in Figure 13. The memory in this figure is shown as one block, but in contemporary computer systems the memory comes in several blocks of varying size and speed. In a contemporary desktop computer 64 or 128 MB of RAM are common. The RAM or main memory in a computer system, however, is slow and several CPU clock cycles are required to access data from it. It is common to have a high-speed memory between the CPU and the main memory so that data can be accessed quickly. This high-speed memory is known as *cache memory*. The cache memory is expensive compared with main memory, so the size of the cache memory is normally limited (typically 1 MB or less). The system attempts to keep the most-often-used instructions and data in cache, thus optimizing the system performance. When data is required that is not in cache (called a *cache miss*), then it must be obtained from main memory. The interface between cache and main memory is often designed such that large pieces of memory, called pages, can be moved from main memory to cache when a cache miss occurs.

Contemporary microprocessors have separate cache memories for instructions and for data, and these cache memories are typically integrated on the microprocessor chip. Many computer systems have two levels of cache between the CPU and main memory, with the second level off-chip (not integrated with the microprocessor). The advantage of two levels of cache is that the second level can be larger than the first because it is not on the microprocessor chip. However, it is also slower than the first level for the same reason.

Complex Instruction Set Computer (CISC)

Most contemporary microprocessors are based on one of two architectures: complex-instruction-set computer (CISC) and reduced-instruction-set computer (RISC). A CISC architecture can have a few hundred instructions and the hard-

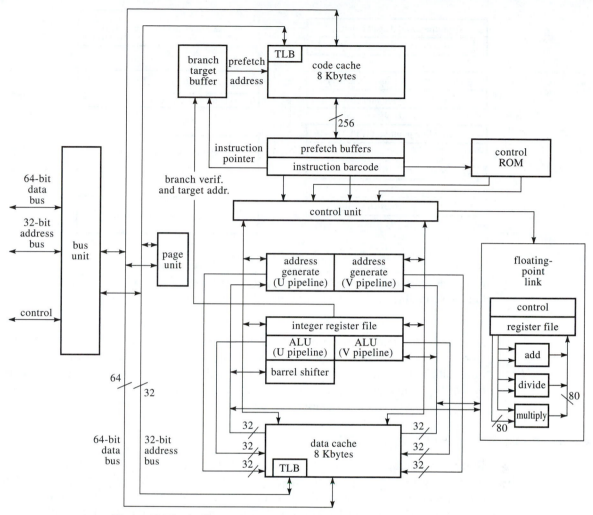

Figure 19 Block diagram of the Pentium microprocessor. (© Intel Corporation, reprinted with permission)

ware contains many features for direct execution of complex instructions. Microprocessors based on a CISC architecture were prevalent in the early and mid-1980s and are still found early in this century. The Intel Corporation's x86 family of microprocessors (including the Pentium series) is based on a CISC architecture. The block diagram of the Pentium microprocessor is shown in Figure 19.

The Pentium microprocessor has a 32-bit address bus and a 64-bit data bus. Most instructions are 32 bits, so most of the time the microprocessor receives two instructions with one instruction fetch from memory. There are two instruction pipelines (U and V), so both instructions can be decoded and executed at the same time (provided one does not depend on the other). The high-level-language compiler must do a good job of organizing the assembly code to permit the maximum use of parallel instruction processing. The Pentium has separate instruction (code) and data caches that are 8 kbytes each. There is a floating-point unit separate from the two instruction pipelines

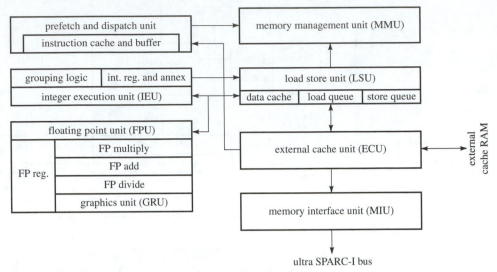

Figure 20 Block diagram of the Ultra-SPARC RISC microprocessor. (© 1997 SUN Microsystems, Inc. All rights reserved. Used by permission.)

for handling difficult floating-point operations. Fixed-point (integer) multiply and divide operations are also handled by the floating-point unit. Notice the address generation units in the instruction pipeline. In a CISC architecture it is common for arithmetic operations to involve data contained in memory. After instruction decode, the address of any operand of the instruction must be determined (or generated).

Reduced Instruction Set Computer (RISC)

Research conducted in the 1980s revealed that most programs utilized only a small percentage (about 20 percent) of the instructions of a CISC microprocessor most (about 90 percent) of the time. The control unit could be greatly simplified by minimizing the number of instructions, and simplification of the control unit would allow much higher clock speeds. Also, simplification of the control unit means extra space on the chip for parallel instruction pipelines, data registers, and cache memory.

RISC microprocessors appeared on the market in the late 1980s and are dominant today in UNIX-based workstations. Examples include the MIPS R4000, the SPARC from Sun Microsystems, the Alpha from Digital Equipment Corporation, the RS/6000 from IBM, the Intel i860/960, the Motorola 88000, and the Hewlett-Packard PA RISC. Early RISC microprocessors were distinguished from CISC microprocessors by instruction pipelines and separate data and instruction caches in addition to a small instruction set. Today the boundary between RISC and CISC architectures is blurred, but the difference in instruction-set complexity still exists. Similar to modern RISC architectures, most modern CISC architectures have multiple-instruction pipelines and separate data and instruction caches.

The block diagram of the Sun Microsystems UltraSPARC microprocessor is shown in Figure 20. It supports 64-bit data words, has a 41-bit physical address, and has a peak execution of four instructions per clock cycle with a maximum clock frequency of 250 MHz. The UltraSPARC can maintain the peak rate of four instructions per cycle even in the presence of conditional branch instructions. The rate of execution is slowed only by data dependencies between operations. It has 16-kbyte instruction and data caches, two integer ALUs, and floating-point and graphics execution units. All memory accesses for data are handled by load/store instructions.

5 MICROCONTROLLER ARCHITECTURES

The architectures of microcontrollers and microprocessors differ in ways dictated by their intended applications. Microcontrollers are designed for real-time applications, those in which it is critical that the system respond in real time. An example of a real-time system is an aircraft radar system, which must respond to objects in its sensor space so that the pilot can react appropriately. On the other hand, a personal computer is not real-time, since the only consequence of a slow computer is the user's frustration. Other examples of real-time systems are communication devices such as two-way voice communication devices (formerly known as telephones) and one-way video and audio devices (formerly known as TVs and radios). If the audio or video signal of such a system is broken, it renders the system unusable.

Microcontrollers have features specific to the implementation of real-time systems. It is common for microcontrollers to have integrated programmable read-only memory (PROM), analog-to-digital converters, and high-priority interrupts. An interrupt is a hardware or software signal that causes the instruction sequence to branch to a specific location in memory where special instructions exist. The special instructions are designed to service a particular request of the system. For example, a control system that monitors the temperature of an oven may be interrupted if the temperature goes above a critical level. The response of the system may be to shut down the oven immediately to avoid the severe consequences of an elevated temperature.

The integrated PROM is useful for code to start up the system, and in some applications can contain all of the instruction code. Analog-to-digital converters are convenient for conversion of analog signals from sensors such as temperature sensors. The block diagram of the Intel 8051 microcontroller is shown in Figure 21.

There are several blocks in a microcontroller that make it distinguishable from a microprocessor: the interrupt controller, ROM, RAM, I/O ports, and a serial port. The internal ROM and RAM may be used for critical code and data segments that require very fast operation to honor a real-time constraint. The interrupt controller manages several interrupts with different priorities to determine which requests should be handled and in what order they should be handled. For example, in a factory, an interrupt signal from a carbon monoxide sensor would be given higher priority than an interrupt signal from a time-of-day clock that is used to adjust thermostat settings in the building. Microcontrollers vary significantly in computing power; 8-bit microcontrollers for low-cost applications (home climate control) and 32-bit microcontrollers for demanding applications (radar signal processing) are still on the market. In contrast, there is not much use today for an 8-bit microprocessor, since all general-purpose computers are expected to run state-of-the-art operating systems and software.

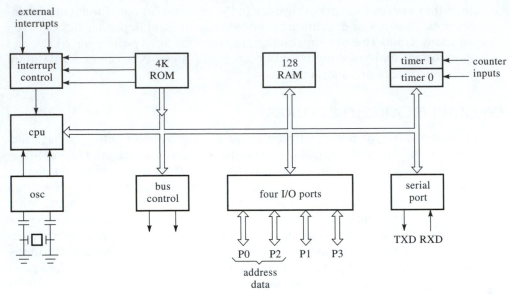

Figure 21 The block diagram of the 8051 microcontroller. (© Philips Semiconductors, reprinted with permission)

CHAPTER SUMMARY AND REVIEW

The subject of computer architecture is vast and requires several semesters of study for the depth and breadth required for a computer engineer. The intent here is to bridge the gap between digital circuit design and large-scale computer design, and to provide an excellent foundation to the study of computer architecture. The following topics were covered:

- Control and datapath units of a processor
- Design of a simple processor: serial multiplier
- Basic stored-program computer
- Fetch, decode, and execute cycle
- Central processing unit (CPU)
- Design of a simple CPU
- Register transfer language (RTL) notation
- Control unit implementations:

  - Hard-wired control unit
  - Micro-programmed control unit

- Contemporary microprocessor architectures
- Instruction pipelining
- Parallel hardware units
- Memory hierarchy
- Complex instruction set computer (CISC)
- Reduced instruction set computer (RISC)
- Microcontroller architectures

PROBLEMS

1 Redesign the datapath of the add-and-shift multiplier in section 1 so that the add and shift operations can be executed in one clock cycle.

2 Redesign the control unit of the add-and-shift multiplier such that if the multiplier bit is 0, the processing unit skips the add step altogether. That is, the datapath performs two shift operations in consecutive clock cycles.

3 Write an ABEL description of the add-and-shift multiplier described in section 1.

4 A processor is needed that computes the difference between two successive 8-bit inputs to a system. The processor receives a stream of 8-bit values, $x(t)$, and produces the result $y(t) = x(t-1) - x(t)$. That is, the present output is the previous input minus the present input. An output should be produced every clock cycle. Write an ABEL description for this processor.

5 Design a processor that accepts a continuous sequence of 8-bit values and computes the average of the most recent eight values. The output is $y(t) = [x(t) + x(t-1) + \cdots + x(t-7)]/8$. Write an ABEL description of this processor.

6 Combine the processors of Problems 4 and 5 to create a system with a single-bit output that is asserted logic 1 whenever the difference between the two most recent inputs exceeds half of the average of the most recent eight inputs.

7 Design the control and datapath units of a processor that accumulates four 8-bit numbers provided sequentially. In four consecutive clock cycles, four 8-bit numbers are provided at the inputs and the processor should accumulate their values in a register. The processor automatically starts again after the last of the four inputs is received. Sketch a block diagram of the datapath unit, and write ABEL descriptions of the control and datapath units of the processor.

8 The function $y(t) = x(t) + 5x(t-1) + 2x(t-2)$ needs to be computed for implementation of a digital filter. Design a processor that implements this function and write its ABEL description. Is it wise to use a multiplier to implement this function?

9 Specify the instructions and datapath for a CPU that has only three instructions and can be used to implement arbitrary programs. (Consider the example in section 2 and determine which of the four instructions is redundant.)

10 Write an ABEL description of the simple CPU in Figure 9; include both datapath and control units.

11 In the ABEL description of Figure 14, describe how an ABEL compiler optimizes the output expressions. For several states, some outputs are not explicitly defined. How does the ABEL synthesis tool treat these? Rewrite the description such that the ABEL synthesis tool takes maximum advantage of the don't-care conditions.

12 Design an 8-bit processor that can perform the arithmetic operation addition and the logical operations AND, Exclusive-OR, and NOT. Assume the instruction code and input data values are stored in registers, and the result of the operation is stored in a fourth register. Write an ABEL description of the processor. Draw a block diagram of its datapath and control units. Synthesize the ABEL description using the Xilinx software.

13 What dictates the maximum clock frequency of a microprocessor? (There may be more than one factor. Consider all of them.)

14 The maximum number of instructions that can be executed in a single clock cycle is determined by the microprocessor architecture. However, the throughputs of several pieces of the architecture have to be matched in order to achieve it. Explain what this means. Give an example in which a system has parallel execution paths, but the maximum number of instructions executed per cycle is limited.

15 For each of the following pairs of program statements, determine whether or not the statements have data dependencies. In other words, can their order be changed without affecting the result of the computation?

    a. $A = B + C, A = C - D$
    b. $A = B + C, D = B + E$

    c. $A = B + C, B = D + C$
    d. $A = B + C, D = A + C$

16 The following sequence of high-level-language statements is to be executed on a micro-processor that has two parallel instruction pipelines. Reorder the statements so that the shortest number of cycles is required for their execution and the same result of computation is achieved. Assume each instruction requires the same number of cycles for fetch, decode, and execute.

$$W = A + B$$
$$X = W + D$$
$$Y = Z + C$$
$$Z = A + D$$

17 For the answer to Problem 16, draw a reservation table for the two instruction pipelines of the CPU. Assume all of the variables are register locations, so no memory accesses are needed to get the data, and assume that fetch, decode, and execute all require one clock cycle. How many clock cycles are saved by reordering the statements?

18 What range of values can be represented in two's complement format by the Pentium and UltraSPARC microprocessors?

19 How many memory locations are directly addressable by the Pentium microprocessor? How many by the UltraSPARC microprocessor?

20 On average, which microprocessor can execute more instructions per second, a 400MHz Pentium or a 250MHz UltraSPARC?

21 Describe an asynchronous interface between a processor and a memory system. Draw state diagrams that characterize the control in each part (processor and memory) of the system.

Appendix

MOSFETs and Bipolar Junction Transistors

This appendix constitutes a brief discussion of the basics of some aspects of electronics, sufficient to give you a working familiarity with the terminology and general operation of electronic logic circuits. It is by no means complete, but it might help you to understand the terminology found in logic circuit manufacturers' data books.

The transistor, the building block from which all logic gates are constructed, was invented over a half-century ago. It is made of *semiconductor* materials that, as the name implies, have properties of electrical conductivity lying between those of good conductors (copper) and insulators (rubber). Examples of semiconductors are *silicon* and *germanium,* both of which lie in the fourth column of the periodic table and, hence, have a valence of 4. The most common semiconductor used today is silicon.

The conductivity of pure semiconductors can be modified and controlled by introducing atoms of materials from adjacent columns in the periodic table into their crystal structure. This is called *doping.* Doping silicon with atoms of valence 5 (e.g., antimony, arsenic) will increase the number of negative charge carriers (electrons). The result is called an *n-type* semiconductor. Doping with atoms of valence 3 (e.g., boron, gallium) will increase the number of positive charge carriers (*holes*), resulting in a *p-type* semiconductor. The material as a whole is electrically neutral in both cases.

A *pn junction* is formed by interfacing a *p*-type semiconductor with an *n*-type, as shown in stylized form in Figure 1.[1] The *bias* across the junction is related to a voltage resulting from external connections. If this voltage is positive from the *p* to the *n* side, the junction is said to be *forward-biased;* if negative, it is *reverse-biased.*

A device consisting of a *pn* junction is called a *diode.* The (forward) current in a forward-biased diode is very low until the voltage reaches about 0.6 V; it then increases rapidly but almost linearly with a further increase in volt-

---

[1]In actual construction, the junction is formed in a single silicon crystal by distributing the doping of the impurity atom so as to create an abrupt change in carrier densities.

Figure 1 Semiconductor diode.

age, as if the diode were a low-valued resistor. When the diode is reverse-biased, the *reverse* (or *leakage*) current is extremely low, several orders of magnitude lower than the forward current. (If the forward current is in mA, the reverse current will be in μA or less.) However, if the reverse voltage increases to what is called the *reverse breakdown* value (typical values are 40–50 V), the reverse current increases precipitously — hence the term *breakdown voltage*. (There are applications in which this property is utilized, but not in logic circuits.)

The only way of controlling the current in a diode is by means of the voltage across the junction. It would be very useful if there were another means of controlling the current. A way of doing this is to form a combination of two junctions that are "sandwiched": *pnp* or *npn*. There are several ways to accomplish this in an integrated circuit. The beginning step is to start with a "slab" of silicon and dope it to form a bulk *n*-type or *p*-type material. After that, different procedures lead to different varieties. Each variety carries the generic name *transistor*. Various types of transistors have made their appearance over time; some types have been entirely discarded when other varieties based on slightly different principles came along. Others still find use in applications different from logic circuits. Two varieties will be discussed here.

MOSFET

For logic devices, two types of transistors are now in use. One is a MOSFET (MOS for metal-oxide semiconductor and FET for field-effect transistor). MOSFETs themselves come in two varieties; the *enhancement-mode* version is the one widely used for logic circuits, so that's the only one we will describe here.[2] A stylized version that can be viewed as a cross-section is shown in Figure 2. What is labeled the *bulk* material is also often called the *substrate*.

Although only three external terminals are shown in Figure 2, the substrate constitutes another terminal that is almost always connected to ground in digi-

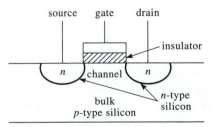

Figure 2 Enhancement-mode MOSFET transistor cross-section.

---

[2]The other version is the *depletion-mode* MOSFET.

tal circuit applications. That is, the enhancement-mode MOSFET is a four-terminal device. The gate terminal—a conductor—is separated from the rest of the device by an insulator (shown crosshatched). That means there is no conduction of current from the gate to any other parts of the MOSFET.[3] The p region lying between the two n regions (drain and source) is called the *channel* because that is the path in which current flows under appropriate bias voltages. During current conduction the carrier concentration in the channel is inverted: electrons are attracted there by the electric field induced by the positive gate-to-source voltage. The transistor in Figure 2 is therefore called an *n-channel* or *n-type* MOSFET. When the gate-to-source voltage is 0, there is no current in the channel. However, for a sufficient positive gate-to-source voltage, there is current flow from drain to source that does not depend on the precise value of the drain-to-source voltage. The MOSFET acts as a switch under these circumstances. This is the manner in which it is used in logic circuit applications.[4]

Common symbols used for the MOSFET in circuits are shown in Figure 3. The most generic symbol is the one in Figure 3*a*. In logic circuits, since the bulk is almost always connected to ground,[5] the bulk terminal is often omitted, as in Figure 3*b*. In addition, since the MOSFET is a physically symmetric device with respect to the source and drain terminals, it is common to use the symbol in Figure 3*c,* where the arrow is removed and the source and drain terminals are not distinguished—except by our labels!

If the p and n regions are interchanged, a variation of the structure in Figure 2 is formed. The result is called (you guessed it!) a *p-channel* or *p-type* MOSFET. Now it is a *negative* gate-to-source voltage that results in a source-to-drain current. Circuit symbols for the p-channel MOSFET are shown in Figure 4. The only distinction is the direction of the arrows in the first two parts and the bubble on the gate in Figure 4*c* compared with Figure 3*c*.

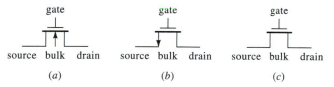

Figure 3 Common circuit symbols for the n-channel MOSFET.

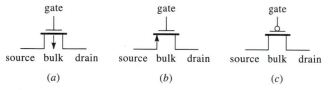

Figure 4 Common circuit symbols for the p-channel MOSFET.

---

[3]Strictly speaking, some (capacitive) current exists, but this is small enough to be called zero.

[4]For lower voltages, the current is proportional to the voltage; in this case the device acts as a resistor. This mode of operation is useful in other applications, but not in logic circuits.

[5]The bulk is always connected to the lowest supply voltage in the circuit; this is almost always ground.

From the preceding description, it is evident that, using appropriate gate-to-source voltages, the MOSFET behaves as a voltage-controlled switch: a positive gate-to-source voltage closes the switch in an *n*-type and a negative voltage does so in a *p*-type. In both types, there is a negligible current at the gate in both conditions. Between source and drain there is (a) a very low impedance (ideally zero) when the transistor is conducting (switch is *on*) and (b) a very high impedance (ideally infinite) when the transistor is not conducting (switch is *off*).

Bipolar Junction Transistor

Another approach to forming two back-to-back junctions is illustrated in Figure 5. This is called a *bipolar junction transistor* (BJT). This version is an *npn* transistor. The physical width of the base is very small. The emitter region is more heavily doped than the bulk *n*-region, while the base is lightly doped. For logic circuits it is again necessary that the transistor operate approximately as a switch. If both junctions are reverse-biased, the transistor is said to be in the *cutoff* region. This requires that the base–emitter voltage (v_{BE}) as well as the base–collector voltage (v_{BC}) be negative. Under these conditions no current (actually, very little) will flow, as in an open switch. However, if both those junctions are forward-biased ($v_{BE}, v_{BC} > 0$), the transistor is said to be in the *saturation* region. The maximum possible current flow will then occur.[6]

It doesn't take much imagination to recognize that the counterpart to an *npn* transistor is a *pnp* transistor, in which the bulk is *p*-type silicon. All the characteristics are complementary to those described for the *npn* transistor. The circuit symbols for both types are given in Figure 6.

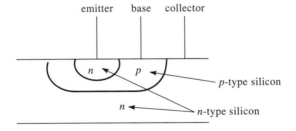

Figure 5 Cross-section of a bipolar junction transistor.

(a)   (b)

Figure 6 Circuit symbols for bipolar transistors. (a) *npn* transistor. (b) *pnp* transistor.

---

[6]There are two other regions in which the transistor can operate: one junction forward-biased and one reverse-biased. The case where $v_{BE} > 0$ and $v_{BC} < 0$ is called the *forward-active* region. In this region the device operates as an *amplifier*, with the collector current being β times the base current, where β can be as high as 100. In this region the transistor is not useful for logic circuits, so the conditions on v_{BE} and v_{BC} are never attained.

In the design of logic circuits with BJTs, the transistors play the role of current-controlled switches; the controlling current being the base current. With zero base current, the switch is open; with a base current greater than some threshold value, the transistor is driven into saturation and the switch is closed.

Bibliography

Historically Important References

George Boole, *An Investigation of the Laws of Thought,* New York: Dover, 1954.

Richard W. Hamming, "Error-Detecting and Error-Correcting Codes," *Bell Syst Tech J* 29, 147–160 (April 1950).

E. V. Huntington, "Sets of Independent Postulates for the Algebra of Logic," *Trans Am Math Soc* 5, 288–309 (1904).

Claude E. Shannon, "A Symbolic Analysis of Relay and Switching Circuits," *Trans AIEE* 57, 713–723 (1938).

Claude E. Shannon, "The Synthesis of Two-Terminal Switching Circuits," *Bell Syst Tech J* 28, 59–98 (1949).

Significant Earlier References

Books

S. H. Caldwell, *Switching Circuits and Logic Design*, New York: Wiley, 1958.

J. Hartmannis and R. E. Stearns, *Algebraic Structure Theory of Sequential Machines,* Englewood Cliffs, NJ: Prentice-Hall, 1966.

Zvi Kohavi, *Switching and Finite Automata Theory,* New York: McGraw-Hill, 1978.

Edward J. McCluskey, *Introduction to the Theory of Switching Circuits,* New York: McGraw-Hill, 1965.

Carver Mead and Lynn Conway, *Introduction to VLSI Systems,* Reading, MA: Addison-Wesley, 1980.

Stephen H. Unger, *Asynchronous Sequential Switching Circuits,* New York: Wiley, 1969.

Articles

David A. Huffman, "The Synthesis of Sequential Switching Circuits," *J Franklin Inst* 257 (March 1954); 257–303 (April 1954).

David A. Huffman, "A Study of the Memory Requirements of Sequential Switching Circuits," MIT Research Lab for Electronics Tech Report 293 (April 1955).

Maurice Karnaugh, "A Map Method for Synthesis of Combinational Logic Circuits," *Trans AIEE* 73(9), 593–599 (November 1953) (part I).

Edward J. McCluskey, "Minimization of Boolean Functions," *Bell Syst Tech J* 35(6), 1417–1444 (November 1956).

G. H. Mealy, "A Method for Synthesizing Sequential Circuits," *Bell Syst Tech J* 34, 1045–1079 (September 1955).

E. F. Moore, "Gedanken Experiments on Sequential Machines," *Automata Studies,* Annals of Math Studies 34, Princeton, NJ: Princeton University Press, 1956, 120–153.

W. V. Quine, "The Problem of Simplifying Truth Functions," *Am Math Monthly* 59(8), 521–531 (October 1952).

John von Neumann, "Probabilistic Logic and the Synthesis of Reliable Organisms from Unreliable Components," *Automata Studies,* Annals of Math Studies 34, Princeton, NJ: Princeton University Press, 1956, 43–49.

Contemporary References (Since 1980s)

S. Brown and Z. Vranesic, *Fundamentals of Digital Logic with VHDL Design,* New York: McGraw-Hill, 2000.

D. J. Comer, *Digital Logic and State Machine Design,* 3rd ed., Philadelphia: Saunders, 1995.

D. D. Gajski, *Principles of Digital Design,* Englewood Cliffs, NJ: Prentice Hall, 1997.

R. H. Katz, *Contemporary Logic Design,* Redwood City, CA: Benjamin-Cummings, 1994.

M. M. Mano and C. R. Kime, *Logic and Computer Design Fundamentals,* 2nd ed., Englewood Cliffs, NJ: Prentice Hall, 2000.

C. H. Roth, *Fundamentals of Logic Design,* 4th ed., St. Paul, MN: West, 1992.

J. F. Wakerley, *Digital Design: Principles and Practices*, 3rd ed., Englewood Cliffs, NJ: Prentice Hall, 2000.

Index